# Handbook of Modern Earth Buildings: Materials, Engineering, Constructions and Applications

# Handbook of Modern Earth Buildings: Materials, Engineering, Constructions and Applications

## Contributors

**Jessica Giro-Paloma,Refat Al-Shannaq and  Mohammed M. Farid et al.**

**AURIS**
Reference

www.aurisreference.com

## Handbook of Modern Earth Buildings: Materials, Engineering, Constructions and Applications

Contributors: Jessica Giro-Paloma,Refat Al-Shannaq and Mohammed M. Farid et al.

**Published by Auris Reference Limited**

www.aurisreference.com

United Kingdom

**Copyright 2016**
**Printed in 2017 for Sale in the Indian Subcontinent**

**Notice**

**Handbook of Modern Earth Buildings: Materials, Engineering, Constructions and Applications**

ISBN: 978-1-78154-828-8

British Library Cataloguing in Publication Data
A CIP record for this book is available from the British Library

Printed in the United Kingdom
Exclusively distributed by CBS Publishers & Distributors Pvt. Ltd.

Sales & Distribution Rights only for India, Pakistan, Bangladesh, Sri Lanka, Nepal and Bhutan.This book is not to be sold outside these territories.

# Contents

# List of Abbreviations

| | |
|---|---|
| ADP | Abiotic depletion potential |
| AHP | Analytical hierarchy process |
| AAC | Autoclaved aerated concrete |
| BOQ | Bill of quantities |
| BIM | Building information model |
| BIM | Building information modeling |
| BM | Building materials |
| BRE | Building research establishment |
| CVA | Change vector analysis |
| CSW | Compressed shopper waste |
| CFD | Computational fluid dynamics |
| CSM | Continuous stiffness measurement |
| DSI | Depth-sensing indentation |
| EUE | End-use of energy |
| EIR | Environmental impact report |
| FLIS | Fuzzy logic inference systems |
| GSA | General services administration |
| GWP | Global warming potential |
| HR | Human resources |
| LA | Lauric acid |
| LOD | Level of detail |
| LCA | Life cycle assessment |
| LCIA | Life cycle impact assessment |
| LCI | Life-cycle inventory |
| LECA | Light expanded clay aggregate |
| LOS | Line of sight |
| MCI | Marginal cost increase |
| MPCM | Microencapsulated phase change materials |
| NCM | National calculation methodology |
| PCM | Phase change material |
| PCM | Phase change materials |
| PCMs | Phase change materials |
| POCP | Photochemical oxidant creation |
| POCP | Photochemical ozone creation potential |
| PCA | Principle component analysis |
| SBS | Sick building syndrome |
| SA | Stearic acid |
| SVM | Support vector machine |
| SPD | Suspended particle devices |
| SD | Sustainable development |

| | |
|---|---|
| TES | Thermal energy storage |
| TIF | Thermal integrity factor |
| TP | True positives |

# List of Contributors

**Jessica Giro-Paloma**
Department of Materials Science and Metallurgical Engineering, Faculty of Chemistry, Universitat de Barcelona, C/Martí i Franquès, Barcelona 1. 08028, Spain

**Refat Al-Shannaq**
Department of Chemical and Materials Engineering, University of Auckland, Private Bag 92019, 20 Symonds Street, Auckland 1142, New Zealand

**Ana Inés Fernández**
Department of Materials Science and Metallurgical Engineering, Faculty of Chemistry, Universitat de Barcelona, C/Martí i Franquès, Barcelona 1. 08028, Spain

**Mohammed M. Farid**
Department of Chemical and Materials Engineering, University of Auckland, Private Bag 92019, 20 Symonds Street, Auckland 1142, New Zealand

**Rongda Ye**
Key Laboratory of Enhanced Heat Transfer and Energy Conservation, the Ministry of Education, School of Chemistry and Chemical Engineering, South China University of Technology, Guangzhou 510640, China

**Xiaoming Fang**
Key Laboratory of Enhanced Heat Transfer and Energy Conservation, the Ministry of Education, School of Chemistry and Chemical Engineering, South China University of Technology, Guangzhou 510640, China

**Zhengguo Zhang**
Key Laboratory of Enhanced Heat Transfer and Energy Conservation, the Ministry of Education, School of Chemistry and Chemical Engineering, South China University of Technology, Guangzhou 510640, China

**Xuenong Gao**
Key Laboratory of Enhanced Heat Transfer and Energy Conservation, the Ministry of Education, School of Chemistry and Chemical Engineering, South China University of Technology, Guangzhou 510640, China

**Paul Joseph**
The Built Environment Research Institute, School of the Built Environment, University of Ulster, Newtownabbey, BT37 0QB, Northern Ireland, UK

**Svetlana Tretsiakova-McNally**
The Built Environment Research Institute, School of the Built Environment, University of Ulster, Newtownabbey, BT37 0QB, Northern Ireland, UK

**Antti Ruuska**
VTT Technical Research Centre of Finland, Tekniikantie 4, 02044 VTT Finland

**Tarja Häkkinen**
VTT Technical Research Centre of Finland, Tekniikantie 4, 02044 VTT Finland

**Sungwoo Lee**
Architectural Engineering, Hanyang University, Sa 3-dong, Sangrok-gu Ansan 426-791, Korea

**Sungho Tae**
School of Architecture & Architectural Engineering, Hanyang University, Sa 3-dong, Sangrok-gu Ansan 426-791, Korea

**Seungjun Roh**
Architectural Engineering, Hanyang University, Sa 3-dong, Sangrok-gu Ansan 426-791, Korea

**Taehyung Kim**
Architectural Engineering, Hanyang University, Sa 3-dong, Sangrok-gu Ansan 426-791, Korea

**Kuang-Sheng Liu**
Department of Interior Design, Tung Fang Design University, No.110 Dongfang Road, Hunei Distract, Kaohsiung City 82941, Taiwan

**Sung-Lin Hsueh**
Graduate Institute of Cultural and Creative Design, Tung Fang Design University, No.110 Dongfang Road, Hunei Distract, Kaohsiung City 82941, Taiwan

**Wen-Chen Wu**
Graduate Institute of Cultural and Creative Design, Tung Fang Design University, No.110 Dongfang Road, Hunei Distract, Kaohsiung City 82941, Taiwan

**Yu-Lung Chen**
Graduate Institute of Cultural and Creative Design, Tung Fang Design University, No.110 Dongfang Road, Hunei Distract, Kaohsiung City 82941, Taiwan

**Miguel A. Gómez**
Industrial Engineering School, University of Vigo, Lagoas-Marcosende s/n 36310 Vigo, Spain

**Miguel A. Álvarez Feijoo**
Industrial Engineering School, University of Vigo, Lagoas-Marcosende s/n 36310 Vigo, Spain

**Roberto Comesaña**
Industrial Engineering School, University of Vigo, Lagoas-Marcosende s/n 36310 Vigo, Spain

**Pablo Eguía, José L. Míguez**
Industrial Engineering School, University of Vigo, Lagoas-Marcosende s/n 36310 Vigo, Spain

**Jacobo Porteiro**
Industrial Engineering School, University of Vigo, Lagoas-Marcosende s/n 36310 Vigo, Spain

**Shaghayegh Mohammad**
The University of Bath, Department of Architecture and Civil Engineering, Bath BA2 7AY, UK

**Andrew Shea**
The University of Bath, Department of Architecture and Civil Engineering, Bath BA2 7AY, UK

**Sudan Xu**
Department of Earth Observation Science, Faculty ITC, University of Twente, 7500 AE Enschede, the Netherlands

**George Vosselman**
Department of Earth Observation Science, Faculty ITC, University of Twente, 7500 AE Enschede, the Netherlands

**Sander Oude Elberink**
Department of Earth Observation Science, Faculty ITC, University of Twente, 7500 AE Enschede, the Netherlands

**Akubue Jideofor Anselm**
Architecture Department, University of Nigeria, Nigeria

**Paolo Frattini**
Department of Earth and Environmental Sciences, Università degli Studi di Milano-Bicocca, p.zza della Scienza 4, 20126 Milan, Italy

**Giovanni B. Crosta**
Department of Earth and Environmental Sciences, Università degli Studi di Milano-Bicocca, p.zza della Scienza 4, 20126 Milan, Italy

**Jacopo Allievi**
Tele-Rilevamento Europa T.R.E., Ripa di Porta Ticinese 79, 20143 Milan, Italy

**Hasan Kaplan**
Pamukkale University, Department of Civil Engineering Turkey

**Salih Yılmaz**
Pamukkale University, Department of Civil Engineering Turkey

**Peng Xingqian**
College of Civil Engineering, Huaqiao University, Quanzhou China

**Liu Chunyan**
College of Civil Engineering, Huaqiao University, Quanzhou China

**Chen Yanhong**
College of Civil Engineering, Huaqiao University, Quanzhou China

# Preface

The construction of earth buildings has been taking place worldwide for centuries. With the improved energy efficiency, high level of structural integrity and aesthetically pleasing finishes achieved in modern earth construction, it is now one of the leading choices for sustainable, low-energy building. The text *Handbook of Modern Earth Buildings: Materials, Engineering, Constructions and Applications* provides an essential exploration of the materials and techniques key to the design, development and construction of modern earth buildings. The purpose of first chapter is to develop, prepare, characterize, study, and compare thermal and mechanical properties of microcapsules containing organic phase change materials (PCM) in order to assess their suitability for use in buildings and other applications. The goal of second chapter is to demonstrate the feasibility of using EP-based composite PCM in cement boards to increase their thermal inertia and to reduce the energy demand of the building. Third chapter aims to analyze recent advances in the area of non-metallic building materials (BM) and outlines future prospects and challenges. Fourth chapter outlines and draws conclusions about different aspects of the material efficiency of buildings and assesses the significance of different building materials on the material efficiency. Fifth chapter develops a template for evaluating the embodied environmental impact of using a building information modeling (BIM) design tool as part of BIM-based building life-cycle assessment (LCA) technology development. In sixth chapter, we explain the Delphi method as a group decision-making technique, including its uses, underlying assumptions, strengths and limitations, potential benefits to qualitative higher education research, and key considerations in its use. In seventh chapter, a CFD-based model has been proposed to analyze the effect of phase change materials (PCMs) on the thermal behavior of the walls of a cubicle exposed to the environment and on the resistance of the walls to climate changes. In eighth chapter, we evaluate the thermal performance of a range of modern wall constructions used in the residential buildings of Tehran in order to find the most appropriate alternative to the traditional un-fired clay and brick materials. Ninth chapter presents a method for detecting and classifying changes to buildings by using classified and well registered (strip difference <10 cm) laser data from several epochs. A review of energy conservation properties in earth sheltered housing has been presented in tenth chapter. The aim of last chapter is to analyze large slope movements in conjunction with radar interferometry and damage data in order to investigate the state of the activity of such phenomena and to describe the resulting level of damage as a function of the ground surface rate of movement.

# Chapter 1

# PREPARATION AND CHARACTERIZATION OF MICROENCAPSULATED PHASE CHANGE MATERIALS FOR USE IN BUILDING APPLICATIONS

Jessica Giro-Paloma[1], Refat Al-Shannaq[2], Ana Inés Fernández[1], and Mohammed M. Farid[2]

[1]Department of Materials Science and Metallurgical Engineering, Faculty of Chemistry, Universitat de Barcelona, C/Martí i Franquès, Barcelona 1. 08028, Spain

[2]Department of Chemical and Materials Engineering, University of Auckland, Private Bag 92019, 20 Symonds Street, Auckland 1142, New Zealand

## ABSTRACT

A method for preparing and characterizing microencapsulated phase change materials (MPCM) was developed. A comparison with a commercial MPCM is also presented. Both MPCM contained paraffin wax as PCM with acrylic shell. The melting temperature of the PCM was around 21 °C, suitable for building applications. The M-2 (our laboratory made sample) and Micronal® DS 5008 X (BASF) samples were characterized using SEM, DSC, nano-indentation technique, and Gas Chromatography/Mass spectrometry (GC-MS). Both samples presented a 6 μm average size and a spherical shape. Thermal energy storage (TES) capacities were 111.73 $J \cdot g^{-1}$ and 99.3 $J \cdot g^{-1}$ for M-2 and Micronal® DS 5008 X, respectively. Mechanical characterization of the samples was performed by nano-indentation technique in order to determine the elastic modulus ($E$), load at maximum displacement ($P_m$), and displacement at maximum load ($h_m$), concluding that M-2 presented slightly better mechanical properties. Finally, an important parameter for considering use in buildings is the release of volatile organic compounds (VOC's). This characteristic was studied at 65 °C by CG-MS. Both samples showed VOC's emission after 10 min of heating, however peaks intensity of VOC's generated from M-2 microcapsules showed a lower concentration than Micronal® DS 5008 X.

# INTRODUCTION

Thermal energy storage (TES) using phase change materials (PCM) has shown a significant increased attention because of its important role on energy conservation in buildings [1,2,3,4]. PCM can be used for TES in buildings [5] either in passive [6] or active systems [7,8,9], aiming to improve the thermal managements of these buildings. In most of the applications, PCM were used either in macroencapsulated [10,11] or microencapsulated [12,13,14] forms, for heating [15], air-conditioning [16], ventilation [17], refrigeration [18], and heat exchangers [19] for building applications [3,20,21,22].

Microencapsulation process is defined as a technique in which small particles or droplet are enclosed by a coating, or surrounded in a homogeneous or heterogeneous matrix, to provide microcapsules (1–100 μm). For this reason, the microencapsulated phase change materials (MPCM) are composed of PCM as a core and a polymer as a shell used to preserve the spherical shape of the microcapsule and avoid PCM leakage during phase change [12,23]. The use of MPCM in buildings [24,25,26,27] can decrease daily inner temperature fluctuation during summer and winter [28]. The suitability of the shell used in encapsulating specific core PCM is a key issue in order to ensure proper thermal performance of the MPCM [13,29,30], especially in preventing PCM leakage when it melts. Additionally, MPCM can be easily incorporated in gypsum board [31,32], plaster [33], and concrete [34] used in buildings.

The complete characterization of materials used in indoor environments like MPCM is very important. The exposure to chemical compounds could cause health problems (nausea; dry skin; eye, nose or throat irritations; headache; irritated eyes; dizziness; difficulty in concentrating; psychological stress) in indoor environments [35,36,37,38] (buildings, for example [39,40,41,42,43,44,45,46,47]) or outdoor environments [48]. These problems are known as sick building syndrome (SBS) [37,49,50,51]. Volatile organic compounds (VOC's) are defined as any organic compound having an initial boiling point less than or equal to 250 °C at a standard pressure of 101.3 kPa [52]. VOC's are one of the most important groups of trace contaminants in the atmosphere known for its photochemical, toxic, and radioactive effects. For this reason there are some studies, guides [53,54], and database [55,56] related to this effect. Formaldehyde [39,46,57] and benzene [58] are some of the most studied pollutants since they are classified in Group 1 of human carcinogens by the IARC 2004 (International Agency for Research on Cancer). Other chemicals known for their health hazard are acetaldehyde, toluene, and xylenes [59]. By this way, VOC's evaluation of the outdoor and indoor air quality has been evaluated [28,29,30,31] in materials for buildings like gypsum base PCM composites [60] but it has not been reported for building materials containing MPCM. For this reason, the characterization of VOC's of MPCM is an important contribution to the state of the art of the environmental properties of MPCM.

The main purpose of this research is to develop, prepare, characterize, study, and compare thermal and mechanical properties of microcapsules containing organic PCM in order to assess their suitability for use in buildings and other applications. The samples under study are commercial MPCM (Micronal® DS 5008 manufactured by BASF, Berlin, Germany) and a laboratory prepared one by us (M-2). Micronal® DS 5008 sample has been used extensively in concrete, gypsum, lime plaster, and gypsum plaster, without being fully characterized for fire hazards. The comparison includes fire retardancy and gas emission released to environment from upon fire. It is important to establish a characterization methodology, which will include both volatile emission measurements and nano-indentation technique to measure the shell mechanical strength of the microcapsules. This is very important issue in the selection of PCM products, especially for use in building application. PCM microcapsules should have high phase change enthalpy, uniform spherical shape, acceptable thermal stability, good mechanical properties, and low release of hazardous gases in the form of volatile organic compounds.

# MATERIALS AND METHODS

## Materials

The chemical preparation of microcapsules required the following reagents:

- Shell: Methyl methacrylate (MMA) (99%, contains ≤ 30 ppm monomethyl ether hydroquinone (MEHQ) as inhibitor, Sigma Aldrich, Auckland, New Zealand) and pentaerythritol tetraacrylate (PETRA) (contains 350 ppm (MEHQ), Sigma Aldrich, Auckland, New Zealand) were used as a monomer and cross-linking agent respectively in order to obtain proper shells for MPCM.

- Free radical thermal initiator: Luperox® A75, Benzoyl peroxide (BPO) (75%, contains 25% water, Sigma Aldrich, Auckland, New Zealand) was used as free radical thermal initiator.

- Surfactants: Polyvinyl alcohol (PVA) ($M_w$ 85,000–124,000, Sigma Aldrich, Auckland, New Zealand) and sodium dodecyl sulfate (SDS) (BioXtra, 99%, Sigma Aldrich, Auckland, New Zealand) were used as a non-ionic and ionic surfactant, respectively.

- PCM: a commercial paraffinic PCM, Rubitherm® RT 21 ($T_m$ = 21 °C,

$\Delta H_m$ = 135 J·g$^{-1}$, Rubitherm® Technologies GmbH, Berlin, Germany) was used.

The bulk density of M-2 microcapsules is 0.496 g·mL$^{-1}$. The commercial MPCM, Micronal® DS 5008 X (BASF®), was also selected for characterization and was compared with the microcapsules produced in this work. This sample is also composed by an acrylate shell and organic PCM in the core [13], and its bulk density is 0.445 g·mL$^{-1}$.

## Synthesis of PCMs Microcapsules

### Emulsification

A standard procedure was used as reported elsewhere [61]. Wherein, an aqueous solution of surface-active agent (called aqueous phase) and a mixture of MMA, PETRA, BPO, and PCM (called organic phase) were prepared separately. The organic phase was added to the aqueous phase and emulsified mechanically using a high shear mixer (Silverson L5M-A laboratory Mixer, Silverson LTD, East Longmeadow, MA, USA). A stirring rate of 3000 rpm for 5 min was chosen to achieve the required emulsification.

### Polymerization

The produced emulsion was transferred to a 2-L four-neck glass reactor (LR-2.ST laboratory reactor-IKA-Werke, Gmbh@Co.KG, Staufen, Germany) consisting of EUROSTAR 200 control P4, Anchor stirrer LR 2000.1, HBR 4 digital heating bath. The agitation speed was set at approximately 300 rpm, and the temperature of the water bath was maintained at 70 °C for 2 h, and then adjusted to 85 °C for another 4 h. The water bath was then switched off and allowed to cool down naturally to room temperature. After cooling, the suspension of PCM microcapsules was transferred to a clean glass beaker and washed three times with distilled water to remove the unreacted monomers and the non-encapsulated PCM. The separated microcapsules were spread on a tray and placed in an oven at 50 °C for 48 h for drying. The dried microcapsules were then collected for testing.

# CHARACTERIZATION OF MICROCAPSULES

## Scanning Electron Microscopy (SEM)

To study the shape and size of microcapsules SEM was used (a FEI Quanta 200 FEG-Field Emission Gun with an EDS Detector SiLi (Lithium drifted) with a Super Ultra-Thin window, FEI Company, Hillsboro, OR, USA). The sputter coater used was a Quorum Q150RS (FEI Company, Hillsboro, OR, USA), and it is designed to give an appropriate thin, slight metal coating proper for SEM observation, using Pt as a target. The coating thickness and uniformity of the sample depends on different factors: distance between sample and target, topography of the sample, and affinity of the material with the metal coating.

## Differential Scanning Calorimetry (DSC)

Phase change properties of the fabricated PCM microcapsules and the pure PCM (such as melting and solidification temperatures and their phase change enthalpies) were determined using a SHIMADZU DSC-60 differential scanning calorimeter (Shimadzu Company, Kyoto, Japan). The measurements were performed by varying the temperature between $-15$ °C and 40 °C with heating and cooling rate of 3 °C·min$^{-1}$. Each sample was analyzed for three times and the average was taken. Consequently, the percentage PCM encapsulated can be determined using DSC results and the following Equation (1) [62,63]. The mass content obtained by DSC measurements does not provide accurate measure of the core mass content. Equation (1), which was used to estimate the mass content from DSC measurements, does not take in account the sensible heat of coating materials. The TGA method provides more accurate measure of the core material mass content than DSC. In our previous publication [63] the core material mass content of sample M-2 obtained by TGA is 77.5 wt%, which is less than the one obtained by DSC.

$$\% \text{ PCM in microcapsules by mass} = \Delta H_{\text{microcapsules}}/\Delta H_{\text{Pure PCM}} \times 100\% \tag{1}$$

where $\Delta H_{microcapsules}$ (J·g$^{-1}$) is the latent heat of the microcapsule containing PCM; and $\Delta H_{purePCM}$ (J·g$^{-1}$) is the latent heat of pure PCM (before encapsulation). In Equation (1), it is assumed that the latent heat of the microcapsule without PCM is zero, which is true if phase change does not occur in the shell does.

## Nano-Indentation Technique

To characterize the mechanical properties of M-2 and commercial Micronal® DS 5008 X samples, a nano-indentation technique was used. Nano-indentation is identified as an appropriate technique to test the strength of individual microcapsules [64]. MTS Nano Indenter XP (MTS Company, Eden Prairie, MN, USA) was the instrument used. Aluminium stubs of 20 mm height and 30 mm diameter were needed to stick the samples at the top to characterize them using a red glue to stick the samples as shown in Figure 1. The instrument parameters were set the same for the two studied samples for a more accurate comparison.

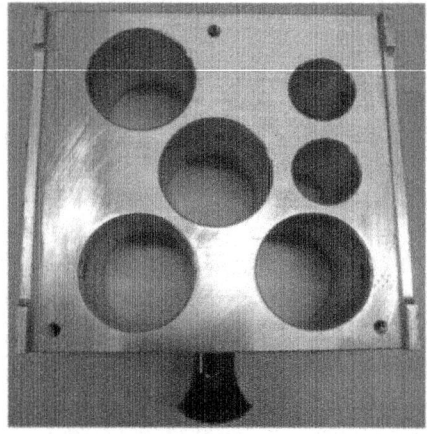 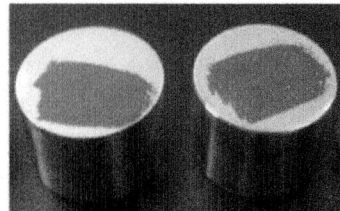

**Figure 1:** Holder and aluminum stubs with the sample over the red glue.

There are some required inputs to set before starting the experiments: strain rate target of 0.05 s$^{-1}$, allowable drift rate of 0.05 nm·s$^{-1}$, a Poissons' ratio of 0.18 for the tip indenter [65,66] a peak hold time of 10 s, a surface speed of 10 nm·s$^{-1}$, 25% of surface approach sensitivity, 90% to unload, an approach distance to store of 1000 nm, a surface approach distance of 1000 nm, and finally, a depth limit of 5000 nm.

To determine the elastic modulus ($E$) of the studied samples, a Berkovich tip TB-13288 (Micro Star Technologies, Huntsville, TX, USA) was used. The use of nano-indentation for the characterization of mechanical properties of materials has been extensively studied by several authors. Oliver and Pharr developed extensively the methodology for characterizing ceramic materials [66,67]. They described the typical load *vs.* displacement curve, where increasing the load ($P$) increases the displacement ($h$) until reaching a maximum load ($P_{max}$) and a maximum displacement ($h_{max}$). Following that, the indenter is removed out of the material (unloading section), the load will be zero, and the final displacement ($h_f$) will be measured. After that, the $E$ value for each sample can be calculated. Hochstetter *et al.* [68] presented results for glassy polymers and Giro-Paloma *et al.* [69] compared both methodologies using continuous stiffness measurement (CSM) by applying a small oscillation to the quasi-static component of loading using different thermoplastics suitable as containers for PCM. They concluded that Loubet's method produce lower values of $H$ and $E$ because it uses a dynamic approach for stiffness measurements and the contact depth is larger due to the contribution of the apparent tip effect. In the light of these findings, it was concluded that Loubet's method should be used only with polymeric materials having a low viscous character ($T_g > T_{measurement}$).

Additionally, in depth-sensing indentation (DSI), which is the mode used in this paper, load is applied as a function of penetration depth during the loading and unloading cycle, as described by Fischer-Cripps [70]:

- Hardness ($H$) is defined as the maximum indentation load divided by the cross-sectional area of the indenter specified at the maximum indentation depth ($A(h_m)$).

- Load ($P_m$) at maximum displacement (mN): It is the load recorded at the maximum load, which occurs when sample fails.

- Elastic modulus ($E$) is evaluated following Equations (2) and (3) from the nano-indentation test using the maximum indentation load ($P_m$) and the depth sensing indentation. Hardness ($H$), elastic work ($W_e$), and total work ($W$) can be calculated by integrating the areas under the indentation unloading. $W_e$ and $W$ are the elastic work and total work, which are equal to the areas under the unloading and loading curves, respectively which is correlated with $E$ and $H$through the function $\psi$ described in [71]. We$W$/ value is independent of the degree of work-hardening behavior [66].

$$H = P_m/A(h_m)$$
(2)

$$E = H/\psi \left( \frac{W_e}{W} \right)$$
(3)

- Displacement ($h_m$) at maximum load (nm) is a measure of the extent the tip penetrates into the material.

## Emission of Volatile Organic Compounds (VOC's)

A GC-17A Gas chromatograph Shimadzu (Shimadzu Corporation, Kyoto, Japan) coupled to GCMS-QP5000 Gas chromatograph/Mass Spectrometer Shimadzu (Shimadzu Corporation, Kyoto, Japan) was used to characterize the VOC's emissions from each sample. A calibration was performed for each pure compound: $C_{14}H_{30}$, $C_{15}H_{32}$, $C_{16}H_{34}$, $C_{17}H_{36}$, $C_{18}H_{38}$, $C_{20}H_{42}$, $C_{22}H_{46}$, and $C_{24}H_{50}$, at different temperatures: 25 °C, 35 °C, 45 °C, 55 °C, and 65 °C. A total of 40 experiments for calibration were executed. When the calibration was completed, the same procedure was performed for the two studied samples: M-2 and Micronal® DS 5008 X.

Each sample was independently located inside a crystal vial HS of 50 mL capacity. The vials were submerged during 30 min in a water bath until reaching the required temperature. Later on, a fibber solid-phase microextraction (SPME) holder with lot number P268618D 57330-U, provided by Supelco (Sigma-Aldrich Corporation, St. Louis, MO, USA), was punctured on the top of the silicone cap. Following that, 10 min desorption was applied. The temperature inside the device was 60 °C during 2 min. After that, a ramp of 15 °C·min$^{-1}$ was programmed.

A HP-5MS (1553434H) (Agilent Technologies, Santa Clara, CA, USA) was the column used. Its thickness, length and diameter were 0.5 μm, 30 m, and 0.32 cm, respectively. Additionally, the injection and interface temperature were 200 °C and 280 °C, respectively inside the gas chromatographer. There were more parameters to take into account, such as inlet pressure: 1 kPa, flow: 1.1 mL·min$^{-1}$, lineal velocity: 38.7 cm·s$^{-1}$, split ratio between peaks of 20, and finally total flow for the He gas of 23.1 mL·min$^{-1}$. On the other hand, the mass spectrometer m/z values are from 35 to 350. Moreover, the solvent cut time was 0.5 min.

# RESULTS AND DISCUSSION

## Characterization of MPCM Shape, Size, and Morphology

SEM images for the two studied samples are shown in Figure 2. M-2 microcapsules morphology appears to be compact and with smooth surface as shown in Figure 2a. Furthermore, their size is around 6 μm and has a spherical shape. On the other hand, commercial Micronal® DS 5008 X looks made of a big sphere (around 150 μm) composed of hundreds of other small microcapsules (the ones which contain the PCM) with 6 μm in size, approximately as shown in Figure 2b. As Figure 2shows, the 6 μm microcapsules of Micronal® DS 5008 X samples are deformed, which is probably due to the process of agglomerating of these microcapsules to form larger particle of 150 μm, which probably has been made for ease of handling. Accelerated thermal cycling experiments of PCM microcapsules (M-2 containing RT-58 sample) were performed in a controlled heating/cooling water bath at temperatures cycling between 2 and 40 °C in our previous publication [63]. The results showed that slight changes in phase transition temperatures of the PCM microcapsules (M-2 containing RT58 sample) after 2000 cycles. The latent heat of M-2 sample (based on DSC measurements) showed only a minor change of 2% after 2000 cycles. Furthermore, SEM images showed that the capsule shape remained spherical and no shell cracks were observed at the end of 2000 cycles.

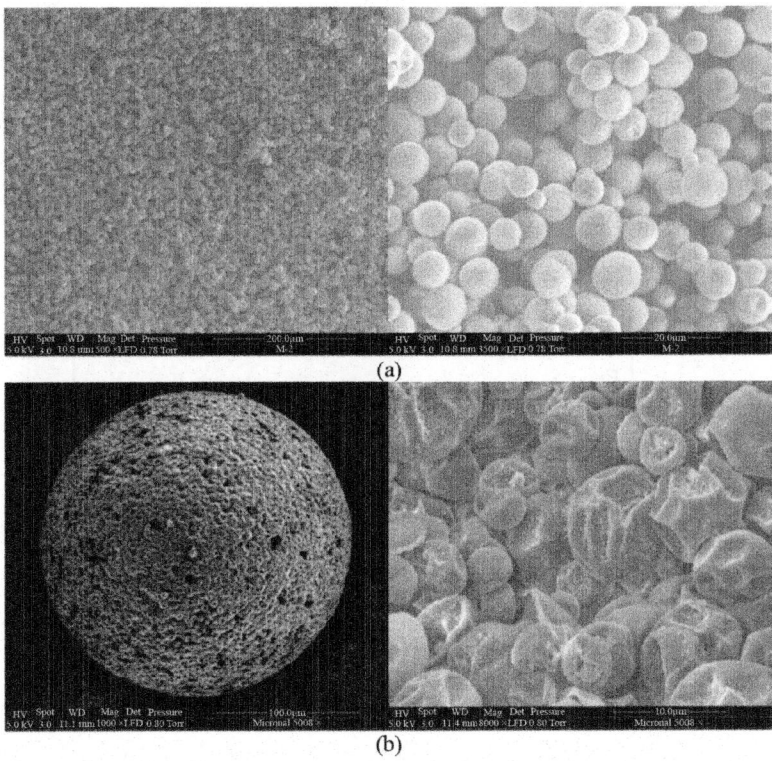

**Figure 2:** SEM images for the studied samples: (**a**) M-2 (magnification: ×500 left and ×3500 right); (**b**) Micronal® DS 5008 X (magnification: ×1000 left and ×8000 right).

## Thermophysical Properties of MPCM

Thermal properties of M-2 and Micronal® DS 5008 X in terms of phase change enthalpies and phase change temperatures were investigated using DSC as shown in Figure 3. Based on DSC measurements, the % mass of the PCM for M-2 sample is 81.7 wt% and for Micronal® DS 5008 X is 78.1 wt%. The DSC results show that the thermal energy storage capacity of the M-2 is 113.8 J·g⁻¹ which corresponds to 85 wt% of RT-21 encapsulation. The melting temperature of the RT-21 in M-2 microcapsules is similar to that of the bulk RT-21. In contrast, the solidification temperature of the PCM microcapsules was about 14 °C lower than that of the bulk RT-21 (super-cooling phenomena) as previously reported [72] (see Table 1). The supercooling of PCM in microcapsules has also been reported by Qiu *et al.* [73]. The increase of the degree of super-cooling of RT-21 microcapsules could be attributed either to the decrease in the amount of RT-21 nuclei inside each microcapsule compared to the bulk RT-21 [74] or due to formation of vacuum pockets space inside the microcapsules [75]. To reduce the supercooling of PCM microcapsules, additives were mixed with the PCM prior encapsulation to act as a nucleating agent. These nucleating agents are usually materials with a similar crystal structure as the solid PCM which allow nucleation at their surface but have a higher melting temperature. Figure 4 shows the DSC curve for the microcapsules containing nucleating agent. Commercial RT-58 (paraffin) with peak melting temperatures of 58 °C was used as nucleating agent. The degree of supercooling has been reduced dramatically as reported in Table 1 and shown in Figure 4.

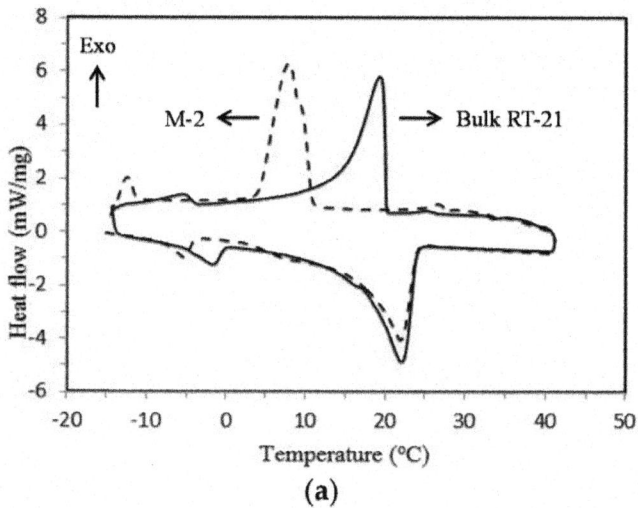

(a)

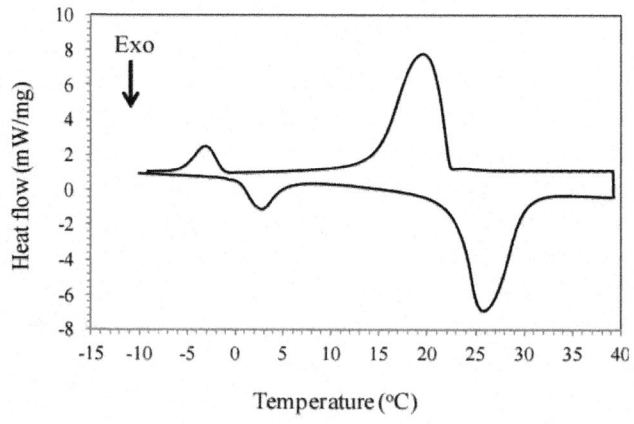

(b)

**Figure 3:** DSC results for: **(a)** M-2; and **(b)** Micronal® DS 5008 X.

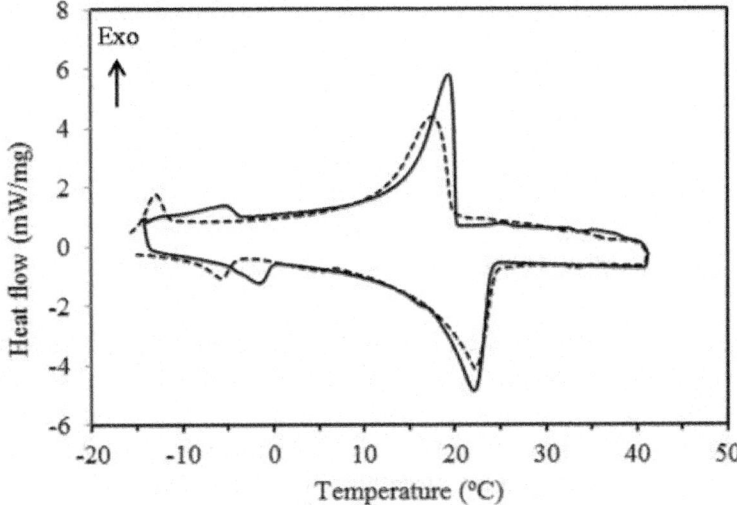

**Figure 4:** DSC curves of the bulk RT-21 (solid line) and M-2 microcapsules containing nucleating agent, RT-58 (dotted line).

**Table 1:** Thermophysical properties of fabricated M-2 and commercial Micronal® DS 5008 X

| Status | Transition Temperatures and Heat of Fusion | M-2 | RT-21 | Micronal® DS 5008 X | M-2 Containing RT-58 |
|---|---|---|---|---|---|
| Heating | $T_{onset}$ (°C) | 16.7 | 16.5 | 22.1 | 16.1 |
| | $T_{peak}$ (°C) | 22.0 | 22.1 | 20.3 | 22.3 |
| | $T_{endset}$ (°C) | 24.0 | 23.9 | 26.3 | 24.3 |
| | $\Delta H$ (J·g$^{-1}$) | 113.9 | 132.0 | 99.9 | 110.4 |
| Cooling | $T_{onset}$ (°C) | 10.9 | 20.2 | 22.5 | 19.8 |
| | $T_{peak}$ (°C) | 7.9 | 19.4 | 24.2 | 17.5 |
| | $T_{endset}$ (°C) | 4.2 | 14.5 | 17.5 | 11.4 |
| | $\Delta H$ (J·g$^{-1}$) | 111.9 | 132.5 | 103.5 | 108.3 |

## Mechanical Properties of MPCM

Results of nano-indentation technique by DSI for the two samples were summarized in Table 2. Ten tests were performed for each sample, but some results were discarded because of the dispersion attributed to indentations in the edge of the microcapsules. For this reason, six results were finally selected. From these measurements the mean value and standard deviation of the elastic modulus ($E$) were calculated for each sample. These results show that M-2 microcapsules are stiffer than Micronal® DS 5008 X ones.

**Table 2:** Elastic modulus results of M-2 and Micronal® DS 5008 X

| Mechanical Property | M-2 | Micronal® DS 5008 X |
|---|---|---|
| | 1.89 | 0.15 |
| | 1.04 | 0.19 |
| E (GPa) | 1.16 | 0.17 |
| | 1.68 | 0.24 |
| | 1.38 | 0.22 |
| | 1.9 | 0.28 |
| Mean | 1.51 | 0.21 |
| Standard Deviation | 0.37 | 0.05 |

Figure 5 is a representation of the *P-h* curves for the M-2 and Micronal® respectively. The measured typical drift behavior can be observed on the plateau at the end of the unloaded curve. It can be concluded that the mechanical response for M-2 is better than Micronal® DS 5008 X as M-2 microcapsules are more rigid and show higher strength. This fact can be attributed to various factors such as the differences in the encapsulation ratio between samples, shell thickness, type of polymers, as well as degree of polymerization of the shells. These results should be compared when mixing these MPCM in a real system, such us mixing them with a cement-based matrix or gypsum.

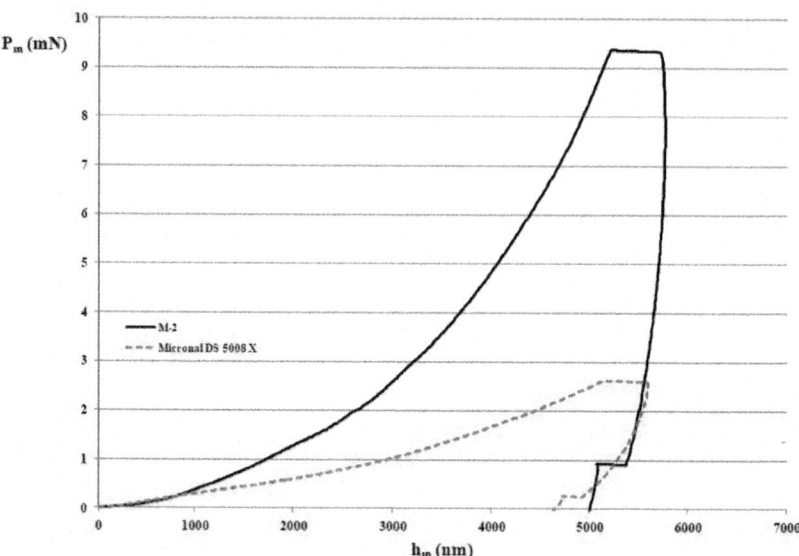

**Figure 5:** $h_m$ vs. $P_m$, comparison of the samples under study.

## Characterization of Volatile Organic Compounds (VOCs)

The results of VOCs' short-term release from the two studied samples are shown in Table 3. Each peak in the GC spectra was given at a certain retention time ($t_r$, in minutes), which was related to a specific compound, according to the calibration conducted in this work.

**Table 3:** VOC emission results for M-2 and Micronal® DS 5008 X samples

| Temperature | M-2 | | Micronal® DS 5008 X | |
|---|---|---|---|---|
| | $t_r$ (min) | Compound | $t_r$ (min) | Compound |
| 25 °C | No signal | | No signal | |
| 35 °C | No signal | | No signal | |
| 45 °C | No signal | | 11.88 | $C_{17}H_{36}$ |
| | | | 12.66 | $C_{18}H_{38}$ |
| 55 °C | No signal | | 11.88 | $C_{17}H_{36}$ |
| | | | 12.65 | $C_{18}H_{38}$ |
| 65 °C | 9.25 | $C_{14}H_{30}$ | - | - |
| | 10.18 | $C_{15}H_{32}$ | - | - |
| | 11.06 | $C_{16}H_{34}$ | - | - |
| | 11.91 | $C_{17}H_{36}$ | 11.85 | $C_{17}H_{36}$ |
| | 12.68 | $C_{18}H_{38}$ | 12.63 | $C_{18}H_{38}$ |

It may be observed from the results that some emissions from Micronal® DS 5008 X sample after 10 min exposure at 45 °C and 55 °C were detected while nothing was detected from M-2 microcapsules. After 10 min exposure at 65 °C, both samples release volatile compounds, but with much lower level for M-2 microcapsules in comparison with Micronal® DS 5008 X. This is confirmed by the high intensity peak for Micronal® DS 5008 X comparing both figures in Figure 6. The multiple peaks for M-2 show the presence of tetradecane, pentacosane, hexadecane, heptadecane, and octadecane in the original PCM, which is RT-21, while only two peaks were observed for the PCM used in Micronal® DS 5008 X, indicating that the latter is of higher purity. Although the results for the short-term emissions are relevant, the long term time-dependent release should be studied in a system after prolonged exhibition to evaluate the VOCs' emissions in real conditions.

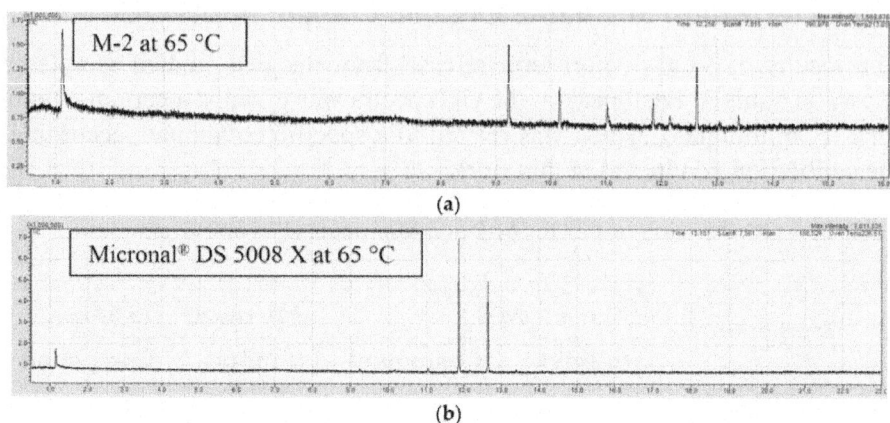

(a)

(b)

**Figure 6:** Revealed peaks for the GC/MS analysis for: (**a**) M-2 and; (**b**) Micronal® DS 5008 X samples at 65 °C.

## CONCLUSIONS

A comparison for the characteristic of two MPCM samples was conducted: one sample was Micronal® DS5008 X while the other was made in our laboratory (M-2). The shell material for both samples is similar in terms of chemical composition. Following SEM observation, it can be concluded that the two samples have similar shape and diameter of about 6 μm, but they have different morphology since Micronal® DS 5008 X capsules were produced as aggregates of many microcapsules. Regarding the thermophysical properties for both samples, their thermal energy storage (TES) capacity were 111.7 J·g⁻¹ and 99.3 J·g⁻¹ for M-2 and Micronal® DS 5008 X, respectively. The mechanical testing was performed by measuring elastic modulus ($E$), load at displacement ($P_m$), and displacement at maximum load ($h_m$) using nano-indentation technique. Different results were obtained for both samples, showing that evaluating the isolated M-2 sample has better mechanical resistance and stiffness. Finally, a comparative evaluation of the VOC's release at 25 °C, 35 °C, 45 °C, 55 °C, and 65 °C was performed in order to study the volatiles emission from these microcapsules. M-2 microcapsules show better stability with less short-term emission of VOC's than Micronal® DS 5008 X.

## ACKNOWLEDGMENTS

Authors acknowledge the support received from the European Union's Seventh Framework Programme (FP7/2007–2013) under grant agreement No: PIRSES-GA-2013-610692 (INNOSTORAGE). Also, we would like to acknowledge the support we have received form Qatar Foundation in supporting the project No: NPRP 5-093-2-034.

## AUTHOR CONTRIBUTIONS

Mohammed M. Farid and Ana Inés Fernández conceived and designed the experiments; Jessica Giro-Paloma and Refat Al-Shannaq performed the experiments; all the authors analyzed the data; Mohammed M. Farid contributed reagents/materials/analysis tools; Jessica Giro-Paloma and Refat Al-Shannaq wrote the paper.

## CONFLICTS OF INTEREST

The authors declare no conflict of interest. The founding sponsors had no role in the design of the study; in the collection, analyses, or interpretation of data; in the writing of the manuscript, and in the decision to publish the results.

## REFERENCES

1.  Zhou, D.; Zhao, C.Y.; Tian, Y. Review on thermal energy storage with phase change materials (PCMs) in building applications. *Appl. Energy* 2012, *92*, 593–605.

2.  Sharma, A.; Tyagi, V.V.; Chen, C.R.; Buddhi, D. Review on thermal energy storage with phase change materials and applications. *Renew. Sustain. Energy Rev.* 2009, *13*, 318–345.

3.  Kuznik, F.; David, D.; Johannes, K.; Roux, J.-J. A review on phase change materials integrated in building walls.*Renew. Sustain. Energy Rev.* 2011, *15*, 379–391.

4.  Farid, M.M.; Khudhair, A.M.; Razack, S.A.K.; Al-Hallaj, S. A review on phase change energy storage: Materials and applications. *Energy Convers. Manag.* 2004, *45*, 1597–1615.

5.  Cabeza, L.F.; Castell, A.; Barreneche, C.; de Gracia, A.; Fernández, A.I. Materials used as PCM in thermal energy storage in buildings: A review. *Renew. Sustain. Energy Rev.* 2011, *15*, 1675–1695.

6.  Soares, N.; Costa, J.J.; Gaspar, A.R.; Santos, P. Review of passive PCM latent heat thermal energy storage systems towards buildings' energy efficiency. *Energy Build.* 2013, *59*, 82–103.

7.  Griffiths, P.W.; Eames, P.C. Performance of chilled ceiling panels using phase change material slurries as the heat transport medium. *Appl. Therm. Eng.* 2007, *27*, 1756–1760.

8.  Huang, L.; Petermann, M.; Doetsch, C. Evaluation of paraffin/water

emulsion as a phase change slurry for cooling applications. *Energy* 2009, *34*, 1145–1155.

9.  Youssef, Z.; Delahaye, A.; Huang, L.; Trinquet, F.; Fournaison, L.; Pollerberg, C.; Doetsch, C. State of the art on phase change material slurries. *Energy Convers. Manag.* 2013, *65*, 120–132.

10. Pendyala, S. Macroencapsulation of Phase Change Materials for Thermal Energy Storage. Master Thesis, University of South Florida, Tampa, FL, USA, 22 June 2012.

11. Calvet, N.; Py, X.; Olivès, R.; Bédécarrats, J.-P.; Dumas, J.-P.; Jay, F. Enhanced performances of macro-encapsulated phase change materials (PCMs) by intensification of the internal effective thermal conductivity. *Energy* 2013, *55*, 956–964.

12. Zhao, C.Y.; Zhang, G.H. Review on microencapsulated phase change materials (MEPCMs): Fabrication, characterization and applications. *Renew. Sustain. Energy Rev.* 2011, *15*, 3813–3832.

13. Giro-Paloma, J.; Oncins, G.; Barreneche, C.; Martínez, M.; Fernández, A.I.; Cabeza, L.F. Physico-chemical and mechanical properties of microencapsulated phase change material. *Appl. Energy* 2013, *109*, 441–448.

14. Al Shannaq, R.; Farid, M.M. Microencapsulation of phase change materials (PCMs) for thermal energy storage systems. In *Advances in Thermal Energy Storage Systems*; Cabeza, L.F., Ed.; Elsevier: Amsterdam, the Netherlands, 2015; pp. 247–284.

15. Alvarado, J.L.; Marsh, C.; Sohn, C.; Phetteplace, G.; Newell, T. Thermal performance of microencapsulated phase change material slurry in turbulent flow under constant heat flux. *Int. J. Heat Mass Transf.* 2007, *50*, 1938–1952.

16. Diaconu, B.M.; Varga, S.; Oliveira, A.C. Experimental study of natural convection heat transfer in a microencapsulated phase change material slurry. *Energy* 2010, *35*, 2688–2693.

17. Zhang, P.; Ma, Z.W.; Wang, R.Z. An overview of phase change material slurries: MPCS and CHS. *Renew. Sustain. Energy Rev.* 2010, *14*, 598–614.

18. Zhang, P.; Ma, Z.W. An overview of fundamental studies and applications of phase change material slurries to secondary loop refrigeration and air conditioning systems. *Renew. Sustain. Energy Rev.* 2012, *16*, 5021–5058.

19. Medrano, M.; Yilmaz, M.O.; Nogués, M.; Martorell, I.; Roca, J.; Cabeza, L.F. Experimental evaluation of commercial heat exchangers for use as

PCM thermal storage systems. *Appl. Energy* 2009, *86*, 2047–2055.

20. Su, J.-F.; Wang, L.-X.; Ren, L. Preparation and characterization of double-MF shell microPCMs used in building materials. *J. Appl. Polym. Sci.* 2005, *97*, 1755–1762.

21. Khudhair, A.M.; Farid, M.M. A review on energy conservation in building applications with thermal storage by latent heat using phase change materials. *Energy Convers. Manag.* 2004, *45*, 263–275.

22. Alawadhi, E.M. Thermal analysis of a building brick containing phase change material. *Energy Build.* 2008, *40*, 351–357.

23. Sarı, A.; Alkan, C.; Karaipekli, A.; Uzun, O. Microencapsulated n-octacosane as phase change material for thermal energy storage. *Sol. Energy* 2009, *83*, 1757–1763.

24. Cabeza, L.F.; Castellón, C.; Nogués, M.; Medrano, M.; Leppers, R.; Zubillaga, O. Use of microencapsulated PCM in concrete walls for energy savings. *Energy Build.* 2007, *39*, 113–119.

25. Castellón, C.; Medrano, M.; Roca, J.; Cabeza, L.F.; Navarro, M.E.; Fernández, A.I.; Lázaro, A.; Zalba, B. Effect of microencapsulated phase change material in sandwich panels. *Renew. Energy* 2010, *35*, 2370–2374.

26. Toppi, T.; Mazzarella, L. Gypsum based composite materials with micro-encapsulated PCM: Experimental correlations for thermal properties estimation on the basis of the composition. *Energy Build.* 2013, *57*, 227–236.

27. Entrop, A.G.; Brouwers, H.J.H.; Reinders, A.H.M.E. Experimental research on the use of micro-encapsulated Phase Change Materials to store solar energy in concrete floors and to save energy in Dutch houses. *Sol. Energy* 2011, *85*, 1007–1020.

28. Tyagi, V.V.; Kaushik, S.C.; Tyagi, S.K.; Akiyama, T. Development of phase change materials based microencapsulated technology for buildings: A review. *Renew. Sustain. Energy Rev.* 2011, *15*, 1373–1391.

29. Zalba, B.; Marín, J.M.; Cabeza, L.F.; Mehling, H. Review on thermal energy storage with phase change: Materials, heat transfer analysis and applications. *Appl. Therm. Eng.* 2003, *23*, 251–283.

30. Alkan, C.; Sari, A. Fatty acid/poly(methyl methacrylate) (PMMA) blends as form-stable phase change materials for latent heat thermal energy storage. *Sol. Energy* 2008, *82*, 118–124.

31. Zhang, H.; Xu, Q.; Zhao, Z.; Zhang, J.; Sun, Y.; Sun, L.; Xu, F.; Sawada, Y. Preparation and thermal performance of gypsum boards incorporated with microencapsulated phase change materials for thermal regulation.

*Sol. Energy Mater. Sol. Cells* 2012, *102*, 93–102.

32.  Borreguero, A.M.; Garrido, I.; Valverde, J.L.; Rodríguez, J.F.; Carmona, M. Development of smart gypsum composites by incorporating thermoregulating microcapsules. *Energy Build.* 2014, *76*, 631–639.

33.  Karkri, M.; Lachheb, M.; Albouchi, F.; Nasrallah, S.B.; Krupa, I. Thermal properties of smart microencapsulated paraffin/plaster composites for the thermal regulation of buildings. *Energy Build.* 2015, *88*, 183–192.

34.  Eddhahak-Ouni, A.; Drissi, S.; Colin, J.; Neji, J.; Care, S. Experimental and multi-scale analysis of the thermal properties of Portland cement concretes embedded with microencapsulated Phase Change Materials (PCMs). *Appl. Therm. Eng.* 2014, *64*, 32–39.

35.  Schlink, U.; Rehwagen, M.; Damm, M.; Richter, M.; Borte, M.; Herbarth, O. Seasonal cycle of indoor-VOCs: Comparison of apartments and cities. *Atmos. Environ.* 2004, *38*, 1181–1190.

36.  Ayoko, G.A. Volatile organic compounds in indoor environments. In *The Handbook of Environmental Chemistry*; Pluschke, P., Ed.; Springer-Verlag: Heidelberg, Germany, 2004; pp. 1–35.

37.  Gallego, E.; Roca, X.; Perales, J.F.; Guardino, X. Determining indoor air quality and identifying the origin of odour episodes in indoor environments. *J. Environ. Sci.* 2009, *21*, 333–339.

38.  Missia, D.A.; Demetriou, E.; Michael, N.; Tolis, E.I.; Bartzis, J.G. Indoor exposure from building materials: A field study. *Atmos. Environ.* 2010, *44*, 4388–4395.

39.  Shin, S.H.; Jo, W.K. Volatile organic compound concentrations, emission rates, and source apportionment in newly-built apartments at pre-occupancy stage. *Chemosphere* 2012, *89*, 569–578.

40.  Huang, H.; Haghighat, F. Building materials VOC emissions—A systematic parametric study. *Build. Environ.* 2003, *38*, 995–1005.

41.  Zhang, Y.; Xu, Y. Characteristics and correlations of VOC emissions from building materials. *Int. J. Heat Mass Transf.* 2003, *46*, 4877–4883.

42.  Kim, S.; Choi, Y.-K.; Park, K.-W.; Kim, J.T. Test methods and reduction of organic pollutant compound emissions from wood-based building and furniture materials. *Bioresour. Technol.* 2010, *101*, 6562–6568.

43.  Lee, C.-S.; Haghighat, F.; Ghaly, W. Conjugate mass transfer modeling for VOC source and sink behavior of porous building materials: When to apply it? *J. Build. Phys.* 2006, *30*, 91–111.

44.  Haghighat, F.; Huang, H. Integrated IAQ model for prediction of VOC emissions from building material. *Build. Environ.* 2003, *38*, 1007–1017.

45. Lee, C.-S.; Haghighat, F.; Ghaly, W.S. A study on VOC source and sink behavior in porous building materials—Analytical model development and assessment. *Indoor Air* 2005, *15*, 183–96.

46. Kim, S.; Kim, H.-J.; Moon, S.-J. Evaluation of VOC emissions from building finishing materials using a small chamber and VOC analyser. *Indoor Built Environ.* 2006, *15*, 511–523.

47. Magee, R.J.; Bodalal, A.; Biesenthal, T.A.; Lusztyk, E.; Brouzes, M.; Shaw, C.Y. Prediction of VOC Concentration Profiles in a Newly Constructed House Using Small Chamber Data and an IAQ Simulation Program. In Proceedings of the 9th International Conference on IAQ and Climate, Monterey, CA, USA, 30 June–5 July 2002; pp. 298–303.

48. Jia, C.; Batterman, S.A.; Relyea, G.E. Variability of indoor and outdoor VOC measurements: An analysis using variance components. *Environ. Pollut.* 2012, *169*, 152–159.

49. Norbäck, D.; Torgén, M.; Edling, C. Volatile organic compounds, respirable dust, and personal factors related to prevalence and incidence of sick building syndrome in primary schools. *Br. J. Ind. Med.* 1990, *47*, 733–741.

50. Redlich, C.A.; Sparer, J.; Cullen, M.R. Sick-building syndrome. *Lancet* 1997, *349*, 1013–1016.

51. Apte, M.G.; Daisey, J.M. VOCs and "Sick Building Syndrome": Application of a New Statistical Approach for SBS Research to US EPA BASE Study Data. In Proceedings of the Indoor Air 99, Edinburgh, Scotland, 30 June–5 July 1999; pp. 2–7.

52. Rufford, T.E.; Zhu, J.; Hulicova-Jurcakova, D. *Green Carbon Materials: Advances and Applications*; Pan Stanford: Boca Raton, FL, USA, 2014; pp. 1–288.

53. *Standard Guide for Small-Scale Environmental Chamber Determinations of Organic Emissions from Indoor Material/Products*; ASTM Standard D5116-97; ASTM International: West Conshohocken, PA, USA, 1997.

54. *Standard Test Method for Determining Formaldehyde Concentrations in Air from Wood Products Using a Small Scale Chamber*; ASTM Standard D6007-96; ASTM International: West Conshohocken, PA, USA, 1996.

55. Won, D.; Magee, R.J.; Lusztyk, E.; Nong, G.; Zhu, J.P.; Zhang, J.S.; Reardon, J.T.; Shaw, C.Y. A Comprehensive VOC Emission Database for Commonly-Used Building Materials. In Proceedings of the 7th International Conference of Healthy Buildings, Singapore, 7–11 December 2003; pp. 1–6.

56. Ouazia, B.; Reardon, J.; Sander, D. Making the Case for Reducing Ventilation Requirements through Selection of Low-Emission Materials. In Proceedings of the 10th International Conference on Indoor Air Quality and Climate, Beijing, China, 4–9 September 2005.

57. Xiong, J.; Zhang, Y.; Huang, S. Characterization of VOC and formaldehyde emission from building materials in static environmental chamber: Model development and application. *Indoor Built Environ.* 2011, *20*, 217–225.

58. Knöppel, H.; Schauenburg, H. Screening of household products for the emission of volatile organic compounds.*Environ. Int.* 1989, *15*, 413–418.

59. Lin, C.-C.; Yu, K.-P.; Zhao, P.; Lee, G. Evaluation of impact factors on VOC emissions and concentrations from wooden flooring based on chamber tests. *Build. Environ.* 2009, *44*, 525–533.

60. Safari, V.; Barreneche, C.; Castell, A.; Basatni, A.; Navarro, L.; Cabeza, L.; Haghighat, F. Volatile Organic Emission from PCM Building Materials. In Proceedings of the Innostock 2012—The 12th International Conference on Energy Storage, Lleida, Spain, 16–18 May 2012.

61. Sánchez, L.; Sánchez, P.; de Lucas, A.; Carmona, M.; Rodríguez, J.F. Microencapsulation of PCMs with a polystyrene shell. *Colloid Polym. Sci.* 2007, *285*, 1377–1385.

62. Sánchez-Silva, L.; Rodríguez, J.F.; Romero, A.; Borreguero, A.M.; Carmona, M.; Sánchez, P. Microencapsulation of PCMs with a styrene-methyl methacrylate copolymer shell by suspension-like polymerisation. *Chem. Eng. J.* 2010, *157*, 216–222.

63. Al-Shannaq, R.; Kurdi, J.; Al-Muhtaseb, S.; Dickinson, M.; Mohammed, F. Supercooling elimination of phase change materials (PCMs) microcapsules. *Energy* 2015, *87*, 654–662.

64. Rahman, A.; Dickinson, M.E.; Farid, M.M. Microencapsulation of a PCM through membrane emulsification and nanocompression-based determination of microcapsule strength. *Mater. Renew. Sustain. Energy* 2012, *1*.

65. Troyon, M.; Huang, L. Correction factor for contact area in nanoindentation measurements. *J. Mater. Res.* 2011, *20*, 610–617.

66. Oliver, W.C.; Pharr, G.M. Measurement of hardness and elastic modulus by instrumented indentation: Advances in understanding and refinements to methodology. *J. Mater. Res.* 2004, *19*, 3–20.

67. Oliver, W.C.; Pharr, G.M. An improved technique for determining hardness and elastic modulus using load and displacement sensing

indentation experiments. *J. Mater. Res.* 1992, *7*, 1564–1580.

68.  Hochstetter, G.; Jimenez, A.; Loubet, J.L. Strain-rate effects on hardness of glassy polymers in the nanoscale range. Comparison between quasi-static and continuous stiffness measurements. *J. Macromol. Sci. Part B* 2006, *38*, 681–692.

69.  Giro-Paloma, J.; Roa, J.J.; Díez-Pascual, A.M.; Rayón, E.; Flores, A.; Martínez, M.; Chimenos, J.M.; Fernández, A.I. Depth-sensing indentation applied to polymers: A comparison between standard methods of analysis in relation to the nature of the materials. *Eur. Polym. J.* 2013, *49*, 4047–4053.

70.  Fischer-Cripps, A.C. *Introduction to Contact Mechanics*, 2nd ed.; Springer: New York, NY, USA, 2007; p. 221.

71.  Ma, D. New relationship between Young's modulus and nonideally sharp indentation parameters. *J. Mater. Res.* 2004, *19*, 2144–2151.

72.  Al-Shannaq, R.; Farid, M.; Al-Muhtaseb, S.; Kurdi, J. Emulsion stability and cross-linking of PMMA microcapsules containing phase change materials. *Sol. Energy Mater. Sol. Cells* 2015, *132*, 311–318.

73.  Qiu, X.; Li, W.; Song, G.; Chu, X.; Tang, G. Fabrication and characterization of microencapsulated n-octadecane with different crosslinked methylmethacrylate-based polymer shells. *Sol. Energy Mater. Sol. Cells* 2012, *98*, 283–293.

74.  Zhang, X.; Fan, Y.; Tao, X.; Yick, K. Crystallization and prevention of supercooling of microencapsulated n-alkanes. *J. Colloid Interface Sci.* 2005, *281*, 299–306.

75.  Chaiyasat, P.; Ogino, Y.; Suzuki, T.; Okubo, M. Influence of water domain formed in hexadecane core inside cross-linked capsule particle on thermal properties for heat storage application. *Colloid Polym. Sci.* 2008, *286*, 753–759.

# Chapter 2

# PREPARATION, MECHANICAL AND THERMAL PROPERTIES OF CEMENT BOARD WITH EXPANDED PERLITE BASED COMPOSITE PHASE CHANGE MATERIAL FOR IMPROVING BUILDINGS THERMAL BEHAVIOR

Rongda Ye, Xiaoming Fang, Zhengguo Zhang, and Xuenong Gao

Key Laboratory of Enhanced Heat Transfer and Energy Conservation, the Ministry of Education, School of Chemistry and Chemical Engineering, South China University of Technology, Guangzhou 510640, China

## ABSTRACT

Here we demonstrate the mechanical properties, thermal conductivity, and thermal energy storage performance of construction elements made of cement and form-stable PCM-Rubitherm® RT 28 HC (RT28)/expanded perlite (EP) composite phase change materials (PCMs). The composite PCMs were prepared by adsorbing RT28 into the pores of EP, in which the mass fraction of RT28 should be limited to be no more than 40 wt %. The adsorbed RT28 is observed to be uniformly confined into the pores of EP. The phase change temperatures of the RT28/EP composite PCMs are very close to that of the pure RT28. The apparent density and compression strength of the composite cubes increase linearly with the mass fraction of RT28. Compared with the thermal conductivity of the boards composed of cement and EP, the thermal conductivities of the composite boards containing RT28 increase by 15%–35% with the mass fraction increasing of RT28. The cubic test rooms that consist of six boards were built to evaluate the thermal energy storage performance, it is found that the maximum temperature different between the outside surface of the top board with the indoor temperature using the composite boards is 13.3 °C higher than that of the boards containing no RT28. The thermal mass increase of the built environment due to the application of composite boards can contribute to improving the indoor thermal comfort and reducing the energy consumption in the buildings.

## INTRODUCTION

Buildings are one of the leading sectors in the energy consumption in the developed countries. Taking the European Union as an example, the buildings sector consumes around 40% of the total fossil energy and produces nearly 40% of the total $CO_2$ emissions. Most of it is due to the increase in the living standard and in occupants' comfort demands, mainly for heating and cooling [1]. Improving the energy efficiency of buildings has a significant benefit for energy-saving and emission-reduction on the earth. In particular, nowadays the trend in the commercial buildings is to decrease the wall thickness to reduce the materials consumed, the transport costs, and the construction time. The main disadvantage of these lightweight buildings is the low thermal mass, resulting in large temperature fluctuations indoors. A phase change material (PCM) can absorb or release a large quantity of latent heat when it changes phase from solid state to liquid state or vice versa, and has been commonly used in thermal energy storage systems [2,3,4,5]. The integration of a PCM in building ceilings, walls, and floors to store significant amounts of thermal energy can compensate for the small storage capacity of the lightweight buildings and thus decrease the frequency of internal air temperature swings, leading to an improvement in the human comfort and a reduction in the energy consumption for the buildings.

Since 1982, the incorporation of PCMs into building fabrics has been investigated as a potential technology for minimizing the energy consumptions in buildings [6,7,8]. The selection of PCMs is almost directed towards the use of organic materials in an effort to avoid some of the problems inherent in inorganic materials, such as the need for special container due to their corrosivity, tendency of super cooling, segregation, *etc.* Up to now, three general methods have been proposed to integrate organic PCMs into construction elements [9,10,11], which are the direction immersion of conventional wallboards into molten PCMs [12,13,14,15,16], the integration of microencapsulated PCMs with ordinary building materials [17,18,19,20,21], and the combination of building materials with a shape-stabilized PCM that is usually prepared by blending an organic PCM with a supporting material [22,23,24]. Although the first one is simple and low cost, the impregnated wallboards are inflammable owing to the diffusion of the liquid PCMs to the surfaces of the wallboards, especially after the PCMs experienced several heating-cooling cycles. The second one suffers from the complicated polymerization processes and the high costs related to the microencapsulation of PCMs. Compared with the two mentioned methods, the third one has the advantages of simple process, low cost, and a large variety of supporting matrices available, such as high density polyethylene, expanded graphite, bentonite, and so on [25,26,27,28,29,30,31,32].

Expanded Perlite (EP), a light-weight (unit volume weight: 0.05–0.30 g/mL), odorless, and heat expanded volcanic mineral, has been commonly used as the ultra-light-weight building material to improve the structure of buildings because of its excellent heat insulation capacity (thermal conductivity rate: 0.03–0.05 kcal/mh °C) [33], environmentally safety, and abundant availability. The properties of high porosity, large surface area, low sound transmission, excellent fire-resistance, and low moisture retention make EP to be a good and cheap supporting matrix for preparing form-stable composite PCMs. Karaipekli and Sari groups [34,35,36] prepared several kinds of EP-based composite PCMs including the capric–myristic acid (CA-MA)/EP, CA/EP, lauric acid (LA)/EP, paraffin/EP and fatty acid esters/EP composites, the thermal properties, thermal reliability and the thermal conductivity of the composite PCMs are determined. The maximum PCMs absorptions of EP are obtained without melted PCMs seepage from the composites, and therefore these mixtures are described as the form-stable composites. Jiao et al. [37] have prepared the binary eutectic of LA-stearic acid (SA)/EP composite PCMs by vacuum impregnation. The structure and properties of the composite PCMs are characterized. The results show that the binary eutectic of fatty acids has been composed with the porous skeleton EP completely in a physical method. Chung et al. [38] prepared and characterized thermal properties and thermal reliability of form-stable composite PCMs composed of n-octadecane, expanded vermiculite, and perlite for thermal energy storage. The results showed that the prepared composite PCMs showed good compatibility between n-octadecane and the expanded vermiculite and pearlite, the thermal conductivities of composites were reduced, the large latent heat capacity and original phase change temperatures were maintained. Zhang et al. [39] prepared LA-palmitic acid (PA)-SA/EP composite PCMs by vacuum impregnation method. The maximum mass ratio of LA-PA-SA retained in EP was found as 55 wt %. They found that the thermal conductivity of LA-PA-SA/EP was increased by 95% by adding 2 wt % expanded graphite, and the thermal energy storage/release rates were also increased. Sun et al. [40] prepared the form-stable PCMs by absorbing paraffin into EP method. Graphite as additive was added into the form-stable PCMs to improve thermal conductivity. The results showed that the thermal conductivity of the form-stable PCMs was increased as much as 192% by graphite with mass fraction of 5%. Ramakrishnan et al. [41] have developed a novel thermal energy storage composite by impregnating paraffin into hydrophobic coated expanded perlite (EPO) granules. They found that no paraffin leakage was observed for novel paraffin/EPO containing 50% by weight of paraffin in the composite. Microstructural and mechanical properties were studied for the compatibility of hydrophobic coated PCM composite in concrete. Peng et al. [42] prepared form-stable composite PCMs for use in wallboards absorbing SA and LA eutectic mixtures into the pores of EP. The microstructure, thermal properties and the thermal reliability of the composite PCMs were characterized. Their results indicated that the maximum SA-LA absorption of the EP was as high as 65 wt % without any melted SA-LA leakage. A gypsum-based building wallboard containing 6 wt % SA-LA/EP had a low density ($0.924$ g/cm$^3$), high mechanical strength ($2.19$ MPa), and remarkable heating preservation performance. Zhang et al.[43] prepared the CA-PA/EP composite PCM and fabricated the thermal-regulated gypsum boards by adding the prepared composite PCM. They found that the higher the composite PCM volume content, the smaller the thermal conductivity of the gypsum board. The bending strength and compressive strength reduced gradually with an increase of the volume fraction of the composite PCM. He et al. [44] prepared the form/stable CA-MA/EP composite PCMs, and paraffin was chosen to encapsulate the composite PCMs avoiding liquid leakage. They also prepared the temperature control mortar and studied the physical mechanical properties. As can be seen in the open literature above, different kinds of composite PCMs by absorbing organic substances into EP were widespreadly prepared, but EP-based composite phase change building materials are a few studied [41,42,43,44] until now.

Cement board that is produced by mixing up Portland cement, EP and water has wide applications in buildings such as creating a smooth surface to walls and thermal insulation structure due to the EP's higher porosity. In the current work, we focus on the incorporation of the RT28/EP composite PCM into ordinary cement to produce thermal energy storage cement board. A systematic experimental study to analyze the important effects of the thermo-mechanical properties of the produced samples, like apparent density and compression behaviors and thermal conductivity was performed. Moreover, the thermal energy storage performances of the cement boards containing different mass percentage of the RT28/EP composite PCM were evaluated. The goal of this study was to demonstrate the feasibility of using EP-based composite PCM in cement boards to increase their thermal inertia and to reduce the energy demand of the building.

# EXPERIMENTAL SECTION

## Preparation and Characterization of RT28/EP Composite Phase Change Materials (PCMs)

RT28, a kind of liquid saturated hydrocarbons, was purchased from Rubitherm GmbH corp. in Berlin, Germany. RT28 as phase change material can be applied in buildings due to its phase change temperature (about 28 °C) in the range of human comfort temperature (between 18 and 28 °C) [19]. EP was used as received. RT28/EP composite PCMs with different mass fractions of RT28 were prepared by adsorbing different amount of RT28 into the pores of EP at 60 °C for 1 h, respectively.

The microstructures of EP and the RT28/EP composite PCMs were observed using a scanning electron micro-scope (SEM, S-3700N, Hitachi, Kyoto, Japan), respectively. The thermal properties of RT28 and the RT28/EP composite PCMs were measured by using a differential scanning calorimeter (DSC, DSC2910, TA Instruments, New Castle, DE, USA) under $N_2$ atmosphere. The measuring temperature ranged from −10 to 60 °C. The temperature rise rate was 5 °C/min.

## Fabrication and Characterization of Cement Boards Containing RT28/EP Composites

Four kinds of slurries were prepared by mixing cement and water with the RT28/EP composite PCMs containing 10, 20, 30 and 40 wt % of RT28, respectively, in which all the mass ratios of the composites to cement to water are 1:1.5:2. The prepared slurries were formed into cubes by using a standard stainless steel mold (size: 70.7 mm × 70.7 mm × 70.7 mm) to measure their mechanical properties and into boards by using a home-made stainless steel mold (size: 100 mm × 100 mm × 10 mm) to measure their thermal conductivity and evaluate their thermal energy storage performance, respectively. Four kinds of the cubes and boards were obtained, in which the mass fractions of RT28 were calculated to be 4%, 8%, 12% and 16%, respectively. For comparison purpose, the EP powder was also mixed with cement and water at the same mass ratio to obtain the blank slurry followed by forming into the cubes and boards using the same fabrication process.

The mechanical properties of the cubes with different mass fractions of RT28 were evaluated by measuring their compressive strength and apparent density, respectively. After being kept under a moist atmosphere for about 24 h, all the fabricated cubes were maintained under water at 20 ± 1 °C for 7 days. Then, some of the cubes used for measuring the compressive strength were kept under a controlled condition (20 ± 2 °C in temperature and 60% ± 3% in humidity) until the tests were performed on a compression-testing machine (5000 A, Jinan Shijin Group Co., Ltd., Jinan, China). Other cubes used for measuring the apparent density were dried at 100 °C until their weight didn't change. The accurate dimensions of the cubes were measured using a vernier caliper with a precision of 0.02 mm, and their weights were measured using an analytical balance with a precision of 0.001 g, respectively. The apparent density (ρ) of the cubes was calculated by the following formula:

$\rho = m/V$, respectively, where $m$ represents the weights of the cubes (unit: g), and $V$ represents the volumes of the cubes (unit: cm$^3$). Conforming to Chinese Standard JGJ/T 70-2009 [45], six same samples were prepared for each compressive strength and apparent density measurements, all measurements were arithmetically averaged.

The thermal conductivity of the fabricated boards was measured at room temperature using a hot disk thermal constant analyzer (Hot Disk TPS2500, Hot Disk AB Company, Uppsala, Sweden). The measurement accuracy of the thermal conductivity was within ±3%. The arithmetically average value of thermal conductivity for three times measurements was chose as the experimental result for each sample.

## Thermal Energy Storage Performance Evaluation of Cement Boards Containing RT28/EP Composites

A sketch of the experimental apparatus for testing the thermal energy storage performance of the composite boards (size: 100 mm × 100 mm × 10 mm) is shown in Figure 1. A small test room (100 mm × 100 mm × 100 mm) that consists of 6 pieces of the boards was set up below a halogen tungsten lamp (500 W) at a distance of 35 cm, which was used as the light source to simulate the sun. The radiation intensity was about 60 mw/cm$^2$ measured by a radiometer (FZ-A type, Beijing Normal University) at the position. Two K-typed thermocouples linked to a data acquisition/switch unit (Agilent 34970A) were used to monitor the temperature variation of the test room. The thermocouples were calibrated before use, the accuracy was ±0.2 °C. One of them was placed at the outside surface of the top board, and the other one was placed in the center of the test room for recording the indoor temperature.

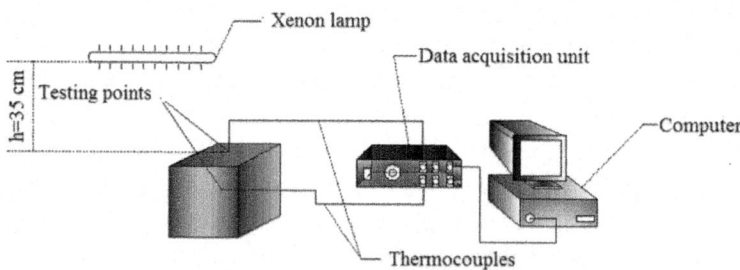

**Figure 1:** An experimental schematic drawing for testing energy conservation performance of composite boards.

When the lamp was switched on, the temperatures at the two spots of the test room started to be monitored. After 1.5 h, the lamp was switched off. The monitoring of the two temperatures continued until the test room cooled down to room temperature. The difference between the two temperatures varying with time was used to evaluate the performance of the boards with different mass fractions of RT28.

# RESULTS AND DISCUSSION

## Characterization of RT28/EP Composite PCMs

In order to obtain form-stable RT28/EP composite PCMs, the maximum absorption ratio of RT28 in the EP powder should be determined. Firstly, the prepared RT28/EP composite PCMs with 40, 45 and 50 wt % of RT28 were placed on three piece of filter paper, respectively, followed by putting them into an oven at 60 °C for 30 min; then, the three pieces of filter paper were taken out from the oven and cooled down to room temperature; finally, the composite PCMs were removed from the three pieces of filter paper, respectively. We checked these pieces of filter paper carefully to make sure if there were any traces of RT28 left in them due to the leakage of RT28 from the pores of EP during the heating. It is shown that there is no trace of RT28 left in the filter paper on which the RT28/EP composite PCM containing 40 wt % of RT28 has been placed. However, as the mass fraction of RT28 is increased to 45%, the trace of RT28 is observed, implying that the maximum adsorption ratio of RT28 in EP is around 40%. Therefore, the RT28/EP composite PCMs containing no more than 40 wt % of RT28 can be considered as the form-stable composite PCMs.

EP possesses a porous structure, which makes it possible to adsorb organic PCMs. Figure 2 shows the SEM images of the RT28/EP PCMs with 30 wt % (a) and 40 wt % (b) of RT28, respectively. We can see that the adsorbed RT28 is uniformly confined into the pores of EP, and the layer of RT28 becomes thicker as its mass fraction is increased from 30 to 40 wt %.

Figure 3 displays the DSC curves of EP and the RT28/EP composite PCMs with 10, 20, 30 and 40 wt % of RT28, respectively. The melting temperatures were measured to be 28.41, 28.43, 28.40 and 28.48 °C for the composite PCMs containing 10, 20, 30 and 40 wt % of RT28, respectively, which are very close to 28.55 °C of RT28. RT28 has a melting latent heat as high as 197.8 kJ/kg. The melting latent heat values of the RT28/EP composite PCMs with 10%, 20%, 30% and 40% of RT28 were measured to be 18.11, 39.11, 58.14 and 77.44 kJ/kg, respectively. Obviously, the measured latent heat values of the composite PCMs are almost equivalent to their calculated latent heat values based on the mass fractions of RT28. It is revealed that the combination of RT28 with EP doesn't change the phase change temperature of RT28, and the maximum latent heat value of the form-stable RT28/EP composite PCMs can reach as high as 77.44 kJ/kg.

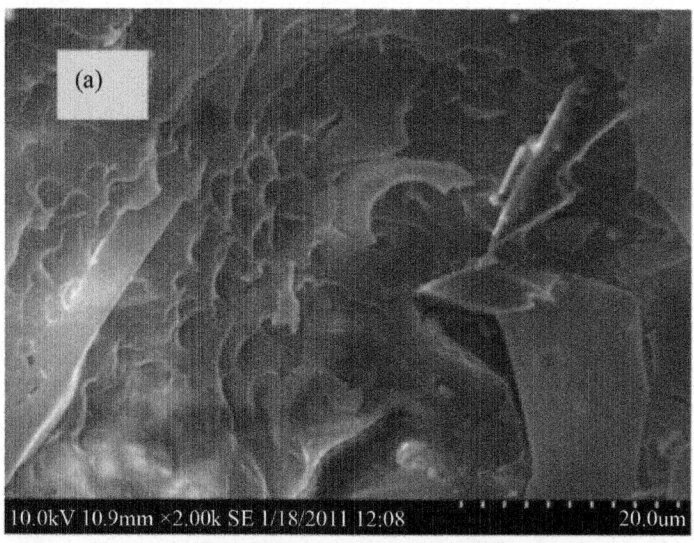

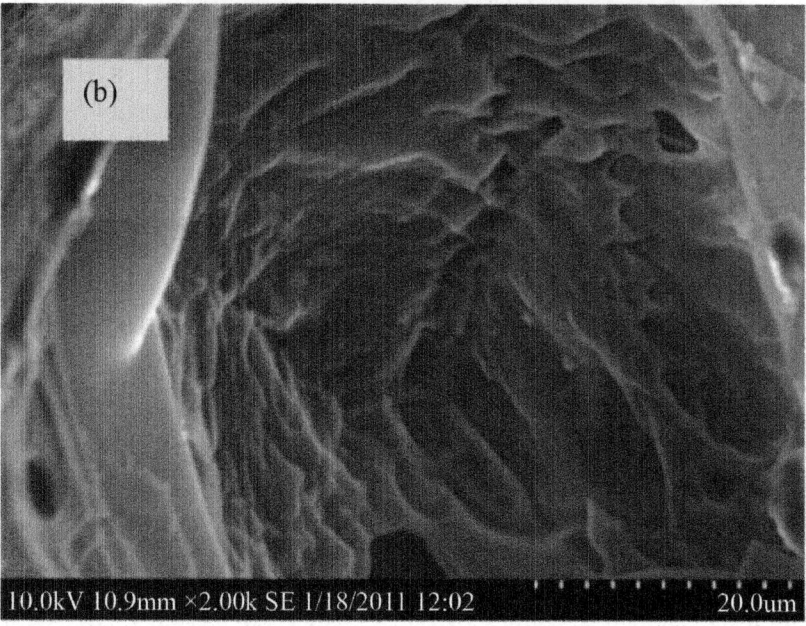

**Figure 2:** SEM images of the RT28/EP composite PCMs with 30 wt % (**a**) and 40 wt % (**b**) of RT28.

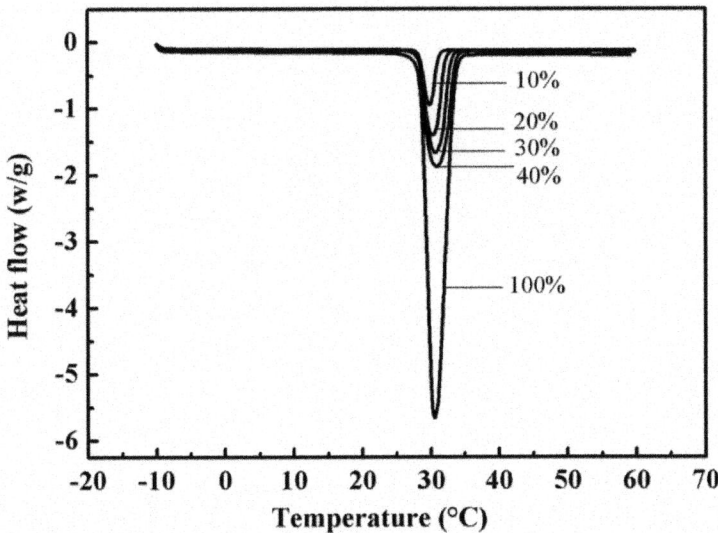

**Figure 3:** DSC curves of RT28 and the RT28/EP composite PCMs with 10, 20, 30 and 40 wt % of RT28, respectively.

## Properties of Cement Cubes and Boards Containing RT28/EP Composite PCMs

Figure 4 displays the apparent density of the four kinds of cubes made of cement and the RT28/EP composite PCMs with 10, 20, 30 and 40 wt % of RT28, respectively, together with that of the cube made of cement and EP. The photograph of the cubes is inserted in this figure. The mass fractions of RT28 in the five kinds of cubes were calculated to be 0%, 4%, 8%, 12% and 16%, respectively. The apparent density of the kind of cubes made of cement and EP was measured to 0.32 g/cm$^3$ on average. For the composite cubes made of cement and the RT28/EP composite PCMs, their apparent density was measured to be 0.35, 0.37, 0.39 and 0.43 g/cm$^3$ as the mass fraction of RT28 in the cubes was 4, 8, 12 and 16 wt %, respectively. The relationship between the apparent density ($\rho$) of the cubes and the mass fraction ($x$) of RT28 can be fitted into a linear equation: $\rho = 0.32 + 0.0065x$. It is revealed that the apparent density of the cubes increases linearly with the mass fraction of RT28. Since RT28 is adsorbed into the pores of EP, the volumes of the RT28/EP composites may not obviously increase with the mass fraction of RT28, but their weight definitely increases accordingly. Therefore, the apparent density of the composite cubes increases with the mass fraction of RT28. Note that the apparent density of the nine kinds of cubes is within the range from 0.3 to 0.8 g/cm$^3$, indicating that all the cubes are up to standard.

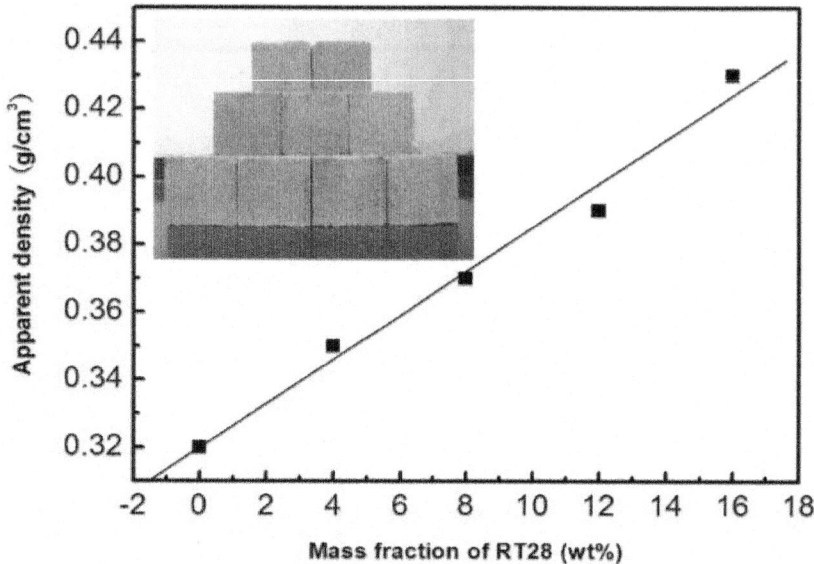

**Figure 4:** Relation between the apparent density of the cubes and the mass fraction of RT28 (the insert is the photograph of the cubes).

The porosity of the composite cubes made of cement and the RT28/EP composite PCMs were estimated from the bulk density, the matrix density assuming that the cement pores are filled of air which remains trapped once the cement has solidified [46]. For the composite cubes made of cement and the RT28/EP composite PCMs, their porosity was calculated to be 0.383, 0.351, 0.339 and 0.313 as the mass fraction of RT28 in the cubes was 4, 8, 12 and 16 wt %, respectively.

Figure 5 shows the compressive strength of the five kinds of cubes containing 0%, 4%, 8%, 12% and 16% of RT28, respectively. The compression strength of the cube containing no RT28 is measured to be 0.35 MPa on average. As the mass fraction of RT28 in the cubes is increased to 4%, 8%, 12% and 16%, their compression strength accordingly increases to 0.39, 0.47, 0.51 and 0.57 MPa, respectively. The relationship between the compression strength ($S$) and the mass fraction ($x$) of RT28 can be also fitted into a linear equation: $S = 0.346 + 0.014x$. It is indicated that the compression strength of the cubes linearly increases with the mass fraction of RT28. Note that all the compression strength of the nine kinds of cubes is more than 0.3 MPa, indicating that all the cubes are up to standard. Obviously, the addition of RT28 not only endows the building materials with the thermal energy storage performance but also improves their mechanical properties.

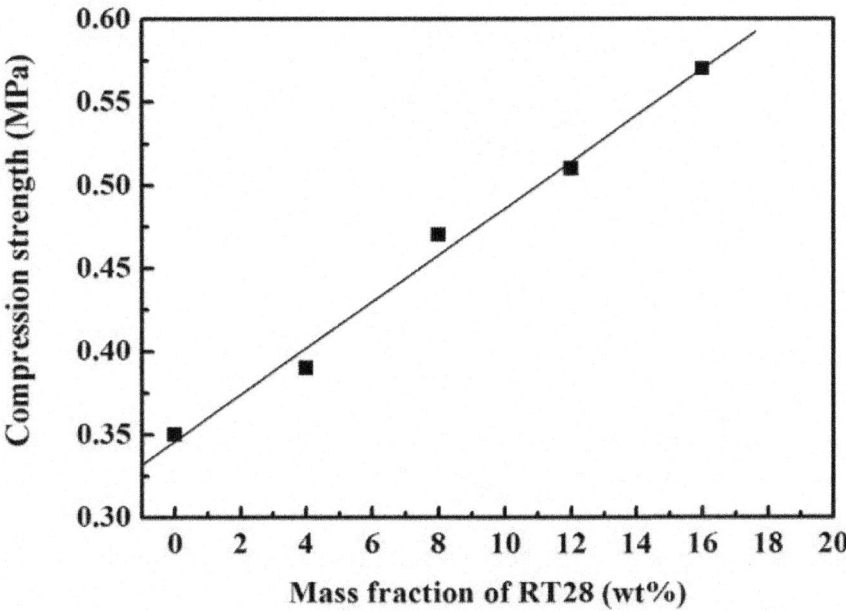

**Figure 5:** Relation between the compression strength of the cubes and the mass fraction of RT28.

Figure 6 shows the thermal conductivity of the five kinds of boards containing 0%, 4%, 8%, 12% and 16% of RT28, respectively. The photograph of the composite boards is inserted in this figure. The thermal conductivity of the board without RT28 is measured to be 0.11 W/m·K. As the mass fraction of RT28 in the boards is increased to 4%, 8%, 12% and 16%, their thermal conductivity increases to 0.126, 0.131, 0.137 and 0.149 W/m·K, respectively. The relationship between the thermal conductivity ($S$) and the mass fraction ($x$) of RT28 can be also fitted into a linear equation: $S = 0.113 + 0.00225x$. As the mass fraction of RT28 in the composite PCMs is increased, the pores of EP occupied by air are gradually filled with RT28. Since thermal conductivity of RT28 (0.276 W/m·K) is far larger than that of air (0.023 W/m·K), the thermal conductivity of the composite boards increases with the mass fraction of RT28.

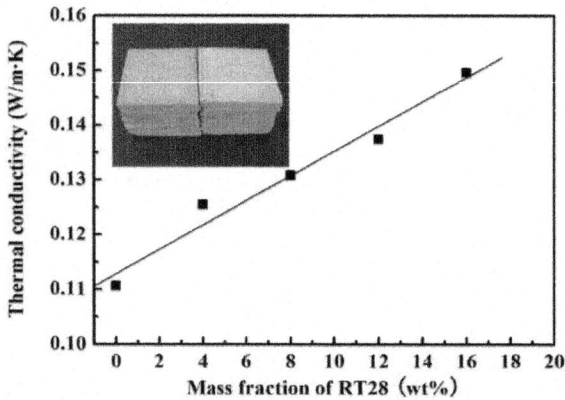

**Figure 6:** Relation between the thermal conductivity of the boards and the mass fraction of RT28 (the insert is the photograph of the boards).

## Thermal Energy Storage Performance of Cement Boards Containing RT28/EP Composites PCMs

To explore the thermal energy storage performance of the five kinds of boards containing 0, 4, 8, 12 and 16 wt % of RT28, respectively, the test room was set up in two ways. In one way, the test room was set up using six identical boards, in which the mass fractions of RT28 were 0, 4, 8, 12 and 16 wt %, respectively. In the other way, except the top board, the other five boards of the test room contained no RT28. The boards with 0, 4, 8, 12 and 16 wt % of RT28 were used as the top board, respectively.

Figure 7 shows the variations of the temperature differences with time for the test rooms set up using six identical boards containing 0, 4, 8, 12 and 16 wt % of RT28, respectively. The temperature difference is defined as the difference of temperature at outside surface of the top board from the indoor temperature of test room. For the test room with the boards containing no RT28, once the light source was switched on, the temperature at the outside surface of the top board increased quickly, resulting in a sharp rise in the temperature difference; 900 s after the light switching on, the temperature difference got stability around 13 °C; after the light was turned off, the temperature difference quickly dropped to zero. The temperature at the outside surface of the top board was always higher than the indoor temperature during the light switching on, and the maximum temperature difference was calculated to be 13.3 °C. It is revealed that the boards made of cement and EP exhibit the heat insulation capacity to prevent the rapid rise in the indoor temperature.

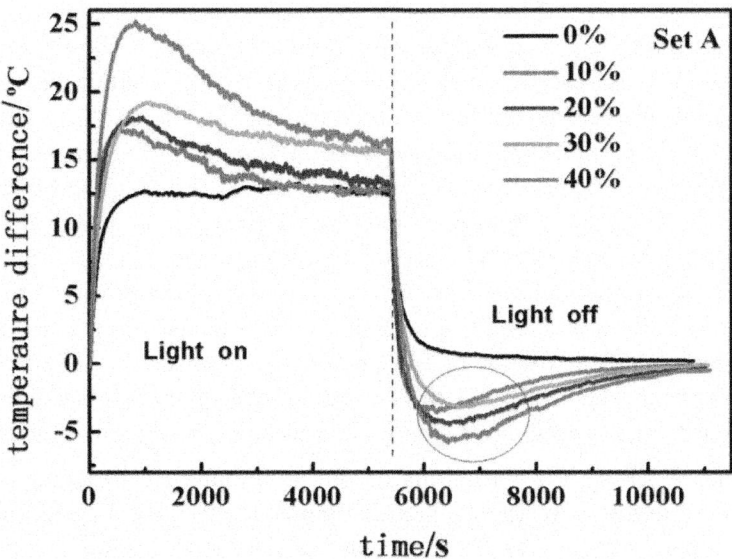

**Figure 7:** Variations of the temperature differences with time for the five test rooms that consist of six identical boards containing 0, 4, 8, 12 and 16 wt % of RT28, respectively (Set A).

It can be also seen from Figure 7 that, the temperature differences in the test rooms that consist of six composite boards are larger than that in the test room with the boards containing no RT28, and the temperature differences increase with the mass fraction of RT28 in the boards. The maximum temperature differences are 17.2, 18.2, 19.3 and 25.2 °C for the test rooms set up using the composite boards containing 4, 8, 12 and 16 wt % of RT28, respectively, which are higher than 13.3 °C of the test room with the boards containing no RT28. The bigger the temperature difference, the lower the room temperature. It is revealed that, the incorporation of RT28 into the boards can improve their heat insulation capacity, and the performance enhances with the increase in the mass fraction of RT28. During the illumination of the light source, RT28 integrated into the boards took the phase change from solid state to liquid state by absorbing some amount of heat coming from the light source. Since some heat has been absorbed by RT28, the indoor temperatures of the test rooms set up using the composite boards should be lower than that of the test room set up using the boards without RT28. The amount of the heat absorbed by RT28 is increased with the mass fraction of RT28 in the boards, resulting in the rise in the temperature difference accordingly. More significantly, as shown in Figure 7, during the cooling, for every test room that consisted of the composite boards, the temperature at the outside surface of the top board is even lower than the indoor temperature of the test room (as shown with a circle in Figure 7). It is revealed that the composite boards containing RT28 have the function of keeping the test room warm during the cooling, whereas the boards containing no RT28 don't possess the function. The function enhances with the increase in the mass fraction of RT28 in the boards. After the light was switched off, RT28 integrated into the boards took the phase change from liquid state to solid state, and simultaneously released some amount of latent heat. It is reasonable that the indoor temperatures of the test rooms that consist of the composite boards are higher than that of the test room with the boards containing no RT28, owing to the latent heat released by RT28. Since the amount of heat released by RT28 is increased with the mass fraction of RT28 in the boards, the function of keeping the rooms warm enhances accordingly.

Figure 8 shows the variations of the temperature differences with time for the test rooms with the top boards containing 0, 4, 8, 12 and 16 wt % of RT28, respectively. It can be seen that the curves in Figure 8 have similar trends to those in Figure 7. During the illumination, the maximum temperature differences for the test rooms with the top boards containing 4, 8, 12 and 16 wt % of RT28 are 15.2, 16.2, 18.2 and 18.7 °C, respectively, higher than 13.3 °C of the test room with the boards containing no RT28.

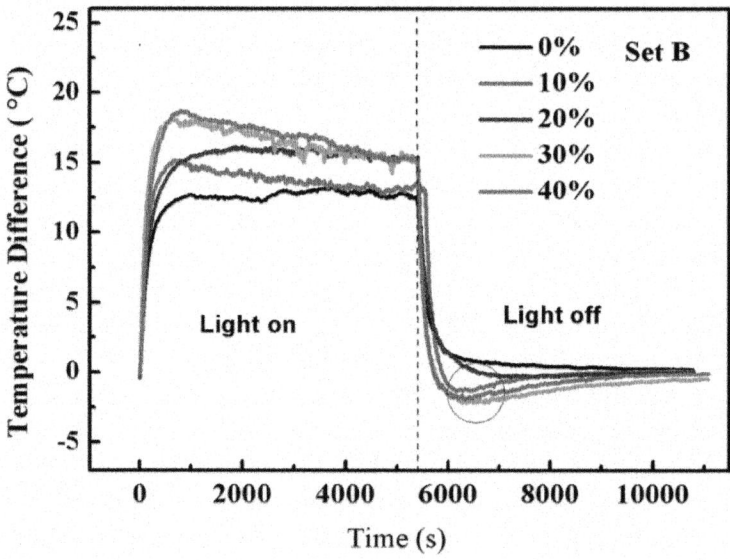

**Figure 8:** Variations of the temperature differences with time for the five test rooms composed of the top boards containing 0, 4, 8, 12 and 16 wt % of RT28, respectively (Set B).

It is suggested that the integration of RT28 into the top board of the test room can also exhibit the increase in thermal mass, and the performance gradually enhances with the increase of the mass fraction of RT28. During the cooling, the temperatures at the outside surfaces of the top boards are even lower than the indoor temperatures, indicating that the top boards containing RT28 also exhibit the function of keeping the rooms warm. It is reasonable that, at the same mass fraction of RT28, the test rooms composed of the six composite boards show superior heat insulation capacity to the test room with only the top board containing RT28. The more amount of the PCM is integrated into a building, the better energy conservation performance can be reached.

# CONCLUSIONS

Form-stable RT28/EP composite PCMs can be obtained by limiting the mass fraction of RT28 no more than 40 wt %. The adsorbed RT28 is observed to be uniformly confined into the pores of EP. The phase change temperatures of the RT28/EP composite PCMs with different mass fractions of RT28 are very close to that of the pure RT28. The latent heat values of the composite PCMs are almost equivalent to the calculated latent heat values based on the mass fractions of RT28. The apparent density and compression strength of the composite cubes increase linearly with the mass fraction of RT28. The thermal conductivity of the composite boards increases with the mass fraction of RT28. The composite boards containing RT28 not only exhibit the increase in thermal inertia during the illumination of a light source but also have the function of keeping the test rooms warm after the light source is turned off, which makes the RT/EP composite PCMs show great promise for improving the human comfort and reducing the energy consumption in the buildings.

# ACKNOWLEDGMENTS

This work is supported by the National Natural Science Foundation of China (No. U1407132), Guangdong Natural Science Foundation (No. 2014A030312009) and the Fundamental Research Funds for the Central Universities (2015ZP006).

# AUTHOR CONTRIBUTIONS

Zhengguo Zhang and Xiaoming Fang conceived and designed the experiments; Rongda Ye performed the experiments; Xuenong Gao analyzed the data; Zhengguo Zhang wrote the paper. All authors read and approved the manuscript.

# REFERENCES

1.  Soares, N.; Costa, J.J.; Gaspar, A.R.; Santos, P. Review of passive PCM latent heat thermal energy storage systems towards buildings' energy efficiency. *Energy Build.* 2013, *59*, 82–103.

2.  Sharma, A.; Tyagi, V.V.; Chen, C.R.; Buddhi, D. Review on thermal energy storage with phase change materials and applications. *Renew. Sust. Energy Rev.* 2009, *13*, 318–345.

3.  Shukla, A.; Buddhi, D.; Sawhney, R.L. Solar water heaters with phase change material thermal energy storage medium: A review. *Renew. Sust. Energy Rev.* 2009, *13*, 2119–2125.

4.  Agyenim, F.; Hewitt, N.; Eames, P.; Smyth, M. A review of materials, heat transfer and phase change problem formulation for latent heat thermal energy storage systems (LHTESS). *Renew. Sust. Energy Rev.* 2010, *14*, 615–628.

5.  Salunkhe, P.B.; Shembekar, P.S. A review on effect of phase change material encapsulation on the thermal performance of a system. *Renew. Sust. Energy Rev.* 2012, *16*, 5603–5616.

6.  Baetens, R.; Jelle, B.P.; Gustavsen, A. Phase change materials for building applications: A state-of-the-art review.*Energy Build.* 2010, *42*, 1361–1368.

7.  Cabezaa, L.F.; Castell, A.; Barrenechea, C.; De Graciaa, A.; Fernández, A.I. Materials used as PCM in thermal energy storage in buildings: A review. *Renew. Sust. Energy Rev.* 2011, *15*, 1675–1695.

8.  Parameshwarana, R.; Kalaiselvamb, S.; Harikrishnanb, S.; Elayaperumala, A. Sustainable thermal energy storage technologies for buildings: A review. *Renew. Sust. Energy Rev.* 2012, *16*, 2394–2433.

9.  Kuznik, F.; David, D.; Johannes, K.; Roux, J.J. A review on phase change materials integrated in building walls. *Renew. Sust. Energy Rev.* 2011, *15*, 379–391.

10. Sá, A.V.; Azenha, M.; de Sousa, H.; Samagaio, A. Thermal enhancement of plastering mortars with phase change materials: Experimental and numerical approach. *Energy Build.* 2012, *49*, 16–27.

11. Borreguero, A.M.; Luz, S.M.; Valverde, J.L.; Carmona, M.; Rodríguez, J.F. Thermal testing and numerical simulation of gypsum wallboards incorporated with different PCMs content. *Appl. Energy* 2011, *88*, 930–937.

12. Hawes, D.W.; Feldman, D. Absorption of phase change materials in concrete. *Sol. Energy Mater. Sol. Cells* 1992, *27*, 91–101.

13. Hadjieva, M.; Stoykov, R.; Filipova, T. Composite salt-hydrate concrete system for building energy storage. *Renew. Energy* 2000, *19*, 111–115.

14. Lee, T.; Hawes, D.W.; Banu, D.; Feldman, D. Control aspects of latent heat storage and recovery in concrete. *Sol. Energy Mater. Sol. Cells* 2000, *62*, 217–237.

15. Athienitis, A.K.; Liu, C.; Hawes, D.; Banu, D.; Feldman, D. Investigation of the thermal performance of a passive solar test-room with wall latent heat storage. *Build. Environ.* 1997, *32*, 405–410.

16. Rudd, A.F. Phase-change material wallboard for distributed thermal storage in buildings. *ASHRAE Trans.* 1993, *99*, 339–346.

17. Schossig, P.; Henning, H.M.; Gschwander, S.; Haussmann, T. Microencapsulated phase-change materials integrated into construction materials. *Sol. Energy Mater. Sol. Cells* 2005, *89*, 297–306.

18. Cabeza, L.F.; Castellón, C.; Nogués, M.; Medrano, M.; Leppers, R.; Zubillaga, O. Use of microencapsulated PCM in concrete walls for energy savings. *Energy Build.* 2007, *39*, 113–119.

19. Tyagi, V.V.; Kaushik, S.C.; Tyagi, S.K.; Akiyama, T. Development of phase change materials based microencapsulated technology for buildings: A review. *Renew. Sust. Energy Rev.* 2011, *15*, 1373–1391.

20. Arce, P.; Castellón, C.; Castell, A.; Cabeza, L.F. Use of microencapsulated PCM in buildings and the effect of adding awnings. *Energy Build.* 2012, *44*, 88–93.

21. Hunger, M.; Entrop, A.G.; Mandilaras, I.; Brouwers, H.J.H.; Founti, M. The behavior of self-compacting concrete containing micro-encapsulated phase change materials. *Cem. Concr. Compos.* 2009, *31*, 731–743.

22. Zhang, Y.P.; Lin, K.P.; Yang, R.; Di, H.F.; Jiang, Y. Preparation, thermal performance and application of shape-stabilized PCM in energy efficient buildings. *Energy Build.* 2006, *38*, 1262–1269.

23. Sari, A.; Biçer, A. Thermal energy storage properties and thermal reliability of some fatty acid esters/building material composites as novel form-stable PCMs. *Sol. Energy Mater. Sol. Cells* 2012, *101*, 114–122.

24. Kenisarin, M.M.; Kenisarina, K.M. Form-stable phase change materials for thermal energy storage. *Renew. Sust. Energy Rev.* 2012, *16*, 1999–2040.

25. Sari, A. Form-stable paraffin/high density polyethylene composites as solid-liquid phase change material for thermal energy storage: preparation and thermal properties. *Energy Convers. Manag.* 2004, *45*, 2033–2042.

26. Zhang, Z.G.; Fang, X.M. Study on paraffin/expanded graphite composite

phase change thermal energy storage material. *Energy Convers. Manag.* 2006, *47*, 303–310.

27.  Fang, X.M.; Zhang, Z.G.; Chen, Z.H. Study on preparation of montmorillonite-based composite phase change materials and their applications in thermal storage building materials. *Energy Convers. Manag.* 2008, *49*, 718–723.

28.  Karaipekl, A.; Sari, A. Capric-myristic acid/vermiculite composite as form-stable phase change material for thermal energy storage. *Sol. Energy* 2009, *83*, 323–332.

29.  Mei, D.D.; Zhang, B.; Liu, R.C.; Zhang, Y.T.; Liu, J.D. Preparation of capric acid/halloysite nanotube composite as form-stable phase change material for thermal energy storage. *Sol. Energy Mater. Sol. Cells* 2011, *95*, 2772–2777.

30.  Karaman, S.; Karaipekli, A.; Sar, A.; Biçer, A. Polyethylene glycol (PEG)/diatomite composite as a novel form-stable phase change material for thermal energy storage. *Sol. Energy Mater. Sol. Cells* 2011, *95*, 1647–1653.

31.  Chen, Z.; Shan, F.; Cao, L.; Fang, G.Y. Preparation and thermal properties of *n*-octadecane/molecular sieve composites as form-stable thermal energy storage materials for buildings. *Energy Build.* 2012, *49*, 423–428.

32.  Wang, Y.; Xia, T.D.; Feng, H.X.; Zhang, H. Stearic acid/polymethylmethacrylate composite as form-stable phase change materials for latent heat thermal energy storage. *Renew. Energy* 2011, *36*, 1814–1820.

33.  Won, J.P.; Kang, H.B.; Lee, S.J.; Lee, S.W.; Kang, J.W. Thermal characteristics of high-strength polymer–cement composites with lightweight aggregates and polypropylene fiber. *Constr. Build. Mater.* 2011, *25*, 3810–3819.

34.  Karaipekli, A.; Sari, A. Capric-myristic acid/expanded perlite composite as form-stable phase change material for latent heat thermal energy storage. *Renew. Energy* 2008, *33*, 2599–2605.

35.  Sari, A.; Karaipekli, A.; Alkan, C. Preparation, characterization and thermal properties of lauric acid/expanded perlite as novel form-stable composite phase change material. *Chem. Eng. J.* 2009, *155*, 899–904.

36.  Karaipekli, A.; Sari, A.; Kaygusuz, K. Thermal characteristics of paraffin/expanded perlite composite for latent heat thermal energy storage. *Energy Sources A* 2009, *31*, 814–823.

37.  Jiao, C.G.; Ji, B.H.; Fang, D. Preparation and properties of lauric acid-

stearic acid/expanded perlite composite as phase change materials for thermal energy storage. *Mater. Lett.* 2012, *67*, 352–354.

38. Chung, O.; Jeong, S.G.; Kim, S. Preparation of energy efficient paraffinic PCMs/expanded vermiculite and perlite composites for energy saving in buildings. *Sol. Energy Mater. Sol. Cells* 2015, *137*, 107–112.

39. Zhang, N.; Yuan, Y.P.; Yuan, Y.G.; Li, T.Y.; Cao, X.L. Lauric-palmitic-stearic acid/expanded perlite composite as form-stable phase change material: Preparation and thermal properties. *Energy Build.* 2014, *82*, 505–511.

40. Sun, D.; Wang, L.J.; Li, C.M. Preparation and thermal properties of paraffin/expanded perlite composite as form-stable phase change material. *Mater. Lett.* 2013, *108*, 247–249.

41. Ramakrishnan, S.; Sanjayan, J.; Wang, X.M.; Alam, M.; Wilson, J. A novel paraffin/expanded perlite composite phase change material for prevention of PCM leakage in cementitious composites. *Appl. Energy* 2015, *157*, 85–94.

42. Peng, K.; Zhang, J.Y.; Yang, H.M.; Ouyang, J. Acid-hybridized expanded perlite as a composite phase-change material in wallboards. *RSC Adv.* 2015, *81*, 66134–66140.

43. Zhang, J.W.; Guan, X.M.; Song, X.X.; Hou, H.H.; Yang, Z.P.; Zhu, J.P. Preparation and properties of gypsum based energy storage materials with capric acid–palmitic acid/expanded perlite composite PCM. *Energy Build.* 2015, *92*, 155–160.

44. He, Y.; Zhang, X.; Zhang, Y.J. Preparation technology of phase change perlite and performance research of phase change and temperature control mortar. *Energy Build.* 2014, *85*, 506–514.

45. *Standard for Test Method of Performance on Building Mortar*; Chinese Standard: JGJ/T 70-2009; China Architecture & Building Press: Beijing, China, 2009.

46. Borreguero, A.M.; Garrido, I.; Valverde, J.L.; Rodríguez, J.F.; Carmona, M. Development of smart gypsum composites by incorporating thermoregulating microcapsules. *Energy Build.* 2014, *76*, 631–639.

# Chapter 3

# SUSTAINABLE NON-METALLIC BUILDING MATERIALS

Paul Joseph, and Svetlana Tretsiakova-McNally
The Built Environment Research Institute, School of the Built Environment, University of Ulster, Newtownabbey, BT37 0QB, Northern Ireland, UK

## ABSTRACT

Buildings are the largest energy consumers and greenhouse gases emitters, both in the developed and developing countries. In continental Europe, the energy use in buildings alone is responsible for up to 50% of carbon dioxide emission. Urgent changes are, therefore, required relating to energy saving, emissions control, production and application of materials, use of renewable resources, and to recycling and reuse of building materials. In addition, the development of new eco-friendly building materials and practices is of prime importance owing to the growing environmental concerns. This review reflects the key tendencies in the sector of sustainable building materials of a non-metallic nature that have occurred over the past decade or so.

## INTRODUCTION

Almost twenty years ago, following the publication of Brundtland's report entitled "Our Common Future" [1] and the 1992 Rio "Earth Summit", the term sustainable development (SD) has gained great attention worldwide. This concept had been defined as the "development that meets the needs of the present without compromising the ability of future generations to meet their own needs" [1]. SD was given a further prominence in the context of the 2002 World Summit on Sustainable Development held in Johannesburg. It became clear that the environment can no longer exist separately from the development of other associated sectors. The idea of SD involves enhancing the quality of life, thus allowing people to live in a healthy environment, with improved social, economic and environmental conditions [2]. In recent years, climate change, air pollution, depletion of natural resources and biodiversity, waste generation, depletion and pollution of water resources and deterioration of the urban environment became global issues that require urgent actions to be taken. Climate change and global warming resulting from carbon dioxide ($CO_2$) and other greenhouse gases (GHG) emissions pose a huge threat to human welfare. To contain that threat, the world needs to cut the emissions by about 50% below current levels by 2050 [3,4].

A great quantity of $CO_2$ is emitted to the atmosphere through the whole life-cycle of a building. This includes the production of building materials (BM), the construction of a building itself, the exploitation, renovation, possible rehabilitation and its final demolition [5]. Construction industry is intensively growing and actively developing worldwide. Only in Europe, construction is the largest industrial employer, accounting for 7% of total employment and 28% of industrial employment at least in fifteen EU countries [6]. On the other hand, this sector is responsible for such environmental burdens as high energy and water consumption, solid waste generation, global GHG emissions, external and internal pollution and depletion of natural resources. Annually, building construction in the world consumes: 25% of the global wood harvest; 40% of stone, sand and gravel; and 16% of water. It generates 50% of global output of GHG and agents of acid rains. Furthermore, almost 3 billion tons of raw materials are turned into foundations, walls, pipes and panels [7]. Generally, energy is used for the extraction, transportation, processing of the BM and assembling of the structures. The $CO_2$ emissions are derived from the combustion of fossil fuels, the land-filling activities and the reactions taking place in the industrial processes [8]. Some authors estimate that almost 50% of total energy costs in the developed countries are a consequence of intensive construction and building practices [4,9,10]. To achieve the goals of SD in building construction, a combination of factors must be considered, such as energy saving methodologies and techniques (use of renewable energy resources), improved use of materials, their further reuse/recycle and emissions control.

A sustainable building is designed, built, renovated, operated or reused in an ecological and resource efficient manner [7]. It has a minimal negative impact on built and natural environment. Sustainable building should meet a number of certain objectives: resource and energy efficiency; $CO_2$ and GHG emissions reduction; pollution prevention; mitigation of noise; improved indoor air quality; harmonization with the environment [11]. "An ideal building would be inexpensive to build, last forever with modest maintenance, but return completely to the earth when abandoned" [12].

One of the most important components of a sustainable building is the material efficiency. Correct selection of BM can be performed by taking into account their complete life time ("from cradle to grave") and by choosing products with the minimal environmental impacts. For instance, González and Navarro estimated that the selection of BM with low environmental impacts can reduce $CO_2$ emissions by up to 30% [5]. The use of renewable and recycled sources is widely encouraged as the life-cycle of a building and its elements can be closed [13]. The other factors that greatly affect the selection of BM are their costs and social requirements such as thermal comfort, good mechanical properties (strength and durability), aesthetic characteristics and an ability to construct quickly. Ideally, the combination of all environmental, economic and social factors can give a clear description of a material, and thus helps in a decision making process regarding the selection of the materials suitable for buildings [14].

The primary aim of this review is to analyze recent advances in the area of non-metallic BM and to outline future prospects and challenges.

## SUSTAINABILITY ASPECTS OF BUILDING MATERIALS

To address the goals of SD the production of materials must use resources and energy from renewable sources instead of non-renewable ones. Sustainable BM are environmentally responsible because their impacts are considered over the complete life time of the products. Sustainable BM should pose no or very minimal environmental and human health risks [15]. They should also satisfy the following criteria: rational use of natural resources; energy efficiency; elimination or reduction of generated waste; low toxicity; water conservation; affordability. Sustainable BM can offer a set of specific benefits to the owner of a building such as reduced maintenance and replacement costs, energy conservation, improved occupant's health and productivity, lower costs associated with changing space configurations, and greater flexibility in design [7].

The major environmental burdens include embodied energy of BM and GHG emissions originated from each stage of their life-cycle. Embodied energy is defined as the amount of energy required to produce a material and supply it to the point of use. It is an important measure of the effectiveness of BM in the environmental terms [14]. Embodied energy consists of: energy required for the manufacturing of BM; energy associated with the transportation of raw materials to the factory and of the finished products to the consumer; energy needed for assembling various BM to form a building [16]. The results presented by Thormark indicate that embodied energy in traditional building can be reduced by approximately 10–15% through the proper selection of BM with low environmental impacts [17]. Although the values of embodied energy can vary widely (sometimes by as much as 100%, depending on the number of factors like country, manufacturing processes, recycling technologies, methodology of analysis, fuel costs and destination, *etc.*), they can be considered as reasonable indicators of an overall environmental impact of BM. Table 1 represents data for embodied energy and embodied carbon collected from UK and EU sources and worldwide averages of BM that were used in the UK (except for wood produced in Canada).

In order for decision makers to select materials suitable for sustainable construction, the assessment of their environmental burdens is necessary. For instance, in the US (California) BM can be considered as sustainable after each material has gone through a three-stage process involving preliminary research, evaluation and selection.

**Table 1:** Embodied energy and embodied carbon of common and alternative BM (taken from [15])

| Type of Material (1 ton) | Embodied Energy (MJ/ton) | Embodied Carbon (kg of $CO_2$/ton) |
|---|---|---|
| Limestone | 240 | 12 |
| Stone/gravel chipping | 300 | 16 |
| Rammed earth | 450 | 24 |
| Soil cement | 850 | 140 |
| Concrete, unreinforced (strength 20 MPa) | 990 | 134 |
| Concrete, steel reinforced | 1,810 | 222 |
| Soft-wood lumber (large dimensions, green)* | 1,971 | 101 |
| Soft-wood lumber (small dimensions, green)* | 2,226 | 132 |
| Portland cement, containing 64–73% of slag | 2,350 | 279 |
| Portland cement, containing 25–35% of fly ashes | 3,450 | 585 |
| Local granite | 5,900 | 317 |
| Engineering brick | 8,200 | 850 |
| Tile | 9,000 | 430 |
| Soft-wood lumber* (small dimensions, kiln dried) | 9,193 | 174 |
| Steel, bar and rod | 19,700 | 1,720 |
| Polypropylene, injection molding | 115,100 | 3,900 |

*Note: System boundary is cradle to average US site.

At first, the research is normally conducted by gathering technical information including material safety data sheets, data tests of indoor air quality, product warranties, source material characteristics, recycled content data, environmental statement, and durability information [7]. After that, further deeper research may be necessary on the issues like building codes, government regulations, building industry articles, *etc.* Secondly, BM must undergo evaluation, which involves confirmation of technical information gathered in the first stage. Life cycle assessment (LCA) is a well-established methodology for evaluating environmental impacts associated with the entire product life time [18]. Although rather simple in principle, LCA can be difficult and expensive [2]. The last step, the selection, often involves the use of tables and matrices to score the specific environmental criteria. The total score of each material will show the product with the best environmental characteristics [7]. However, it is also important for sustainable development to consider social and economic factors as well.

LCA of BM includes an analysis of the following aspects: resource base; embodied pollution; impact during use; final disposal. Certain sources of raw materials are becoming exhausted; therefore, the use of remaining stocks should be treated with great caution. Most rare materials used in construction can be substituted by others, more abundant or renewable. Environment (natural habitat, flora, fauna and landscape), human health and well-being can be severely damaged by the extraction or harvesting of raw materials and by the production and distribution of BM that make up the whole supply chain of the construction industry. There may also be negative effects for the local communities associated with noise, dust, local transport problems or general disruption. Some extraction processes are inherently rational in terms of resources used, whereas others are extremely inefficient, leading to a significant amount of waste. Attention to this aspect has led to a new trend in manufacturing BM from the waste products of various origins [19]. For instance, utilization of waste products had been successfully implemented in countries like Holland and Japan, where "construction industry practically lack raw materials" [20].

Pollution, caused by the processes taking place during the production of BM, has a huge negative environmental impact as well. Highly processed components must be avoided in the future in favor of the less processed ones, which can serve the same purpose. Reduction of the pollution caused by the combustion of fossil fuels and cutting down the costs of energy required for manufacturing of BM are also the main challenges for the producers of highly energy intensive products, like concrete, bricks, plastics and metals. For example, several British brick manufacturers partially use bio-gas as a fuel for the firing [19]. A wide range of traditional building products (wood treatments, foams, chipboards, vinyl flooring, paints and varnishes) contain compounds, which adversely affect the health of occupants of a building. Although some harmful substances are regulated by health and safety policies such as Control of Substances Hazardous to Health (COSHH), in reality great level of health risks comes from the unknown "cocktail" effect of the many chemicals that are present in the buildings. Further detailed research is required to address these concerns and produce information on health risks associated with possible substances combinations [19]. The use of BM sourced locally can help lessen the environmental burdens. This would considerably cut transportation costs and provide support of the local economies. For instance, a good choice of material suitable for sustainable construction can be timber, available from a local source, used in untreated form and designed for long life. It is preferable to employ the vernacular traditions and skills, often connected with a particular regional material that is acceptable to local planning authorities. Also, it is quite important to take into account an inherent durability and quality of BM and increase them as much as possible. In addition, materials and components should have a good recycling potential [19].

Glass *et al.* mentioned that the construction industry is quite conservative, and currently it is underperforming in addressing such issues as sustainability, low replacement rates, lack of innovation, inadequate level of skills and external factors (oil depletion, water pollution and globalization) [21]. According to recent European building regulations on energy efficiency and to the "Code for Sustainable Homes", new standards are established in order to produce "carbon neutral" buildings [21,22]. In 2003 European Commission released the integrated product policy, aiming to identify within the construction sector products with the best environmental performance [23]. This policy takes into consideration the whole life-cycle of the product. There are three main phases in this approach: environmental impact of products; environmental improvement of products; policy implications. Eco-design and the environmental product declarations (EPD) are employed to implement the integrated product policy. Eco-design is a set of techniques that can design a product with low negative environmental impacts during its complete life-cycle. EPD is a communication tool providing customers and international markets with relevant and verified information on the environmental performance of the products [24]. EPD is based on LCA and contain data associated with the acquisition of raw materials, chemical nature of BM, possible air-land, and water-pollutions and waste generation as well [2].

Researchers use different criteria to classify BM. For instance, Asif *et al.* categorize construction materials into six groups: concrete, metals, wood, stone, plastics and ceramics [9]. Classification conducted by Sun *et al.*, on the other hand, is based on materials environmental impact drivers [18]. By means of this method, 16 groups were identified for the families of materials such as glass and ceramics, ferrous metals, non-ferrous metals, paper, polymers and woods. This classification can be suitable at the early stages of product design and development. Calkins used sustainability criterion and defined the following groups of sustainable BM [15]:

- materials that reduce the use of resources;

- materials that minimize environmental impacts;
- materials that pose no or low human health risks;
- materials that assist with sustainable site design strategies;
- materials from companies with sustainable social, environmental and corporate policies.

The following sections of this review focus on the recent advances in the field of the most common (cement/concrete; wood; brick; stone; ceramics; glass; plastics) and alternative (bamboo; cob; adobe) BM that have non-metallic nature.

## CONCRETE AND CEMENT

Concrete as a construction material is widely used for building structural frames, ground-works, floors, roofs, and prefabricated elements [25]. Annually more than 10 billion tons of concrete are produced in the world [26]. Concrete is a durable material with excellent mechanical properties. It is adaptable to different climates, relatively fire resistant, widely available and affordable. Concrete can be molded almost into any shape and can be designed to satisfy almost any performance requirements [26]. It can be reinforced with either steel or fibers. Moreover, recycled materials can be incorporated into the concrete mix, thus reducing consumption of raw materials and disposal of waste products. The use of admixtures—materials added to concrete—becomes very popular as the final composite can have better durability and gains some specific unique properties [15]. Typical composition of concrete is shown in Table 2.

**Table 2:** Typical constituents of concrete (taken from [15])

| Constituent | Average Content, wt.% |
|---|---|
| Portland cement | 9.3 |
| Fly ash | 1.7 |
| Fine aggregate | 26 |
| Coarse aggregate | 41 |
| Water | 16 |
| Air | 6 |

In spite of the advantages mentioned above, concrete unfortunately has an enormous negative impact on the environment. It is estimated that cement and concrete industry generates up to 7% of global anthropogenic $CO_2$ emissions, and it is set to increase dramatically in the coming decades as the Earth's population grows [15]. Apart from the emissions related to the combustion of fossil fuels, there is a release of $CO_2$ associated with unavoidable de-carbonation of limestone (raw material) [28]. Concrete manufacturing is responsible for generating not only carbon dioxide but also other air pollutants like carbon monoxide (CO), sulfur oxides ($SO_x$), nitrogen oxides [(NO)$_x$], hydrogen chloride (HCl), volatile hydrocarbons and particulate matter. Production of concrete causes depletion of non-renewable mineral and water resources required in extremely large quantities. World concrete industry uses 10 billion (in short scale billion) tons of rock and sand, and 1 billion ton of water annually. Although Portland cement composes about 10% of concrete mix (see Table 2), its production accounts for 92% of the total energy demand [15]. Finally, demolition and disposal of concrete structures pose another significant environmental threat [26]. Concrete is estimated to account for up to 70% by weight of construction and demolition waste. At the present moment, concrete industry must take urgent actions in order to reduce the emissions of $CO_2$ and other air pollutants; to reduce the use of energy; to cut down the use of natural resources (including water); and to minimize the amount of waste generated. The environmental impacts of concrete/cement materials are largely discussed in detail elsewhere [15,25,26,28,29,30].

The Cement Sustainability Initiative (CSI) is a serious international effort set by ten leading cement companies to reduce the environmental and human health damage caused by cement manufacturing. The group of eighteen cement producers, accounting for 40% of global cement production, is organized under the World Business Council for Sustainable Development. The purpose of this initiative is to implement the main principles of SD and to identify actions needed to achieve SD in cement industry. CSI prepares guidelines and protocols for addressing such issues as energy and $CO_2$ management, fuel and material use, employee health and safety, reduction of emissions, impacts on land and local communities, and communication [15,28].

## Main Strategies Dealing with Challenges of Modern Cement and Concrete Industries

Improvement of durability, mechanical properties and service life of concrete

One of the effective ways to deal with negative environmental impact of concrete is to reduce the total volume of this material needed for a certain construction process by enhancing its performance. It is important to consider the overall quality of BM, which strongly depends on durability and associated mechanical properties and the life time. Habert and Roussel estimated that in France, the reduction of the concrete volume required for a particular building, by increasing the mechanical strength of the concrete, could lead to the reduction of $CO_2$ emission by approximately 30% [27]. In a report [31] it is claimed that many exposed exterior concrete structures are only in place for half of the designated service time. Premature failure can result in a great amount of resources needed for structures to be fixed, replaced or demolished before the end of their original life time, and therefore causing an extra negative impact. Designing of smaller and thinner concrete sections can also reduce the total amount of materials and energy resources required to produce concrete. However, this implies that the material should have a significant level of strength. There are several solutions for this problem.

The first is the development and application of high performance concrete (HPC). HPC is a type of concrete that has a low water to cement or water to binder ratio, properties of which are improved by the use of super-plasticizers. HPC has a higher level of compressive strength (40–50 MPa) compared to a traditional concrete (15–25 MPa) [30]. It is more economical as the designed structures can be smaller or thinner. It also has a low porosity that makes HPC more resistant to low temperatures and chemical exposure [15].

Secondly, by using self-compacting concrete (SCC), which is defined as "concrete, which without any mechanical action is able to fill a given form without separation" [28]. In 2004 only 1% of European ready-mix construction was SCC. SCC has following economic, social and environmental advantages compared to a traditional concrete:

- less labor involved, thus reducing costs, increasing productivity and allowing to build faster;
- the absence of large voids and inhomogeneities inside SCC results in its improved mechanical characteristics, better performance and longer service life;
- SCC casting requires no additional electrical energy for vibration (as this stage is eliminated);
- low level of noise and the absence of problems normally associated with vibration at the plants and construction sites;
- SCC also has new aesthetic potentials and more complicated geometries can be designed.

SCC technology provides an opportunity to use fine fillers of ground limestone or by-products such as fly-ash, quarry dust *etc.* Ye *et al.* have investigated the behavior of self-compacting cement paste at elevated temperatures [32]. It was found that a dramatic loss of mass was observed in the samples of self-compacting cement paste with addition of limestone filler when temperature is higher than 700 °C. This implies that SCC made by this type of paste will probably have a bigger damage once exposed to the fire. This can be efficiently avoided when polypropylene fibers (*ca.* 0.5 kg/m$^3$) are added to the paste.

In general, the use of polymers in manufacturing more sustainable concrete is continuously growing. They can be used for the following applications: concrete crack injections; repair of mortars for concrete and stone; consolidation of masonry; admixtures; and pure polymer concrete building components. Concrete modified with polymers is a composite material consisting of two phases: the aggregate, which is discontinuously dispersed through the material, and the binder, which itself consists of cementitious and polymer phase [33]. The main issues here are physical and chemical incompatibility of polymers and concrete, mechanical malfunctioning and low durability of the finished composite. Depending on the volume fraction of the polymer, the material shifts from polymer cement concrete to polymer concrete. Polymer impregnated concrete is a special composite, in which polymers are combined with concrete. In this case, low-viscosity monomers are injected into the pores of the hardened concrete and polymerized later. The resultant polymers form a second matrix if the pores are interconnected throughout the material [33].

Thirdly, by applying adequate reinforcing techniques that will enhance the durability of concrete structures. There are two ways of concrete reinforcement: steel reinforcing and fiber reinforcing. The idea of reinforcement concludes in the prevention of cracks developing inside the concrete before or after it happened, and as a result will lead to an improved impermeability, strength, weather and impact-resistance of the material [15]. Application of the first method is less preferable as durability of the finished product can be affected by corrosion, and the production of steel has some serious environmental implications (high energy use, emission of hazardous air pollutants, *etc.*). The second method involves inclusion either synthetic (nylon, glass or polypropylene) or natural fibers (vegetable, hemp [34], flax [35], coir, eucalyptus pulp, residual sisal [36]) in to the concrete mix. In spite of the main drawbacks like low durability performance of concrete and incompatibility issues, the consumption of BM made of biological fiber reinforced cement is increasing rapidly, especially in the developing countries having access to significant sources of cellulose fibers [35]. Two articles [36,37] have appeared in the literature that represent overviews of Australian and Brazilian experiences, respectively, in natural fiber reinforced cement composites. The authors state that it is possible to develop a material with properties suitable for building purposes with adequate mix design and taking into account the mechanical properties of fibers [36]. The study conducted by American researchers on self-healing ("bleeding") concrete has shown that the repair of cracks and filling the voids occurs immediately through the internal release of chemical agents from the fibers or beads embedded into concrete matrix [38]. Fourthly, it is the development of new ultra-high performance cement composites, which have unique structural and aesthetic potential. These are compact reinforced composite (CRC) and Ductal®. CRC is a composite of special fiber reinforced concrete with extremely high compressive strength (150–400 MPa) and reinforcing bars arranged in a particular manner [28,39]. CRC has been used in structural application, mainly for the production of precast elements (balconies and staircases). Ductal® had been developed by three French companies, namely Lafarge, Bouygues and Rhodia [28]. It possesses improved rheological properties and a unique combination of attributes detailed in [40]. In comparison to a traditional concrete, Ductal®'s compressive strength is 6–8 times higher, the flexural strength is 10 times higher, the durability is from 10 to 100 times better. Moreover, Ductal® can deform under excessive loads without rupture and has excellent surface aspects. From an environmental point of view, Ductal® technology requires only 65% of raw materials, 51% of the primary energy and 47% of the overall $CO_2$ emissions of the traditional concrete [28].

Finally, the use of nanomaterials might be very powerful in order to achieve sustainability objectives. Nanoscience of cements is a relatively new discipline with a huge potential to manipulate the nanostructure of calcium silicate hydrate [41]. Because the full environmental and human impacts of nanoparticles are unknown, they might pose some risks by inhalation or skin absorption during their manufacture, use and disposal [15]. Concrete reinforcement with nanofibers, including carbon nanotubes, has a potential to improve strength of concrete significantly, possibly eliminating the need of the reinforcement with steel. Moreover, nanocoatings containing titanium dioxide ($TiO_2$) can make self-cleaning buildings in the future, reducing the amount of harmful cleansers used currently. Molecules of $TiO_2$ have photo-catalytic properties [42]. They release an electric charge when absorbing sunlight that forms reactive radicals, which oxidize the nearby organic (and some inorganic) substances when they exposed to ultraviolet and/or sun rays [15]. The acidic products obtained in this process are washed away by rain or neutralized by alkaline calcium carbonate contained in the concrete. It is reported that nanoparticles of $TiO_2$ can even reduce air pollution by removing nitrogen oxides [43]. Tests showed that road surfaces with incorporated nano-$TiO_2$ reduce concentrations of nitrogen oxides by up to 60%. The use of nanoparticles of Portland cement, silica ($SiO_2$), titanium dioxide ($TiO_2$), and iron oxide ($Fe_2O_3$) can significantly improve compressive and flexural strength of concrete [15]. In addition, nanosensors can be integrated into concrete with the aim to collect performance data such as stress, corrosion of steel, pH levels, moisture, temperature, density shrinkage, *etc.* [44].

Reduction of the cement content in the concrete mix by increasing the application of supplementary cementitious materials (SCM)

Reduction of cement use in a concrete mix is most easily achieved through the substitution of Portland cement with other pozzolanic or hydraulic materials [15]. Depending on physical characteristics (grading curve or size), chemical composition and properties of SCM, they can perform either a function of ordinary filler (*i.e.*, they would fill the porosity of the material and thus increase elasticity modulus and improve its mechanical strength) or work as a binding agent (*i.e.*, they would react with water or with clinker hydration products and form stable hydrates). The most common SCM include fly ash (by-product from coal fired power plants), ground granulated blast furnace slag (GGBFS: by-product of steel industry), and silica fume (by-product of semi-conductor industry) [15,26,28,45,46]. Following the increasing popularity of SCM, Meyer in a recent review discusses properties, optimum levels of cement replacement, benefits and disadvantages in the application of each type of substituents [26]. Other SCM that can be used for cement replacement belong to a family of natural pozzolans such as calcined clay, calcined shale and metakaolin [15].

The utilization of industrial waste products as SCM definitely has a positive environmental impact because otherwise they would be land-filled. Moreover, they improve durability and mechanical properties of concrete, reduce thermal stress and cracking. However, in some cases longer set and curing times are required. Damtoft *et al.* stated that in reality the reduction of $CO_2$ emission is limited when SCM used for the Portland cement production. This is primarily due to the low content of calcium oxide (CaO) in the majority of SCM (except GGBFS). In practice, level of limestone replacement by GGBFS constitutes only 10%. When reductions in fuel consumption are taken into consideration as well, the total reductions of $CO_2$ emission theoretically do not exceed more than 25%. Also, the availability of GGBFS in the near future is most likely to decrease as existing steel plants are due to be replaced by more efficient electric arc furnaces. Fly ashes (class C) enriched with CaO can also be used to replace limestone in clinker production; however, the availability of this product is limited as well. Therefore, 100% utilization of current world sources of blast furnace slag and class C fly ashes would result in $CO_2$ emission reductions only by 10% [28]. Habert and Roussel evaluated that in France $CO_2$ emissions can be cut by 15% by increasing level of clinker substitution on mineral additives, both industrial by-products and natural pozzolans [27]. The authors also stressed that this could be achieved in a medium term perspective, *i.e.*, by 2020. However, in the long time perspective, *i.e.*, by 2050, the authors recommend considering other options, e.g., developing new types of clinker such as sulfoaluminate or belite activated sulfoaluminate clinker that has low

$CO_2$ emission. Increase in the use of recycled materials in place of natural non-renewable resources

Since aggregate constitutes the largest volume fraction of the concrete mix, the substitution of natural aggregates with recycled products can result in reducing the consumption of raw materials in manufacturing process, in reducing the exploitation of quarries, and can thus result in the minimization of the land areas for disposal. The products that can replace fine and coarse aggregates are: recycled concrete aggregate (RCA); crushed blast furnace slag; sand; brick; glass; granulated plastics; waste fiberglass; mineralized wood shavings; *etc.* This strategy gains a great importance and has been discussed by many authors [26,47,48,49,50,51,52]. For instance, Sani *et al.* studied leaching and mechanical behavior of concrete containing RCA [47]. It was indicated that the use of recycled aggregate as a 100% replacement of natural aggregate causes an increase in total porosity of concrete, although the leaching rate of sodium, potassium and calcium ions becomes lower as it is directly related to the percentage of macro/meso-pores. Mechanical strength in the presence of RCA drops by approximately 40% compared to a traditional concrete. However, this loss can be contained by adding fly ash. In spite of these negative effects, the authors still suggest that application of RCA would be acceptable as more environmentally sustainable. Compressive strength of RCA can be increased to adequate values of traditional concrete (30–35 MPa) by adding to the mixture SCM (fly ash or silica fume) with the aid of an acrylic-based super-plasticizer and at the same time by decreasing the water/cement ratio [53]. RCA also causes a reduction of elasticity modulus (by 35%), an increase in creep and shrinkage deformations, as well as a higher permeability of concrete, which decreases its durability [26]. A variety of contaminants (soil, plaster, wood, gypsum, asphalt, and rubber) found in recycled concrete can also be an important issue as their presence degrades the durability and strength. Another limitation of using RCA is a larger water consumption compared to an ordinary concrete. Nevertheless, RCA is quite acceptable for many applications. For instance, the successful case is the renewal of Denver's Stapleton Airport in the US, where 6.5 million tons of concrete had been recycled or reused [54].

Post-consumer glass bottles and post-industrial float glass cullet are offered as suitable aggregates for concrete [26,28,55,56,57,58]. Recycled glass has zero water absorption, high hardness, good abrasion resistance, excellent durability and chemical resistance. All these characteristics can improve the overall performance of concrete and impart color and aesthetic properties to it. The only technical problem here is an alkali silica reaction that can occur with coarse glass aggregate (less with fine aggregate). This reaction leads to the formation of a gel, which swells in the presence of water, causing cracks and damage in concrete [15]. Bignozzi *et al.* investigated a new application of matt waste, derived from the purification of cullet by separated collection, as a filler and as a partial Portland cement replacement (up to 50%) for newly blended cement [55]. When it is used as a filler, the resulting composite material showed higher compressive strength (up to 23%) and lower water absorption than self-compacting concrete. These significant improvements of mechanical properties are due to the good pozzolanic activity of the glass. Matt waste of amounts up to 25% was shown also to be very effective in new cement formulations. Kralj proposed a method of recycling of lightweight concrete with aggregate containing expanded glass [56]. Although the values of compressive strength, density and thermal conductivity for a new product are similar to the ones for lightweight concrete containing only aggregate of expanded glass, this technology is necessary for a production of cheaper and more environmentally friendly material. Guerra *et al.* studied an effect of recycled porcelain materials on the mechanical properties of concrete [59]. The substitution of natural aggregate with ceramic debris from sanitary ware waste does not improve significantly the mechanical properties of the new material compared to an ordinary concrete. However, it provides a good opportunity for the recycling of construction industry residues.

There are a number of publications related to the replacement of natural aggregate with the wastes from wood processing activity [60,61,62,63,64,65]. The research conducted by Becchio *et al.* focuses on the possibility of the mineralized wood concrete production by incorporating wooden waste, which was pre-treated with silica fume [65]. The inclusion of wood aggregate into concrete leads to a decrease of material density and final material becomes lighter. It also improves thermal insulation, although mechanical properties of the composite drop.

There are reports that describe preparation of rubberized concrete composites by replacing fine (up to 10%) and coarse (up to 20%) natural aggregate with waste tire rubber [66,67]. The most common ways of using recycled tires in cement concrete composite are shredding, chipping or grounding the rubber to the particles with sizes ranging from 450 mm to 75 μm [26]. The main drawbacks of this method are a significant decrease of the compressive and tensile strength as well as a reduction of stiffness of the composite product with the increasing amount of rubber in the mix. This can also lead to the earlier developing of cracks and the overall failure of concrete matrix. In order to improve mechanical behavior of concrete, Bignozzi and Sandrolini suggested using a self-compacting technology that helps binding rubber phase with cement matrix [66]. On the other hand, owing to the presence of rubber particles, the concrete can gain extra ductility and energy absorption [26]. Other potential advantages of rubberized concrete are good sound absorption capacity as well as excellent thermal properties. However, the incompatible Young's moduli of rubber and concrete often lead to inadequate mechanical properties of materials.

Recycled waste plastic is not generally available widely [15]. A major obstacle is the poor adhesion of plastic particles with cement matrix, which can also considerably reduce mechanical performance of concrete [26]. This problem can be solved by combining 10–15% of waste plastics with other materials like fly ash, thus leading to the production of lightweight structures and blocks that increase the deformation characteristics of concrete without failure [15].

## WOOD-BASED BUILDING MATERIALS

Wood is one of the most common and oldest forms of BM. It is easy to work with, structurally strong construction material suitable for numerous applications, e.g., framing, flooring, roofing and lining. There are different varieties and sources of timber. Sun *et al.* classified 82 types of wood into four groups on the basis of magnitude of their environmental impacts (eco-indicators) [18].

Up until recently such negative factors as deforestation, destruction of natural habitats, acidic rains, high rate of wood consumption, extensive use of toxic preservatives have resulted in wood being viewed as un-sustainable material. A representative consumer survey conducted in Germany has aimed to explore the image of timber as a construction material in general and timber framed houses in particular [68]. The study found that although timber has a positive association with such values as well-being, aesthetics and eco-friendliness, prejudice regarding high combustibility, low durability and extensive maintenance still persists in the minds of consumers. These barriers constitute a real challenge that producers of timber houses would have to face by optimizing processes and products.

An increased use of wood in construction is a rather controversial topic in recent literature. On one hand, forests purify the air and sequester carbon, even after being harvested and processed into lumber products. Furthermore, trees need mainly solar energy to grow and manufacture of wooden materials requires fewer amounts of fossil fuels, and emits less GHG over their life-cycle than other common BM. On the other hand, some wood harvesting practices and techniques have caused the global problems such as clearing large expanses of forests; loss of biological diversity; water and soil pollution due to the liberal use of fertilizers and pesticides; generation of waste that was land-filled [15]. In addition, some wood finishes can release volatile organic compounds, negatively affecting air quality and human health in general. Wood can be considered as a renewable material, and have a huge potential to be sustainable in the future, given the strategy of sustainable forest management and harvesting practices monitored by forest certification programs are in place. Forests, as a source of wood, play a vital role in the Earth's carbon cycle. Photosynthesis, which occurs in forests, provides an efficient mechanism for the removal of carbon dioxide from the atmosphere and the release of oxygen back to it. This process is the most productive in the new forests where rapid tree grow takes place [15]. Forests, wood and wood products store carbon until its eventual release through burning, bacterial or fungal decay, or consumption by insects [69]. If trees are replanted, carbon sinks would be added to the carbon cycle [70]. In contrast, deforestation leads to an imbalance of carbon flows by the removal of trees that can sequester carbon [15]. It is estimated that that 17.3% of carbon dioxide emissions, caused by humans, are related to deforestation, biomass decay, *etc.* [4]. A sustainable balance can be achieved if the annual harvest level is equal to, or below, the annual forest growth increment [4,71], and when intensive forest management regime is employed [72,73]. Consortium for Research on Renewable Industrial Materials (CORRIM) found that growing wood on the shorter rotations as opposed to longer intervals between harvesting can sequester more total carbon over time [70]. An accumulation of carbon in wood products, by an increase of their consumption or by using long-lived products, can positively benefit the environment, but only in the short or medium term [74]. Some authors believe that, in the long term, a greater use of wood in buildings at the expense of energy-intensive BM and substitution effect of avoiding fossil fuel emissions are more important than carbon stored in wood [69].

It is also important to consider how wood is treated at the end of its life cycle [15]. In the EU, land-filling both combustible and organic waste has been banned [75]. Land-filling is still a common practice in North America and in the developing countries. Residues, resulting from the harvesting of forests and the manufacture of wood products, should ideally be completely utilized to replace fossil fuels [71]. Reclaiming and reusing of wood are also widely encouraged. While untreated wood can be recycled into other products (such as mulch or compost), the treated wood would pose a more significant problem for disposal as it may contain harmful compounds [15]. Some amount of carbon from land-filled wood will return to the soil, but another fraction may decompose into methane, which has a much higher global warming potential than $CO_2$. Although part of the methane gas can be recovered and used as bio-fuel, the rest of it will be emitted to the atmosphere [74].

The use of timber in construction gains more and more support, especially in the regions with vast forest resources, because it can reduce both the energy demands of the buildings and the concentration of GHG in the atmosphere. Generally, this can be achieved by making use of wood instead of either fossil fuels (fuel or direct substitution) or non-wood materials, such as steel, aluminum and concrete (material or indirect substitution). There is a large potential to increase wood substitution in Europe. Gustavsson *et al.* point at a rather low level of timber use in Western and Central Europe, excluding Scandinavia [76].

In general, there are a large number of recent publications regarding the substitution between timber products and other BM [8,69,70,71,72,73,74,76, 77,78,79,80,81,82,83,84,85]. For instance, in the case study by Borjesson and Gustavsson on the multi-storey building in the south of Sweden, the primary energy use and the emissions of $CO_2$ and methane have been calculated and compared for two design options (either wooden or concrete frame) from life-cycle and forest land-use perspectives [77]. They evaluated that the primary energy input was about 60–80% higher when concrete frames were used. The authors suggested that the net GHG balance is strongly affected by the method, in which wood is being utilized after the demolition of the building. The net GHG emissions estimated to be clearly positive if all of the demolition wood is land-filled and slightly positive if all wood from this building is used instead of fossil fuel. GHG emissions can be improved and even may be negative if demolition wood is re-used. The comparison of timber and concrete design options of the same building was performed by Lenzen and Treloar by employing an Australian environmentally extended input-output framework in a tiered hybrid LCA, and in structural path analysis instead of process analysis [78]. Although the authors of this study reported that values of the energy use are twice as large as in similar study conducted by Borjesson and Gustavsson, the fundamental result, that the concrete-framed building causes higher level of emissions and uses more energy, has been confirmed.

Cole has provided a detailed examination of the energy and GHG emissions associated with on-site construction of a selection of alternative wood, steel and concrete structural assemblies [85]. Significant differences between the amount of energy and GHG emissions were observed for the construction with these materials, indicating that the use of concrete typically involves an order of magnitude higher quantities.

A great majority of scientists are convinced that using wood products in construction can result in lower fossil energy demands and significant cuts of GHG emissions compared to non-renewable alternatives such as steel and concrete [74]. For instance, Buchanan *et al.* found that a 17% increase in wood content of buildings in New Zealand could lead to a 20% decrease in fossil fuel consumption and to a 20% reduction of atmospheric carbon emissions from the manufacture of all BM [69]. This would account for a reduction of about 1.5% of New Zealand's total emissions. The reduction in emissions is mainly associated with using wood instead of brick and aluminum, and to a lesser extent steel and concrete.

A study conducted by Upton *et al.* focused on the energy requirements and GHG emissions associated with the use of wood-based BM in residential construction in the US [81]. The authors compared houses with similar heating and cooling regimes but using wood-based and non-wood-based construction materials. The differences were estimated over a period of 100 years. The results indicate that houses built with wood-based BM require 10–15% less total energy for non-heating/cooling purposes and their net GHG emissions are 20–50% lower than thermally equivalent houses employing steel-or concrete-based BM.

Salazar and Meil discussed the prospects for carbon-neutral housing by greater wood use on the example of single-family residence [83]. This article compared energy and carbon balance of two residential houses: a typical wood-framed home using more traditional materials (brick cladding, vinyl windows, asphalt shingles, and fiberglass insulation) and a wood-intensive house with maximized timber use throughout (cedar shingles, wood windows, and cellulose insulation). The wood-intensive home's life-cycle consumed only 45% of the fossil fuels used in the typical house. Including land-fill methane emissions, the wood-intensive house produced 20 tons of $CO_2$ emission as opposed to 72 tons for typical house. It was estimated that the house with higher wood content can be energy efficient and carbon neutral for 35–68 years in Ottawa region. The authors showed that wood waste can be recovered and used to generate enough energy to completely offset the manufacturing emissions and even partially offsetting the heating or cooling energy demands for this house.

Calkins in a recent book recommended the following strategies for design and specification of sustainable timber [15]:

- use wood resources efficiently, which means: using lowest quality wood for applications; build smaller and durable structures; simplicity in design details; minimal preservative treatments; reduce wood waste; build for disassembly; use engineered wood products;
- use certified wood;
- use reclaimed wood;
- preferential use of natural or low-toxic wood finishes.

The use of residues from agricultural activities can improve conservation of timber stock. Van Dam *et al.* developed an efficient technology to produce boards with high strength and density by processing whole coconut husk without the addition of synthetic binders [86]. The board had excellent mechanical properties, which are similar or even better than commercial wood-based panels. The recycling of wood is encouraged as it effectively addresses the problems of waste management and lack of natural resources, especially in countries like Japan with limited land areas suitable for waste disposal. Obata *et al.* discussed an application of recycled medium density fiberboard to produce a base material for floor heating systems [87].

The industry of wood preservatives and finishes currently focuses on the manufacture of products with improved environmental performance [88]. The comprehensive comparison of various wood preservatives is represented by Calkins [15]. A recent development of micro manufacturing heat treatment with the aid of sodium silicate is at the preliminary stage of testing, but it has a great potential in wood preservation. The use of nanomaterials as wood coatings, preservatives, adhesives, sealants and impregnators has also very promising future [43]. For instance, Calkins mentioned the development of a preservative containing organic insecticide and fungicide embedded in 100 nanometers plastic beads [15]. The suspension of these beads in water had been passed through the wood under pressure. Owing to their nanosize, they were able to disperse completely within the wood fibers. As soon as a finished wood product is well protected from the effect of fungus and insects, it is suitable for application in the exterior structures. Unfortunately, the availability of some new products is limited at the moment.

# BRICK, STONE AND CERAMICS

Brick is one of the major BM in modern construction industry. It has a very good durability and long service life. Bricks are mainly used for the outer and inner walls construction. Primarily bricks are made of non-toxic natural materials like clay and shale. Furthermore, brick manufacturing has a good potential for utilization of the solid wastes, which can be incorporated into the brick and neutralized by firing at high temperatures. The main environmental concerns all over the world for the brick production process are high energy usage and GHG emissions. For instance, LCA conducted by researchers from Greece quantified environmental performance of brick production in that country [89]. It was shown that most of the emissions are directly associated with the burning of fossil fuels. Among other environmental indicators acidification had a highest value (56%), which is explained by the combustion of low-grade fuel with high sulfur content, producing large amounts of $SO_2$ and $(NO)_x$.

All these negative factors encourage researchers to develop new type of masonry materials with improved environmental profile [15,89,90,91]. One of them is unfired clay bricks [91]. The usage of unfired bricks, in place of conventional fired ones, can significantly reduce the energy use and also cut down $CO_2$ emissions. Unfired clay soil (in the form of sun-baked bricks, mortars or plaster) is classified as a traditional BM that was very popular in the past, especially in rural areas. The main disadvantage of using these products is susceptibility to water damage, which can be avoided by stabilizing the clay soil with the addition of small quantity of lime [92]. Although the durability of lime-stabilized soil remains quite low and further improvement is required [93]. The results of several studies [91,94,95] showed that increase in durability is occurred when GGBFS is added to lime-stabilized systems. Oti *et al.* described a new technology of production of unfired clay bricks containing blended binders: lime and GGBFS [91]. The use of only 1.5% of lime in the formulations makes possible to obtain clay masonry units with engineering standards acceptable for wall construction. The price of the final products was relatively low. Furthermore, unfired clay bricks demonstrated excellent environmental performance; their total energy input was estimated of 657 MJ/ton as opposed to 4,187 MJ/ton for the common fired bricks, while an equivalent output of $CO_2$ emission was 41 kg $CO_2$/ton compared to 202 kg $CO_2$/ ton for traditional bricks in mainstream construction. There are also reports in the literature regarding "smart" brick or masonry [96,97]. This approach consists in incorporating sensors that monitor environmental parameters such as force, stress, temperature, tilt and moisture.

Stone can be considered as a low impact BM, if quarried locally, minimally processed and used appropriately. There is a tendency of rehabilitating the use of dry stone for modern sustainable construction [98]. The environmental burdens and potential applications of natural stone and aggregate are extensively discussed by Calkins [15].

According to a LCA of BM conducted by Asif *et al.*, ceramic tiles are quite energy-intensive (32,240 MJ) that accounts for 15% of the total embodied energy in the house [9]. Nicoletti *et al.* presented a comparative LCA carried out for marble and ceramics used as flooring materials [99]. The analysis indicated that marble tiles have better environmental performance compared to ceramic ones. In case of ceramics, due to the composition of raw materials used for the glaze production, the emission of arsenic and lead containing compounds took place.

## GLASS AND PLASTICS

Windows are very important in sustainable construction of residential and commercial buildings. They are responsible for heat transfer, provision of daylight, ventilation, weather protection and acoustic insulation. Due to the high heat conductivity of the glass, an unwanted heat gain or loss takes place between building and surroundings. There are several ways to improve energy conservation and windows sustainability: use of low-emissivity (low-e) coatings; replacement of air with inert gases; adjustment of the gap between glass panes in double or triple glazing. Low-e coatings improve thermal characteristics of the windows by blocking the transmission of rays with wavelengths responsible for solar heat gain [100]. Lampert reported that low-e glass accounts for almost 40% of the insulated glass market in the US [101]. The optimization of the thickness of air layer in glazing cavity [102] or substitution of the air with gases having low thermal conductivity (argon or krypton) [103] can significantly reduce energy losses. A design, incorporating thermal breaks between inner and outer surfaces of frames, can also considerably improve thermal insulation. In addition to this, the material of a window frame, which normally has a higher U-value than the glazing component, must be considered. Generally, wooden frames provide better thermal insulation compared to aluminum or plastic ones. Moreover, the lower values of embodied energy make the timber frames more sustainable than aluminum, uPVC, steel and aluminum-clad timber types of frames [103].

The emergence of chromogenic technology gave a start to a range of new products, often called "smart" or switchable BM [101,104,105,106,107,108]. Electrically switchable chromogenic devices could either change their color or transmittance due to the action of an electric field [105]. These devices can be incorporated into glass or plastic materials. The main advantage of electrochromics (EC) is that the low electric field is required only during the switching operations. Common types of electrically powered technologies are EC, suspended particle devices (SPD), also known as an electrophoretic media, and phase dispersed liquid crystals (PDLC) [101,104,105]. Currently, the most popular products are EC-windows, which change their color upon exposure to ultraviolet and thus can control light, glare and heat entering a building [101,106]. Witter et al. reported about gasochromic windows that could change their transmittance characteristics [108]. In this case, when glass containing a layer of tungsten oxide ($WO_3$) covered with a very thin layer of platinum, is exposed to diluted hydrogen gas, it changes its color due to the reduction of $WO_3$. This process can be reversed by introducing diluted oxygen. The main advantages of these windows are high solar transmittance in the bleached state and simple layer configuration that does not require transparent conducting electrodes. The prototype of SPD windows in "on" and "off" positions is represented by Lampert [101,104]. An active layer of the glazing consists of needle-shaped dipole particles (less than 1 mm long) suspended in a polymer.

In "off" position the particles are randomly arranged, absorbing the light. When an electric field is applied, particles align and transmission is increased. The principle of the optical switching in all systems based on liquid crystals (LC) is the reorientation or twist of their molecules, which causes the change in materials transmittance. PDLC can be obtained either by embedding LC droplets into polymer matrix or when LC fills the voids of polymer network. In "off" position PDLCs are translucent due to the light scattering effect. When an electric field is applied ("on" position), the reorientation of LC directors changes refractive index of LC domains and devices become transparent. Composites that combine electro-optical and chromogenic response change their transmittance within milliseconds [105].

Numerous products used in construction are made of plastic, e.g., pipes and drainage systems, composite lumber, panels and fences, *etc.* Plastics impart such properties to the BM as water-and decay-resistance, durability, flexibility, relatively light weight, integrated color and low maintenance. They can incorporate a substantial amount of recycled products or can be recyclable themselves. Nevertheless, plastics can have negative effects on the environment including high consumption of fossil fuels required for their productions; release of toxic by-products like heavy metals and furans during their manufacture, use and disposal; generation of large amounts of waste. The most common plastics used in building construction are: high-density polyethylene; cross-linked polyethylene; polypropylene (PP); polyvinyl chloride; polystyrene; polyacrylonitrile; *etc.* Characteristics, associated risks and benefits of these materials are discussed by Calkins [15]. In recent literature there are accounts of two main tools that improve environmental impacts of plastic BM: reuse/recycle and development of new materials with better sustainable properties [15,109,110]. The interest in the field of reinforced plastic composites is rapidly growing [110,111,112,113]. The finished composites, which could contain a certain amount of different additives (aluminum, steel, glass, ceramics, nanoclays, natural or synthetic fibers), have better mechanical properties and higher potential for further recycling. Xu *et al.* presented a LCA study carried out in New Zealand for wood-fiber-reinforced PP composite and compared it to a traditional PP [110]. Composite pre-forms, containing natural fibers in the amount of 10%, 30% or 50% by weight, were produced by compression molding. The authors mentioned the following advantages of bio-fibers over the synthetic glass ones: low costs, low density, renewability, excellent chemical resistance, good strength and significant processing benefits. They also concluded that environmental performance of the composite improves due to its lower density than original PP. Some researchers express concerns about the structural integrity of composite materials [38,111,114]. This indicates that the cracks can develop inside the material upon loading. Further repair of these micro-cracks is difficult or sometimes even impossible. Therefore, the development of self-repairing composites is very important. Unlike a conventional repair, self-repair would occur with the aid of materials contained within damaged structure. The process begins as soon as damage has happened without affecting an overall performance of the structure [11,111]. In this case, the principles of biomimetics and biological self-healing are applied [11]. An ability to use this technique in fiber reinforced polymer composites has been demonstrated by several authors [111,114,115,116,117]. For instance, Pang and Bond have developed novel hollow fiber reinforced polymer composite [111]. The release and infiltration of UV fluorescent dye occur from fractured hollow fibers into the damaged parts of the composite. This method can also visually highlight the damage on its surface [111]. The researchers from the University of Illinois (USA) have developed a material with an encapsulated healing agent and a solid catalyst dispersed in the matrix of polymer. Once a crack is formed, these microcapsules rupture and release healing agent into the damaged area. Its subsequent exposure to the catalyst initiates polymerization resulting in the filling the crack. A good level of mechanical strength recovery is observed. The negative effects of moisture swelling and destruction of the composite are also significantly mitigated [116,117].

## ALTERNATIVE BUILDING MATERIALS

Due to the exhaustion of non-renewable resources in the near future, there will be a shift in construction towards BM with low embodied energy and that are preferentially available locally [118]. Although it must be acknowledged that in some cases transportation costs can be compensated if non-local materials with better overall performance can be found. Compared to the common BM (concrete, steel, wood and plastics), these materials have a range of beneficial properties such as low toxicity, durability, low level of GHG and other pollutants emissions, high recycling potential and minimal processing requirements. Many of them are biodegradable and do not produce hazardous by-products. Examples of these products include bio-based [15,19,119,120,121,122,123,124] and earthen [15,19,125,126,127,128] BM.

Bio-based BM are generally originated from renewable organic constituents of plants and animals. The resources for these products could be agricultural crops and residues, animal wastes, forest materials and post-consumer biological waste [15]. The examples represented in recent literature include bamboo, straw bales, fiber crops, agricultural residues and plant seed oils. Bamboo, as a sustainable alternative to traditional BM, has attracted attention of many researchers [119,120,121,122,123,124]. Bamboo, which is a member of giant grasses, is an abundant material in tropical regions of

Latin America, Asia and Africa. Thanks to its excellent mechanical properties, light weight, flexibility, high growing rate and relatively low costs, bamboo has many opportunities as sustainable BM, especially in the areas where it occurs naturally [120,129]. The use of bamboo is growing rapidly, particularly in the sector of house interior, for production of laminate flooring, panels, chipboards and fireboards [15]. van der Lugt *et al.* discussed the possibilities of using bamboo as a building material in Western Europe [119]. In this study the suitably of bamboo culms for construction of supporting structures was assessed from the environmental and economical points of view. It was shown that bamboo has a very low environmental impact (20 times better) compared to other more conventional BM. The authors also mentioned the problems associated with the application of bamboo, such as difficulties in joining techniques due to its hollow round form. This issue can be solved by laminating the material. The financial assessment of a bridge in the Amsterdam Woods showed that among other BM bamboo is the least expensive, even with included costs for its transportation from Costa Rica.

De Flander and Rovers have presented quantitative analysis of a laminated bamboo-framed house [121]. It was shown that bamboo has a great advantage in annual yield per forest area compared to a traditional wood. It was demonstrated that one laminated bamboo-framed house can be build from one hectare of bamboo forest. Calkins mentioned the other barriers reducing bamboo overall performance, leading to its short service life: developing cracks for minimally processed culms; slippery outer surface when its wet; susceptibility to the attack of insects, fungi and microbes; deterioration of durability upon the exposure to adverse weather [15]. Obviously, some time and efforts will be required to overcome these issues and make bamboo strong and competitive building material that meets the standards of modern construction.

Materials for earthen construction such as hydrated lime, clay, cob, adobe (mud bricks), compressed earth blocks and rammed earth have been known and used for many years all over the world. Currently, there is a growing interest in these BM as sustainable alternatives to traditional concrete, brick and wood. Many recent publications raise questions related to the soil characterization, manufacturing process and materials testing [130,131,132,133,134,135]. Earthen BM normally contain soil with some percentage of clay (less than 20%) and water.

Collet *et al.* demonstrated that pre-fabricated cob blocks can be used in modern construction [127]. It was shown that the thermal behavior of south facing 50 cm thick cob wall is about the same as the one for concrete block wall with 7–9 cm of insulation. An additional 5 cm of insulation for the cob wall makes it equal to the dense concrete block with the insulation layer of

15 cm. In another study, Kouakou and Morel examined an effect of clay as a natural binder on the mechanical properties of adobes [126]. Traditional adobe and pressed adobe blocks were studied. Once the adobes dried, they were subjected to a compression testing. The results showed that the mechanical strength depends on the manufacturing process and the content of water in the adobes. Pressed adobe blocks were more homogeneous than traditional adobes and had a higher compressive strength with the gain of approximately 50%.

Loss of strength when saturated with water, erosion due to the wind or driving rain, low dimensional stability are the main problems that have to be eliminated in order to provide successful future application of earthen BM. Nevertheless, it should be noted that the use of bamboo, straw bales, cob, adobe, rammed earth is growing in popularity. Case studies and modern examples of buildings are widely represented in publications [19,128].

## CONCLUSIONS

Nowadays, principles of sustainability have become mandatory in order to tackle global warming and the associated climate change. Governments of several countries have adequate policies in place with a view to controlling and improving the current state of construction industry. The major actions include minimization of energy consumption in the buildings, rational use of natural resources and stricter control of the emissions. All these measures should be systematically applied during the selection of materials suitable for sustainable buildings and construction activities. General issues on the selection BM are sourcing, performance, maintenance and cost.

The present review has analyzed the recent innovations, techniques, tools and strategies in the sector of non-metallic BM spanning over a decade. The main approaches could be summarized into the following:

- Use of renewable energy resources for extraction of raw materials, for manufacturing, processing, finishing and transportation of BM
- Use of materials originated from renewable sources
- Reduce the consumption of disproportional amount of natural resources
- Emphasis on BM available locally and affordable even for poor communities. Although, in some cases, when non-local materials produced on a larger scale than non-local, the transportation of them for long distances can be more beneficial.
- Rehabilitation and application of some vernacular building skills and techniques
- Elimination of energy, water or materials wastage by using manufacturing

processes with closed cycle

- Increase the use of waste or recycled products as raw materials or additives to design composite BM with improved environmental performance
- Increase the potential for reuse or recycle of BM or structures
- Increase durability, strength and total service life of traditional and alternative BM
- Design and use composite BM, combining materials with different properties, to achieve improved standards of performance
- Design non-polluting BM required very low maintenance and repair.

With the aim to achieve sustainable BM in the near future, one of the main strategies would be to improve the functionality and environmental performance of traditional materials through the use of more sustainable technologies, for example, utilizing nanotechnology. Producers of sustainable BM will also have to make them more affordable and harmless for human health and the environment.

## REFERENCES

1.  Report of the World Commission on Environment and Development: Our Common Future. Available online: http://www.un-documents.net/wced-ocf.htm (accessed on 30 September 2009).

2.  Ortiz, O.; Castells, F.; Sonnemann, G. Sustainability in the construction industry: A review of recent developments based on LCA. *Constr. Build. Mater.* 2009, *23*, 28–39.

3.  *Building a Low-Carbon Economy—The UK's Contribution to Tackling Climate Change*; Committee on Climate Change: London, UK, 2008. Available online: http://www.theccc.org.uk/pdf/TSO-ClimateChange. pdf (accessed on 30 September 2009).

4.  International Panel on Climate Change (IPCC). *Climate Change 2007: Mitigation of Climate Change*, IPCC Fourth Assessment Report (AR 4). Available online: http://www.ipcc.ch/publications_and_data/publications_ipcc_fourth_assessment_report_wg3_report_mitigation_of_climate_change.htm (accessed on 8 December 2009).

5.  González, M.J.; Navarro, J.G. Assessment of the decrease of $CO_2$ emissions in the construction field through the selection of materials: Practical case study of three houses of low environmental impact. *Build. Environ.* 2006, *41*, 902–909.

6.  European Commission Enterprise & Industry. Construction. *Overview,*

Available online: http://ec.europa.eu/enterprise/construction/index_
en.htm (accessed on 30 September 2009).

7.  Green Building Home Page. Available online: http://www.ciwmb.ca.gov/
    GreenBuilding/ (accessed on 1 October 2009).

8.  Gustavsson, L.; Sathre, R. Variability in energy and carbon dioxide
    balances of wood and concrete building materials.*Build. Environ.* 2006,
    *41*, 940–951.

9.  Asif, M.; Muneer, T.; Kelly, R. Life cycle assessment: A case study of a
    dwelling home in Scotland. *Build. Environ.* 2007,*42*, 1391–1394.

10. Dimoudi, A.; Tompa, C. Energy and environmental indicators related
    to construction of office buildings. *Resour. Conserv. Recycl.* 2008, *53*,
    86–95.

11. John, G.; Clements-Croome, D.; Jeronimidis, G. Sustainable building
    solutions: A review of lessons from natural world. *Build. Environ.* 2005,
    *40*, 319–328.

12. Bainbridge, D.A. Sustainable building as appropriate technology. In
    *Building without Borders: Sustainable Construction for the Global
    Village*; Kennedy, J., Ed.; New Society Publishers: Gabriola Island,
    Canada, 2004; pp. 55–84.

13. Chwieduk, D. Towards sustainable-energy buildings. *Appl. Energ.* 2003,
    *76*, 211–217.

14. Abeysundara, U.G.; Babel, S.; Gheewala, S. A matrix in life cycle
    perspective for selecting sustainable materials for buildings in Sri Lanka.
    *Build. Environ.* 2009, *44*, 997–1004.

15. Calkins, M. *Materials for Sustainable Sites: A Complete Guide to the
    Evaluation, Selection, and Use of Sustainable Construction Materials*;
    John Wiley & Sons: Hoboken, NJ, USA, 2009.

16. Venkatarama-Reddy, B.V.; Jagadish, K.S. Embodied energy of common
    and alternative building materials and technologies. *Energ. Bldg.* 2003,
    *35*, 129–137.

17. Thormark, C. The effect of material choice on the total energy need and
    recycling potential of a building. *Build. Environ.* 2006, *41*, 1019–1026.

18. Sun, M.; Rydh, C.J.; Kaebernick, H. Material grouping for simplified
    product life cycle assessment. *J. Sustain. Product Des.* 2003, *3*, 45–58.

19. Halliday, S. *Sustainable Construction*, 1st ed.; Butterworth-Heinemann:
    Oxford, UK, 2008.

20. Peris-Mora, E. Life cycle, sustainability and the transcendent quality of
    building materials. *Build. Environ.* 2007, *42*, 1329–1334.

21. Glass, J.; Dainty, A.R.J.; Gibb, A.G.F. New build: Materials, techniques, skills and innovation. *Energ. Policy* 2008, *36*, 4534–4538.

22. Communities and Local Government. *Code for Sustainable Homes;* 2008. Available online: http://www.communities.gov.uk/planningandbuilding/ buildingregulations/legislation/codesustainable/ (accessed on 1 October 2009).

23. European Commission. Environment. *Identifying Products with Greatest Potential for Environmental Improvement*, 2003. Available online: http:// ec.europa.eu/environment/ipp/identifying.htm (accessed on 2 October 2009).

24. Environmental Product Declaration (EPD). *The International EPD System*, Available online: http://www.environdec.com/pageId. asp?id=200 (accessed on 2 October 2009).

25. Pulselli, R.M.; Simoncini, E.; Ridolfi, R.; Bastianoni, S. Specific emergy of cement and concrete: An energy-based appraisal of building materials and their transport. *Ecol. Indic.* 2008, *8*, 647–656.

26. Meyer, C. The greening of the concrete industry. *Cem. Concr. Compos.* 2009, *31*, 601–605.

27. Habert, G.; Roussel, N. Study of two concrete mix-design strategies to reach carbon mitigation objectives. *Cem. Concr. Compos.* 2009, *31*, 397–402.

28. Damtoft, J.S.; Lukasik, J.; Herfort, D.; Sorrentio, D.; Gartner, E.M. Sustainable development and climate change initiatives. *Cem. Concr. Res.* 2008, *38*, 115–127.

29. Pade, C.; Guimaraes, M. The $CO_2$ uptake in a 100 year perspective. *Cem. Concr. Res.* 2007, *37*, 1348–1356.

30. Aïtcin, P.C. Cements of yesterday and today. Concrete of tomorrow. *Cem. Concr. Res.* 2000, *30*, 1349–1359.

31. Mehta, P.K. Greening of the concrete industry for the sustainable development. *Concr. Int.* 2002, *24*, 23–28.

32. Ye, G.; Liu, X.; de Schutter, G.; Taerwe, L.; Vandevelde, P. Phase distribution and microstructural changes of self-compacting cement paste at elevated temperature. *Cem. Concr. Res.* 2007, *37*, 978–987.

33. van Gemert, D.; Czarnecki, L.; Maultzsch, M.; Schorn, H.; Beeldens, A.; Lukowski, P.; Knapen, E. Cement concrete and concrete-polymer composites: Two merging worlds. A report from 11th ICPIC Congress in Berlin, 2004. *Cem. Concr. Compos.* 2005, *27*, 926–933.

34. de Bruijn, P.B.; Jeppsson, K.H.; Sandin, K.; Nilsson, C. Mechanical

properties of lime-hemp concrete containing shives and fibres. *Biosyst. Eng.* 2009, *103*, 474–479.

35. Fernandez, J.E. Flax fiber reinforced concrete—A natural fiber biocomposite for sustainable building materials. *High Perform. Struct. Mater.* 2002, *4*, 193–207.

36. Agopyan, V.; Savastano, H., Jr.; John, V.M.; Cincotto, M.A. Development of vegetable fibre-cement based materials in São Paulo, Brazil: An overview. *Cem. Concr. Compos.* 2005, *27*, 527–536.

37. Coutts, R.S.P. A review of Australian research into natural fibre cement composites. *Cem. Concr. Compos.* 2005, *27*, 518–526.

38. Dry, C.M. Three designs for the internal release of sealants, adhesives, and waterproof chemicals into concrete to reduce permeability. *Cem. Concr. Res.* 2000, *30*, 1969–1977.

39. Compact Reinforced Composite (CRC) Home Page. Available online: http://www.crc-tech.com/ (accessed on 2 October 2009).

40. Ductal® Home Page. Available online: http://www.ductal-lafarge.com/wps/portal/Ductal/ (accessed on 2 October 2009).

41. Beaudoin, J.J.; Raki, L.; Alizadeh, R. A $^{29}$Si MAS NMR study of modified C-S-H nanostructures. *Cem. Concr. Compos.* 2009, *31*, 585–590.

42. Chen, J.; Poon, C.S. Photocatalytic construction and building materials: From fundamentals to applications. *Build. Environ.* 2009, *44*, 1899–1906.

43. Green Technology Forum. *Nanotechnology for Green Building*, 2007. Available online: http://www.greentechforum.net/greenbuild/ (accessed on 2 October 2009).

44. Martinez, I.; Andrade, C. Examples of reinforcement corrosion monitoring by embedded sensors in concrete structures. *Cem. Concr. Compos.* 2009, *31*, 545–554.

45. Papadakis, V.G.; Tsimas, S. Greek supplementary cementing materials and their incorporation in concrete. *Cem. Concr. Compos.* 2005, *27*, 223–230.

46. Paya, J.; Monzo, J.; Borrachero, M.V.; Peris-Mora, E.; Amahjour, F. Mechanical treatment of fly ashes. Part IV. Strength development of ground fly ash-cement mortars cured at different temperatures. *Cem. Concr. Res.* 2000, *30*, 543–551.

47. Sani, D.; Moriconi, G.; Fava, G.; Corinaldesi, V. Leaching and mechanical behavior of concrete manufactured with recycled aggregates. *Waste Manage.* 2005, *25*, 177–182.

48. Evangelista, L.; de Brito, J. Mechanical behavior of concrete made with

fine recycled concrete aggregate. *Cem. Concr. Compos.* 2007, *29*, 397–401.

49.  Casuccio, M.; Torrijos, M.C.; Giaccio, G.; Zerbino, R. Failure mechanism of recycled aggregate concrete. *Constr. Build. Mater.* 2008, *22*, 1500–1506.

50.  Uchikawa, H. Approaches to ecologically benign system in cement and concrete industry. *J. Mater. Civ. Eng.* 2000, *12*, 320–329.

51.  Mymrin, V.; Correa, S.M. New construction material from concrete production and demolition wastes and lime production waste. *Constr. Build. Mater.* 2007, *21*, 578–582.

52.  Achtemichuk, S.; Hubbard, J.; Sluce, R.; Shehata, M.H. The utilization of recycled concrete aggregate to produce controlled low-strength materials without using Portland cement. *Cem. Concr. Compos.* 2009, *31*, 564–569.

53.  Corinaldesi, V.; Moriconi, G. Influence of mineral additions on the performance of 100% recycled aggregate concrete.*Constr. Build. Mater.* 2009, *23*, 2869–2876.

54.  Yelton, R. Concrete recycling takes off: The renewal of Denver's Stapleton Airport showcases concrete's place as a sustainable material. *Concr. Prod.* 2004, *22*, 28–31.

55.  Bignozzi, M.C.; Saccani, A.; Sandrolini, F. Matt waste from glass separated collection: An eco-sustainable addition for new building materials. *Waste Manage.* 2009, *29*, 329–334.

56.  Kralj, D. Experimental study of recycling lightweight concrete with aggregates containing expanded glass. *Process Saf. Eviron. Prot.* 2009, *87*, 267–273.

57.  Shao, Y.; Lefort, T.; Moras, S.; Rodriguez, D. Studies on concrete containing ground waste glass. *Cem. Concr. Res.* 2000,*30*, 91–100.

58.  Federico, L.M.; Chidiac, S.E. Waste glass as a supplementary cementitious material in concrete—Critical review of treatment methods. *Cem. Concr. Compos.* 2009, *31*, 606–610.

59.  Guerra, I.; Vivar, I.; Llamas, B.; Juan, A.; Moran, J. Eco-efficient concretes: The effects of using recycled ceramic material from sanitary installations on the mechanical properties of concrete. *Waste Manage.* 2009, *29*, 643–646.

60.  Al Rim, K.; Ledhem, A.; Douzane, O.; Dheilly, R.M.; Queneudec, M. Influence of the proportion of wood on the thermal and mechanical performances of clay-cement-wood composites. *Cem. Concr. Compos.* 1999, *21*, 269–276.

61. Bouguerra, A.; Sallee, H.; de Barquin, F.; Dheilly, R.M.; Queneudec, M. Isothermal moisture properties of wood-cementitious composites. *Cem. Concr. Res.* 1999, *29*, 339–347.

62. Bederina, M.; Marmoret, L.; Mezreb, K.; Khenfer, M.M.; Bali, A.; Queneudec, M. Effect of the addition of wood shavings on thermal conductivity of sand concretes: Experimental study and modeling. *Constr. Build. Mater.* 2007, *21*, 662–668.

63. Bederina, M.; Laidoudi, B.; Goullieux, A.; Khenfer, M.M.; Bali, A.; Queneudec, M. Effect of the treatment of wood shavings on the physic-mechanical characteristics of wood sand concretes. *Constr. Build. Mater.* 2009, *23*, 1311–1315.

64. Turgut, P. Cement composites with limestone dust and different grades of wood sawdust. *Build. Environ.* 2007, *42*, 3801–3807.

65. Becchio, C.; Corgnati, S.P.; Kindinis, A.; Pagliolico, S. Improving environmental sustainability of concrete products: Investigation on MWC thermal and mechanical properties. *Energ. Bldg.* 2009, *41*, 1127–1134.

66. Bignozzi, M.C.; Sandrolini, F. Tyre rubber waste recycling in self-compacting concrete. *Cem. Concr. Res.* 2006, *36*, 735–739.

67. Hernandez-Olivares, F.; Barluenga, G.; Bollati, M.; Witoszek, B. Static and dynamic behavior of recycled tyre rubber-filled concrete. *Cem. Concr. Res.* 2002, *32*, 1587–1596.

68. Gold, S.; Rubik, F. Consumer attitudes towards timber as a construction material and towards timber frame houses—Selected findings of a representative survey among the German population. *J. Cleaner Prod.* 2009, *17*, 303–309.

69. Buchanan, A.H.; Levine, S.B. Wood-based building materials and atmospheric carbon emissions. *Environ. Sci. Policy* 1999, *2*, 427–437.

70. Lippke, B.; Wilson, J.; Perez-Garcia, J.; Bowyer, J.; Meil, J. CORRIM: Life-cycle environmental performance of renewable building materials. *Forest Prod. J.* 2004, *54*, 8–19.

71. Gustavsson, L.; Pingoud, K.; Sathre, R. Carbon dioxide balance of wood substitution: Comparing concrete and wood-framed buildings. *Mitig. Adapt. Strat. Global Change* 2006, *11*, 667–691.

72. Eriksson, E.; Gillespie, A.R.; Gustavsson, L.; Langvall, O.; Olsson, M.; Sathre, R.; Stendahl, J. Integrated carbon analysis of forest management practices and wood substitution. *Can. J. For. Res.* 2007, *37*, 671–681.

73. Perez-Garcia, J.; Lippke, B.; Comnick, J.; Manriquez, C. An assessment of carbon pools, storage, and wood products market substitution using

life-cycle analysis results. *Wood Fiber Sci.* 2005, *37*, 140–148.

74. Sathre, R.; O'Connor, J. *A Synthesis of Research on Wood Products and Greenhouse Gas Impacts*; Technical Report No. TR-19. FPInnovations, Forintek Division: Vancouver, BC, Canada, 2008. Available online: http://www.forintek.ca/public/pdf/    Public_Information/technical_rpt/ TR19%20Complete%20Pub-web.pdf (accessed on 11 January 2010).

75. Commission of the European Communities. *On the National Strategies for the Reduction of Biodegradable Waste Going to Landfills Pursuant to Article 5(1) of Directive 1999/31/EC on Landfill of Waste*; Report from the Commission to the Council and European Parliament: Brussels, Belgium, 2005. Available online: http://eur-lex.europa.eu/LexUriServ/ LexUriServ.do?uri=COM:2005:0105:FIN:EN:PDF (accessed on 11 January 2010).

76. Gustavsson, L.; Madlener, R.; Hoen, H.F.; Jungmeier, G.; Karjalainen, T.; Klöhn, S.; Mahapatra, K.; Pohjola, J.; Solberg, B.; Spelter, H. The role of wood material for greenhouse gas mitigation. *Mitig. Adapt. Strat. Global Change* 2006, *11*, 1097–1127.

77. Börjesson, P.; Gustavsson, L. Greenhouse gas balances in building construction: Wood *versus* concrete from life-cycle and forest land-use perspectives. *Energ. Policy* 2000, *28*, 575–588.

78. Lenzen, M.; Treloar, G. Embodied energy in buildings: Wood *versus* concrete—Reply to Börjesson and Gustavsson.*Energ. Policy* 2002, *30*, 249–255.

79. Gustavsson, L.; Joelsson, A.; Sathre, R. Life cycle primary energy use and carbon emission of eight-storey wood-framed apartment building. *Energ. Bldg.* 2009. (in press).

80. Sathre, R.; Gustavsson, L. Using wood products to mitigate climate change: External costs and structural change.*Appl. Energ.* 2009, *86*, 251–257.

81. Upton, B.; Miner, R.; Spinney, M.; Heath, L.S. The greenhouse gas and energy impacts of using wood instead of alternatives in residential construction in the United States. *Biomass Bioenerg.* 2008, *32*, 1–10.

82. Petersen, A.K.; Solberg, B. Environmental and economic impacts of substitution between wood products and alternative materials: A review of micro-level analysis from Norway and Sweden. *Forest Policy Econ.* 2005, *7*, 249–259.

83. Salazar, J.; Meil, J. Prospects for carbon-neutral housing: The influence of greater wood use on the carbon footprint of single-family residence. *J. Cleaner Prod.* 2009, *17*, 1563–1571.

84. Petersen, A.K.; Solberg, B. Greenhouse gas emissions, life-cycle inventory and cost-efficiency of using laminated wood instead of steel construction. Case: Beams at Gardermoen airport. *Environ. Sci. Policy* 2002, *5*, 169–182.

85. Cole, R.J. Energy and greenhouse gas emissions associated with the construction of alternative structural systems.*Build. Environ.* 1999, *34*, 335–348.

86. van Dam, J.E.G.; van den Oever, M.J.A.; Keijsers, E.R.P. Production process for high density high performance binderless boards from whole coconut husk. *Ind. Crops Prod.* 2004, *20*, 97–101.

87. Obata, Y.; Takeuchi, K.; Soma, N.; Kanayama, K. Recycling of wood waste as sustainable industrial resources—Design of energy saving wood-based board for floor heating systems. *Energy* 2006, *31*, 2341–2349.

88. *Green Is the Colour*, 2009. Available online: http://www.timber-building. com/news/categoryfront.php/id/59/Spring.html (accessed on 3 June 2009).

89. Koroneos, C.; Dompros, A. Environmental assessment of brick production in Greece. *Build. Environ.* 2007, *42*, 2114–2123.

90. Roth, M. Sustained brick construction. *ZI, Ziegelindustrie International/ Brick and Tile Industry International* 2004, *5*, 50–52.

91. Oti, J.E.; Kinuthia, J.M.; Bai, J. Engineering properties of unfired clay masonry bricks. *Eng. Geol.* 2009, *107*, 130–139.

92. Mckinley, J.D.; Thomas, H.R.; Williams, K.P.; Reid, J.M. Chemical analysis of contaminated soil strengthened by the addition of lime. *Eng. Geol.* 2001, *60*, 181–192.

93. Okagbue, C.O.; Yakubu, J.A. Limestone ash waste as a substitute for lime in soil improvement for engineering construction. *Bull. Eng. Geol. Env.* 2000, *58*, 107–113.

94. Tasong, W.; Wild, S.; Tilley, R.J.D. Mechanisms by which ground granulated blastfurnace slag prevents sulphate attack of lime-stabilised kaolinite. *Cem. Concr. Res.* 1999, *29*, 975–982.

95. Rajasekaran, G. Sulphate attack and ettringite formation in the lime and cement stabilized marine clays. *Ocean Eng.*2005, *32*, 1133–1159.

96. Engel, J.M.; Zhao, L.; Fan, Z.; Chen, J.; Liu, C. Smart brick. *Masonry Constr. World Masonry* 2005, *18*, 39–41.

97. Bastianini, F.; Corradi, M.; Borri, A.; di Tomasso, A. Retrofit and monitoring of a historical building using "Smart" CFRP with embedded fibre optic Brillouin sensors. *Constr. Build. Mater.* 2005, *19*, 525–535.

98.  Villemus, B.; Morel, J.C.; Boutin, C. Experimental assessment of dry stone retaining wall stability on a rigid foundation. *Eng. Struct.* 2007, *29*, 2124–2132.

99.  Nicoletti, G.M.; Notarnicola, B.; Tassielli, G. Comparative Life Cycle Assessment of flooring materials: Ceramic *versus*marble tiles. *J. Cleaner Prod.* 2002, *10*, 283–296.

100. Robinson, P.D.; Hutchins, M.G. Advanced glazing technology for low energy buildings in the UK. *Renewable Energy*1994, *5*, 298–309.

101. Lampert, C.M. Large-area smart glass and integrated photovoltaics. *Sol. Energy Mater. Sol. Cells* 2003, *76*, 489–499.

102. Aydin, O. Determination of optimum air-layer thickness in double-pane windows. *Energ. Bldg.* 2000, *32*, 303–308.

103. Menzies, G.F.; Wherrett, J.R. Windows in workplace: Examining issues of environmental sustainability and occupant comfort in the selection of multi-glazed windows. *Energ. Bldg.* 2005, *37*, 623–630.

104. Lampert, C.M. Chromogenic smart materials. *Mater. Today* 2004, *7*, 28–35.

105. Cupelli, D.; Nicoletta, F.P.; Manfredi, S.; de Filpo, G.; Chidichimo, G. Electrically switchable chromogenic materials for external glazing. *Sol. Energy Mater. Sol. Cells* 2009, *93*, 329–333.

106. Sottile, G.M. 2004 Survey of United States architects on the subject of switchable glazings. *Mater. Sci. Eng. B* 2005, *119*, 240–245.

107. Smestad, G.P.; Lampert, C.M. Solar power 2006, San José, CA. *Sol. Energy Mater. Sol. Cells* 2007, *91*, 440–444.

108. Witter, V.; Datz, M.; Ell, J.; Georg, A.; Graf, W.; Walze, G. Gasochromic windows. *Sol. Energy Mater. Sol. Cells* 2004, *84*, 305–314.

109. Ross, S.; Evans, D. The environmental effect of reusing and recycling a plastic-based packaging system. *J. Cleaner Prod.*2003, *11*, 561–571.

110. Xu, X.; Jayaraman, K.; Morin, C.; Pecqueux, N. Life cycle assessment of wood-fibre-reinforced polypropylene composites. *J. Mater. Process Technol.* 2008, *198*, 168–177.

111. Pang, J.W.C.; Bond, I.P. A hollow fibre reinforced polymer composite encompassing self-healing and enhanced damage visibility. *Compos. Sci. Technol.* 2005, *65*, 1791–1799.

112. Corbière-Nicollier, T.; Laban, B.G.; Lundquist, L.; Leterrier, Y.; Månson, J.A.E.; Jolliet, O. Life cycle assessment of biofibres replacing glass fibres as reinforcement in plastics. *Resour. Conserv. Recycl.* 2001, *33*, 267–287.

113. Pervaiz, M.; Sain, M.M. Carbon storage potential in natural fiber

composites. *Resour. Conserv. Recycl.* 2003, *39*, 325–340.

114. Motuku, M.; Vaidya, U.K.; Janowski, G.M. Parametric studies on self-repairing approaches for resin infused composites subjected to low velocity impact. *Smart Mater. Struct.* 1999, *8*, 623–638.

115. Bleay, S.M.; Loader, C.B.; Hawyes, V.J.; Humberstone, L.; Curtis, P.T. A smart repair system for polymer matrix composites. *Compos. Part A* 2001, *32*, 1767–1776.

116. Kessler, M.R.; White, S.R. Self-activated healing of delamination damage in woven composites. *Compos. Part A* 2001, *32*, 683–699.

117. Kessler, M.R.; Sottos, N.R.; White, S.R. Self-healing structural composite materials. *Compos. Part A* 2003, *34*, 743–753.

118. Morel, J.C.; Mesbah, A.; Oggero, M.; Walker, P. Building houses with local materials: means to drastically reduce the environmental impact of construction. *Build. Environ.* 2001, *36*, 1119–1126.

119. van der Lugt, P.; van den Dobbelsteen, A.A.J.F.; Janssen, J.J.A. An environmental, economic and practical assessment of bamboo as a building material for supporting structure. *Constr. Build. Mater.* 2006, *20*, 648–656.

120. Utama, A.; Gheewala, S.H. Influence of material selection on energy demand in residential houses. *Mater. Des.* 2009, *30*, 2173–2180.

121. de Flander, K.; Rovers, R. One laminated bamboo-frame house per hectare per year. *Constr. Build. Mater.* 2009, *23*, 210–218.

122. Hoang, C.P.; Kinney, K.A.; Corsi, R.L. Ozone removal by green building materials. *Build. Environ.* 2009, *44*, 1627–1633.

123. Jayanetti, L.; Follet, P. Building with sustainable forest products. *Struct. Eng.* 2003, *81*, 14–17.

124. Paudel, S.K.; Lobovikov, M. Bamboo housing: Market potential for low-income groups. *J. Bamboo Rattan* 2003, *2*, 381–396.

125. Isik, B.; Tulbentci, T. Sustainable housing in island conditions using Alker-gypsum-stabilized earth: A case study from northern Cyprus. *Build. Environ.* 2008, *43*, 1426–1432.

126. Kouakou, C.H.; Morel, J.C. Strength and elasto-plastic properties of non-industrial building materials manufactured with clay as a natural binder. *Appl. Clay Sci.* 2009, *44*, 27–34.

127. Collet, F.; Serres, L.; Miriel, J.; Bart, M. Study of thermal behaviour of clay wall facing south. *Build. Environ.* 2006, *41*, 307–315.

128. *Building without Borders: Sustainable Construction for the Global*

*Village*; Kennedy, J., Ed.; New Society Publishers: Gabriola Island, Canada, 2004.

129. Singh, M.K.; Mahapatra, S.; Atreya, S.K. Bioclimatism and vernacular architecture of north-east India. *Build. Environ.*2009, *44*, 878–888.

130. Hall, M.; Allison, D. Assessing the moisture-content-dependent parameters of stabilised earth materials using the cyclic-response admittance method. *Energ. Bldg.* 2008, *40*, 2044–2051.

131. Jayasinghe, C.; Kamaladasa, N. Compressive strength characteristics of cement stabilized rammed earth walls. *Constr. Build Mater.* 2007, *21*, 1971–1976.

132. Bui, Q.B.; Morel, J.C.; Venkatarama Reddy, B.V.; Ghayad, W. Durability of rammed earth walls exposed for 20 years to natural weathering. *Build. Environ.* 2009, *44*, 912–919.

133. Morel, J.C.; Pkla, A.; Walker, P. Compressive strength testing of compressed earth blocks. *Constr. Build. Mater.* 2007,*21*, 303–309.

134. Venkatarama-Reddy, B.V.; Gupta, A. Influence of sand grading on the characteristics of mortars and soil-cement block masonry. *Constr. Build. Mater.* 2008, *22*, 1614–1623.

135. Maniatidis, V.; Walker, P. Structural capacity of rammed earth in compression. *J. Mater. Civ. Eng.* 2008, *20*, 230–238.

# Chapter 4

# MATERIAL EFFICIENCY OF BUILDING CONSTRUCTION

Antti Ruuska, and Tarja Häkkinen
VTT Technical Research Centre of Finland, Tekniikantie 4, 02044 VTT Finland

## ABSTRACT

Better construction and use of buildings in the European Union would influence 42% of final energy consumption, about 35% of our greenhouse gas emissions and more than 50% of all extracted materials. It could also help to save up to 30% of water consumption. This paper outlines and draws conclusions about different aspects of the material efficiency of buildings and assesses the significance of different building materials on the material efficiency. The research uses an extensive literature study and a case-study in order to assess: should the depletion of materials be ignored in the environmental or sustainability assessment of buildings, are the related effects on land use, energy use and/or harmful emissions significant, should related indicators (such as GHGs) be used to indicate the material efficiency of buildings, and what is the significance of scarce materials, compared to the use of other building materials. This research suggests that the material efficiency should focus on the significant global impacts of material efficiency; not on the individual factors of it. At present global warming and greenhouse gas emissions are among the biggest global problems on which material efficiency has a direct impact on. Therefore, this paper suggests that greenhouse gas emissions could be used as an indicator for material efficiency in building.

## INTRODUCTION

Resource efficiency means efficient use of energy, natural resources, and materials, in order to create products and services with lesser resources and environmental impacts. It is based on life-cycle thinking and comprises of energy efficiency and material efficiency. Whereas the energy efficiency considers sparing use of energy, and ratio of energy use and production, material efficiency is about sparing use of natural material resources, effective management of side-streams, reduction of waste, and recycling [1].

Natural resources underpin the functioning of the European and global economies and the quality of life. These resources include raw materials, such as fuels, minerals and metals, as well as food, soil, water, air, biomass, and ecosystems [2]. A roadmap to a resource-efficient Europe [3] highlights the buildings sector as one of the three key sectors for improvements. Better construction and use of buildings in the European Union would influence 42% of final energy consumption, about 35% of our greenhouse gas emissions and more than 50% of all extracted materials. It could also help to save up to 30% of water consumption.

The importance of material efficiency and the need to improve it can be studied from several perspectives. Limited availability or scarcity of materials may lead to threats to the economy, and the production processes of materials can have significant environmental impacts. The extraction of raw materials and the production of materials may also be energy and/or labor intensive and very costly, and the extraction of materials may lead to land use changes and related impacts.

This article presents an overview of the different aspects of resource and material efficiency in building construction. The paper also presents the results of a case study and analyses the significance of building materials in terms of material scarcity.

## Classification of Resources (and Aspects of Scarcity)

Natural resources can be divided into renewable and non-renewable resources. Non-renewable resources are those that can only be harvested once. These are often referred to as stocks (e.g., iron ore) or resources that form extremely slowly (e.g., crude oil) [4]. Azapagic [5] divides the minerals industry into energy minerals (e.g., coal, oil), metallic minerals (e.g., iron, copper and zinc), construction minerals (e.g., natural stone, aggregates, sand, gravel, gypsum), and industrial minerals (e.g., borates, calcium carbonates, kaolin, plastic clays, talc).

A reserve is defined as that part of the reserve base that could be economically extracted or produced at the time of determination (in accordance with the terminology used by the European Commission [6]). The reserves of the most common building materials (aggregates, clay, lime and stone, gypsum, and quartz) are either large or very large [4]. However, buildings also consume materials whose reserves are more limited, for example, coal, oil, and metallic minerals.

The usability of resources depends specifically on the economy and the available technology. Resources that have previously been uneconomical to extract may become usable because of rising values and improved extraction technologies. Political situations and the effects of extraction on the landscape and environment may also affect the usability of resources. Scarcity always has a time dimension: it can be interpreted as a change in availability over time [7]. Steen [8] claims that many life cycle impact assessment (LCIA) approaches mix scarcity with issues such as difficulty of extraction. This can be viewed as double counting, as the effects thereof, such as high energy demand, are accounted for in other categories. Metals in use can also be seen as a global inventory of available metals. Virgin metal is added when necessary to this inventory [9]. Future backup technologies will probably require significantly less energy and other resources than the extraction of virgin metal.

Meadows *et al.* [10] identify that the increasing cost of resources is becoming a major problem for societies. As resources become scarcer, this may influence the quality of life in some parts of society. This, in turn, may have negative impacts on human health as a specific area of protection [11]. It may therefore be important not to separate the environmental and economic aspects. Yellyshetty *et al.* [12] argue that resource depletion needs to be considered in LCAs from the perspective of time, environmental and economic aspects of mineral extraction, and future consequences of decreased availability of mineral resources for a region. Steen [8] highlights three issues that should be considered when drawing conclusions about the inclusion of resource depletion in LCAs: (1) the time perspective when evaluating impacts on abiotic resources; (2) the separation of environmental and economic aspects; and (3) whether the consequences of decreased availability should form part of the LCI or the LCIA. The socio-economic value of mineral extraction can be significant in some regions, and changes in the extraction industry can have important social consequences [13]. Söderholm and Tilton argue that economic depletion will occur long before physical depletion [14].

Another way of looking at the issue of mineral resource scarcity is the surplus cost method, which assumes that future increases in mining volume will lead to increasing production costs per metal or mineral extracted. This is defined as the marginal cost increase (MCI). When the MCI is multiplied by future resource demand, the future costs to society can be determined [15,16].

## Indicators for Resource Efficiency and Material Efficiency in the Building Industry

Resource efficiency can be defined with a number of indicators. Each indicator has a specific definition, which contains only certain aspects of the issue.

Resource efficiency may be defined, for example, in terms of land area that an economy requires [17], human impacts on natural processes [18], impacts on land use [19], amount of material use [20] or related environmental impacts [21], ratio of GDP to material use [3], or national monetary input-output tables expanded with environmental information [22].

When moving from the level of economies to the level of technologies or products, other life-cycle related indicators are more common. The indicators are typically not correlated, so a wide range of environmental indicators are needed [23]. For example, life cycle assessment (LCA) methodology assesses the harmful impacts of buildings in terms of global warming, ozone depletion, acidification of soil and water, eutrophication, photochemical ozone creation, and depletion of abiotic resources (elements and fossil fuels) [24,25].

The impacts from resource use, often referred to as, resource depletion, is a prominent impact category in LCA [26]. LCA methodology addresses abiotic, or non-living, resources in terms of their availability for present and future generations. The depletion of such resources can be studied from the perspective of amounts of deposits, extraction rates, future ore extractions, or exergy consumption [27].

The use of natural raw materials in building can be decreased by using lightweight structures, minimizing loss, improving durability and service life, using secondary materials and improving appropriate flexibility [28,29]. Improved space efficiency also contributes to better material efficiency when assessing it in terms of functional units (a building that fulfils the required performance).

The following equation shows how these different aspects of material efficiency relate to the wider concept of resource efficiency. Equation (1) defines the total impacts associated with the production and processing of a specific material as (adopted from [30]):

$$I = D \times M \times Y \times E$$

(1)

In Equation (1), the impacts (I) are due to the demand (D) for products containing material, the average mass of material per product (M), the yield ratio of supplied material *versus* material in the final product (Y), and the average emissions per unit of material (E). The impacts of material efficiency extend to all the factors, D, M, Y, and E. In the context of buildings, the demand for new buildings is influenced by their durability, service life and flexibility. The use of lightweight structures impacts the average mass per product, and the yield ratio is affected by material losses during processes. Finally, the use of secondary materials impacts—in addition to the use of natural material resources—the average emissions, as reuse and recycling are typically significantly less energy intensive than primary production [30]. Instead of viewing material efficiency through the multiple viewpoints presented above, this research focuses on their total impacts. This research outlines the related impacts as follows: (1) depletion of natural raw materials; (2) impacts of material-related harmful emissions; (3) impacts due to material-related land use; and (4) life cycle costs due to the use of materials. The following sections discuss the importance of these different impacts, on the basis of literature.

## AIM AND SCOPE

The objectives of the research were as follows:

- to outline and draw conclusions about different aspects of the material efficiency of buildings;
- to assess the significance of different building materials on the material efficiency of buildings.
- The study was founded on the premise that the importance of material efficiency is based on one or more of the following impacts:
- the depletion of raw materials and its long-term socio-economic impacts;
- land use change due to the extraction of raw materials and its environmental impacts, and impacts on the landscape and future recreational use;
- the use of energy in production processes of materials and depletion of non-renewable energy;
- harmful emissions from production processes of materials and their local and/or global environmental impacts;
- material cost impacts due to the limited availability of raw materials or a higher need for energy and/or labor in the different phases of production processes.

The different aspects of the material efficiency of buildings were outlined and analyzed with the help of a literature study. The importance of the different groups of building materials and the significance of building materials compared with the use of energy resources was studied with the help of a case study. The Abiotic Depletion Potential (ADP) was calculated in terms of ADP elements and ADP fossil, and the significance of different building materials was assessed.

With regard to the building sector, the research questions of interest are as follows: (1) "As the global availability of the main building materials is very good, should the depletion of materials be ignored in the environmental or sustainability assessment of buildings?"; (2) "Although the availability is good, are the related effects on land use, energy use and/or harmful emissions significant and should related indicators (such as GHGs) be used to indicate the material efficiency of buildings?"; (3) "Although the availability of the main building materials is very good, what proportion of buildings use scarce materials and what is the significance of these compared with the use of other materials in buildings?".

# LITERATURE REVIEW

This section presents the literature review, which answers the research questions of this paper on a general level. It also points out the gaps in literature and gives reasoning for the selected case-study approach, which is presented later in this paper.

The literature review examined the impacts of material efficiency on: (1) depletion of natural raw materials; (2) impacts of material-related harmful emissions; (3) impacts due to material-related land use; and (4) life cycle costs due to the use of materials. It aimed to identify and fill potential gaps in the current knowledge and point out needs for more detailed studies.

## Scarcity and Availability of Abiotic Building Materials

Material efficiency is a way to reduce the demand of abiotic building materials. Whereas the importance of material scarcity is growing in general, the issue is not as clear for building materials. Common building materials, such as metals and ceramics, are derived from ores. Some of the minerals are approaching their production peaks and some have already passed their peak [31]. There is also a continuous decrease in ore grade at which some materials are being mined [32]. The inevitability of peaking of oil is generally acknowledged, although, it is still under debate, whether or not the peak has already passed [33]. Oil is needed, for example, for production of polymer-based building materials.

The building industry uses large amounts of materials, equating to approximately 50% of European resource extraction [3], but the most common building materials are also common in nature. Aggregates, for example, are the key component of many building elements but are generally not a scarce resource [34]. However, due to their heavy and bulky nature, aggregates need to be sourced close to their markets. Viable sources may be constrained at regional and local level [35], for example in rapidly growing developing countries [36], if their viable local supply is not strategically planned [6]. Relating to these problems, approaches which account for local resources have been proposed in literature [37].

The buildings also require metallic minerals for the production of, for example, concrete reinforcements and structural steel in the building frames, roofs, façades, windows and doors of the building envelope and pipes, ducts and wirings of building systems. Despite of dependence on the import of metallic minerals in some countries [34] these resources are not considered scarce, as their global availability is good [6]. However, mining of these minerals may become critical in terms of social impacts that mining activities cause locally on land and ecosystems [38].

When buildings become more energy efficient and building systems more advanced and complex, the demand for scarcer resources may increase. Some of the components of advanced, energy-efficient building systems, such as wind turbine magnets, high-capacity batteries, energy-efficient lighting and photovoltaic cells require rare earths and critical natural resources in their production [39]. However, the exact selection and weighting of factors, which make a raw material critical or scarce, are still open research questions [40]. Raw materials may be considered critical, for instance, if they have national significance for economies and their current or future supply is at risk [39]. Other sources of criticality may rise from specific ecological, social, or political considerations [6].

## Greenhouse Gases

The building sector is the single largest contributor to global greenhouse gas emissions. On the other hand, it also has a substantial emission saving potential. Material efficiency extends to all the underlying factors of resource efficiency, making it a significant contributor to resulting impacts from materials. Considering these viewpoints, material efficiency has a significant role in reducing the global GHG emissions from buildings.

The greenhouse gas emissions from buildings are related to the embodied energy of building materials and the emissions from operational energy use and the role of materials is becoming increasingly important. The research and policies have focused only on the operational energy use until recently [41,42,43]. This can be explained by the fact that, the role of embodied energy has been relatively low, at some 10%–20% [44,45], but development towards more energy efficient buildings increases the importance of materials. In low-energy buildings the role of materials can be as high as 50% [41] and ultimately, at zero-energy-level, all the energy-consumption, and related greenhouse gas emissions come from the embodied energy of building materials [42]. Due to this development, the embodied energy and related emissions cannot be omitted in life cycle assessments. In addition to initial material consumption, the buildings also need materials for their lifetime renovations. The energy consumption of interior renovations over the lifetime of a building can account for some 20% to 30% of the initial embodied energy [46]. The need of this recurrent embodied energy can be almost halved, with the use of materials with longer service life [47].

When looking at the issue from the level of residential areas, also transport needs to be considered. Significant greenhouse gas savings can be achieved in all, embodied, operational and transport energy needs when planning residential areas [48]. From sector-level, the most important factors affecting the greenhouse gas emissions are housing size, style and location [49].

Another viewpoint to the issue is the temporal perspective of emissions from building. The initial GHG emissions emitted over a short period of time in the construction phase may compromise the greenhouse gas mitigation goals in short and medium term [50]. Therefore, the greenhouse gas emission targets cannot be achieved with energy-efficient new buildings alone.

Ruuska and Häkkinen [28] assess the total greenhouse gas emissions of a multi-storey residential building in Finland with the help of a parametric study. The results show for a concrete building case that material-related emission account for some 40% of 50-year lifetime total GHG-emissions for a passive-level building in Southern Finland. Furthermore, if soil stabilization of a building site is included in the figures, the role of materials rises to over 50% of lifetime totals.

## Land Use

Construction causes irreversible land changes. Use of land means consumption of resources, in terms of changing the potential end-use and the consumption of soil materials. Buildings use land directly by occupying the land under their footprints and through their embodied land use, relating to their raw material and energy use throughout the building's value chain. An impact because of land use occurs when the land properties are modified (transformation) and also when the current man-made properties are maintained (occupation) [51]. Changes in land use can have wide-ranging environmental consequences, including biodiversity loss, changes in emissions of gases affecting climate change, changes in hydrology, and soil degradation [52].

Buildings and other construction assets cause soil sealing as land remains below constructions. Artificial sealing is generally extensive and permanent [53]. When vegetated soils are replaced with impermeable surfaces, it results in the increase of overland flow, reduction of infiltration and bypass of natural storage [54]. Although the global availability of the main building materials is good, the consideration of land use may affect the importance of material efficiency with regard to buildings. However, an LCA-based case study analysis [55] indicates that when only non-renewable material resources are considered, the land occupied by buildings is more important that the land use due to the extraction of raw materials used for buildings. However, when wood is used as a building material, the land use (in terms of occupied land area) required for the production of building materials becomes more significant than the land occupancy of the building itself.

The extraction of aggregate materials also affects the landscape and the natural geological and biological conditions. In addition to this, in Finland, the extraction of gravel affects the quality of the groundwater because the extraction increases the variety in the quality and pollution risk of the groundwater [56]. In addition to the impacts on groundwater and surface water, the production of aggregates causes local impacts, such as vibration, and noise and dust emissions.

## Cost and Productivity

Material efficiency has an important effect on construction cost efficiency. The positive impacts on cost and productivity can be seen as a natural driver towards material efficiency in the building industry.

The importance of materials in relation to the investment costs of construction varies. The approximate magnitude has been estimated at 15%–40% of the investment cost (including the cost of design, interfaces, labor costs, site overheads, taxes and the contractor's profits) [29]. Minimizing the loss of materials has a direct impact on the investment costs. On the other hand, better and appropriate flexibility in the design of spaces can also have a significant impact on the life cycle costs, especially in the context of retail and office buildings.

Goodrum et al. [57] studied the relationship between changes in material technology and productivity in construction. The results show that changes in material technology correlate with improvements in both labor and partial factor productivity (physical output per material cost + equipment cost + labor cost). The authors found that the relationship between changes in material technology and construction productivity was weaker for labor productivity than for partial factor productivity. The strongest relationship between changes in material technology and labor productivity was also found among changes in the unit weight of materials followed by modularity, curability, and installation.

## Existing Standards and Regulation

The current European regulation, as well as the work done for the development of assessment standards, reflects the stated policy targets to consider and improve the material efficiency and the overall resource efficiency of societies. However, unlike energy performance, which is defined by European Directives [58,59], material efficiency is not tightly controlled or regulated. Also, contrary to the energy efficiency of building and renovation [60,61], there are no fiscal instruments or incentives in place for improvements in material efficiency of buildings.

In Europe, the Construction Product Regulation [62] gives basic requirements for construction products. Construction works as a whole and in their separate parts must be fit for their intended use, throughout the life cycle of the works and fill the basic requirements. Sustainable use of resources is included in the requirements, and the CPR states that construction works must be designed, built and demolished in such a way that the use of natural resources is sustainable. Especially the following is highlighted: (1) re-use or recyclability of the construction works, their materials and parts after demolition; (2) durability of the construction works; and (3) use of environmentally compatible raw and secondary materials in the construction works. Even though the Construction Product Regulation emphasizes the importance of material efficiency, it does not give normative rules for it, or dictate mandatory information about material efficiency.

Assessments of resource depletion and comparisons of buildings and building products are supported by international and European standards. The current standardization and guidelines suggest using two separate impact categories for resource depletion: ADP elements for all non-renewable abiotic materials and ADP fossil fuels for all fossil resources [24,25,63]. Previously, both these items were assessed in terms of antimony equivalents [64]. However, as the two contribute towards the decrease of different resources, their ADP is characterized by different units [65]. The unit of measurement for the depletion of natural resources is the antimony equivalent (kg Sb eq) and for the depletion of natural fossil energy the resources, their net calorific value (MJ). Despite of its established status through the current standardization and guidelines, the calculation of ADP has some shortcomings. For example, the characterization factors for its calculation do not exist for many of the common building materials. The basic problem behind this is that such factors cannot be defined for many of the common building materials, such as gypsum, silica sand, construction sand, clays, limestone, and such, due to lack of data on material configurations, reserves, reserve bases, and ultimate reserves for these materials [65].

The status of ADP calculations in standardization and the identified shortcomings in the calculation method, give a basis for the case-study of this research. The literature study was unable to identify detailed ADP calculations, which would show the importance of different building materials. The case-study aims to create new knowledge on the importance of different building materials, in terms of their ADP. It also aims to compare the material-related ADP to the ADP from lifetime operational energy use. Finally, it aims to give more information on the significance of the use of different scarce materials in buildings.

# QUANTIFYING THE ABIOTIC DEPLETION POTENTIAL (ADP) OF BUILDINGS

The case-study aims to add to the existing knowledge by showing the importance of different building materials, in terms of their abiotic resource depletion potential (ADP). It also studies the importance of building materials, in relation to operational energy use and the role of advanced building systems. Finally, the case-study offers new information on the current calculation method for ADP, together with its limitations. These issues were selected as the focus of the case-study, based on the gaps in the existing literature.

This section presents the case-study building, and explains the calculation method and main data sources used in the study. This case-study assesses the resource depletion of a case-building, by using impact categories of ADP elements and ADP fossil, recommended by current standardization and guidelines. The following subsections go through the calculation method, principles of the used life cycle assessment method, material quantities used in the assessment, calculation of energy consumption and, especially, calculation of ADP elements and ADP fossil.

## Calculation Method

This research used life cycle assessment to determine the ADP of a case building. The calculation was carried out by using the bill of quantities (BOQ) of a real world building and assigning each of the materials with a specific characterization factor for their ADP (elements). For ADP fossil, the energy consumption associated with the materials of BOQ was completed with lifetime energy consumption information.

### *Life Cycle Assessment*

Life cycle assessment means compiling and evaluating inputs, outputs and environmental impacts of a product system throughout its life cycle [66]. It is widely accepted as one of the best tools for environmental assessment of a variety of products and processes [67]. This research uses a process-based analysis, which is generally recognized as more accurate, but more labor, and time-intensive than, for example, input-output analysis [68,69]. The selected method and its limitations and benefits are examined in more detail in the 'discussion'-section. The life cycle assessment is limited to the abiotic depletion (ADP) of non-renewable raw materials and fossil fuels. The assessment does not aim to be exhaustive, but it aims to define the ADP of building materials with sufficient accuracy. The specific focus of this research is on the product stage, but also construction, use and end of life stages are assessed to cover the whole life cycle of the building, following the division of current standardization [24]. The assessment period of this research was 50 years.

## Material Quantities

The material quantities required for the assessments of the product stage were based on the bill of quantities (BOQ) of a real-world case building, which is described in further detail later on. The BOQ was derived from the building's building information model (BIM), hence offering a high level of accuracy in material amounts. For calculation purposes, the materials of the BOQ were categorized under nine identified main material groups, namely: aluminium, concrete, copper, fossil materials, gravel, other mineral resources, steel, wood boards, and other wood-based products.

In addition to the quantities of the BOQ, the lifetime material consumption, including waste during construction stage and material requirements for use stage were also accounted for. The material loss was estimated to be 5% for all the building materials, for both construction and use phase material needs, based on literature [70]. The material needs of the use phase were assessed by estimating replacement and refurbishment needs over the lifetime of the building, for different building parts and components. The material needs of maintenance and repair were estimated to be insignificant and they were not accounted for in the assessment.

The following assumptions were made for the lifetime renovations. Firstly, the load-bearing structures were assumed to last for the whole lifetime of the building. Secondly, the roofing, building systems, windows, doors, glazing, and the surfaces of sanitary spaces were expected to be replaced (or refurbished) once over the 50-year assessment period. Thirdly, the surface finishes, fittings and furniture were expected to require replacement in every 10 years, thus, they were assumed to undergo four renewals over the assessment period.

## Energy Consumption

The energy consumption of the product stage was taken into account by using life cycle inventories (LCI), which included energy consumption from raw material supply, transport and manufacturing from cradle-to-gate. Energy consumption of construction installation process and transportations were taken into account in the construction phase. For the use phase, the assessment included the energy consumption of replacement and refurbishment, transportation of materials and operational energy use of the building. The end of life phase included energy needed for deconstruction and transport of waste from site. The waste processing and disposal stages were excluded for this assessment. Data sources used for these calculation are shown in more detail in the next section.

## Calculation of ADP Elements

The calculation of ADP elements had five steps. Firstly, total material needs over the lifecycle of the building were defined. This was done by combining the information from the original BOQ with the estimates on material losses during construction (5%) and assumptions on replacements and refurbishments. Each of the materials was then categorized under one of the identified nine main material groups.

The second step was to define the total abiotic material inputs for each of the main materials. This was done using the European ELCD database and its LCI data [71] for this purpose.

The third step was to derive the ADP characterization factors for each of the abiotic inputs. The characterization factors used are based on the CLM database's base reserve figures [72], as recommended in current guidelines [64].

Fourthly, after designating the ADP characterization factors for each of the abiotic inputs of the main materials, the average ADP factor for each main material was calculated.

Finally, when all the material amounts, and corresponding ADP characterization factors were defined, the ADP for each material was calculated. After this, the building-level ADP was calculated by adding together the ADPs of all the nine main materials. The results of calculations, together with references to the used data sources are presented in section "ADP Elements".

This research also considers the specific issue of soil stabilization, which may be needed in case of poor ground conditions on building site, as it has been previously found to be significant building factor impacting (GHG) emissions [28] and its main components have high embodied energy. The ADP Elements calculations follow the same methodology as described previously, the only difference being the main materials in stabilization are cement (CEMII) and quicklime (CaO) with a mixing ratio of 1:1. In addition, ADP Fossil is assessed for the soil stabilizations. The assessment results for soil stabilization, along with the data sources, are presented in section "ADP of soil stabilization".

Another specific issue studied by this research is the ADP of advanced building systems of energy-efficient buildings, because such systems typically include rare earth elements and other critical materials [39]. The components selected for study are energy-efficient lighting and PV panels. The ADP Fossil of these is not calculated due to a lack of reliable data. The calculation results and data sources are shown in section "ADP of advanced building systems".

## Calculation of ADP Fossil

The ADP fossil calculations followed a similar methodology to that of the ADP elements. For material-related ADP-fossil, the calculation comprised of three stages.

Firstly, the total material needs over the lifecycle of the building were based on the total masses calculated for ADP elements.

Secondly, the non-renewable energy inputs for each of the main materials were derived from the ELCD database [71] to give a characterization factor for ADP fossil for each of the main materials.

Thirdly, when all the material amounts, and corresponding ADP Fossil characterization factors were defined, the ADP for each material was calculated. After this, the building-level ADP was calculated by adding together the ADPs of all the nine main materials. The results of calculations, together with references to the used data sources are presented in section "ADP Fossil".

In addition to the direct, material-related energy consumption, fossil energy is also consumed in material transportation. The contribution of transportations to the ADP Fossil is calculated by assuming a 50 km transport with a semi-trailer combination to the building site for all the materials and the same 50 km distance for all the materials to cover their transport off the site with earth moving lorry at the end-of-life. The construction installation process, lifetime replacement and refurbishment activities, and deconstruction of the building at the end-of-life also consume fossil energy. These are assessed using values from previous research. The assessment results for transportations, and construction, lifetime renovations and demolition are shown in section, along with the used data sources "ADP Fossil of Material Transportation and Construction Work".

To complete the ADP Fossil calculations, lifetime operational energy use is also assessed. This is done by assessing the operational energy consumption over the lifetime of a building. This research divides the operational energy consumption into three items: space heating, hot water, and electricity. The calculations are based on standard energy consumption of buildings, in terms of end use of energy. The end use of energy is then converted into non-renewable primary energy, based on country-specific energy production profile. Furthermore, as energy production is constantly developing, future energy production scenarios are used to forecast the development of the use of non-renewable primary energy over the life cycle of 50 years. All of these calculations, together with the used calculation data, are presented in section "ADP of the operational energy use".

## Case Study Building

All of the ADP calculations were made for a specific case building, which was located in Southern Finland, and represented a typical Finnish contemporary building. The building under study was a six storey residential building with a basement floor. The gross floor area of the building was 3060 m² and the number of apartments was 28. The structures of the building were passive-level and the heating method was district heating. The load-bearing frame, consisting of internal and external walls, floor slabs and roof, were precast concrete structures. The bill of quantities, extracted from the building information model (BIM) of the case building was used as the basis of the calculations of this research. Material quantities of the case building are not shown here, as they are presented later in this paper, in the result tables for the ADP (Table 1 and Table 2).

# RESULTS

The following subsections present the calculation results of the case-study, along with the references for the used data sources. The ADP elements and ADP fossil for the case building are shown in the first two subsections, followed by results for soil stabilization. After this, the impacts of advanced building systems are assessed, followed by the impacts of transports and construction work. The last result section shows the results for ADP from operational energy use and compares it to the material-related ADP results.

## ADP Elements of Building Materials

This section shows the results for ADP Elements of building materials for the case building. The following table (Table 1) shows that the total need of building materials over a 50-year life cycle for the case building is 4960 t, or 1.62 $t/m^2$. The total material need includes the initial material needs for construction of the building (89%), recurrent material needs for replacements and refurbishments (6%), and material losses (5%). The table also shows that the production of the building materials for the case-building requires a total of 7320 t of abiotic inputs, or 2.39 $t/m^2$. According to the results, the building-level abiotic depletion potential, over the lifetime of the building, is 1.05 kg of Antimony equivalents, or 0.34 $g/m^2$.

In addition to these results, the following Table 1 also includes the ADP characterization factors used in the calculations for each of the main materials. It also shows the noteworthy information on abiotic inputs, which lack an ADP characterization factor, and are therefore not included in the calculation results.

## ADP Fossil of Building Materials

This section shows the results for ADP Fossil of building materials for the case-building. The total material needs presented in the following table (Table 2) match those presented in the previous section. According to the results, the ADP Fossil of the case-building is 15,900 GJ of fossil energy inputs, or 5.2 $GJ/m^2$.

## ADP of Soil Stabilization

This section studies the effect of soil stabilization on the ADP elements and ADP fossil. The total material need for stabilization is 1420 t, including material losses (5%). The following Table 3 shows that the ADP elements value of soil stabilization is 530 g, or 0.17 $g/m^2$, and that the ADP Fossil is 3500 GJ, or 1.14 $GJ/m^2$.

## ADP of Advanced Building Systems

This section assesses the ADP elements of advanced building systems. The ADP Fossil is not assessed, due to lack of reliable data.

### Energy-Efficient Lighting

This section shows the calculation results for ADP Elements of energy-efficient lighting. The selected lamp type is a standard T12-type fluorescent lamp with a rare earth triphosphor coating, with a coating thickness of 5 $mg/cm^2$ [73] and a total of 7 grams of phosphorous coating. In addition to this, the lamp has low-pressure mercury vapor, with an estimated amount of 25 mg per lamp [74]. Assuming a service life of 10 years for the lamps, four replacements are required over the 50-year life cycle. The case building has a total of 355 lamps.

The following Table 4 shows that ADP Elements for energy-efficient lighting is 0.12 kg of Antimony equivalents, or 0.38 g/m².

**Table 1:** Total mass of materials, abiotic material inputs per ton material ton, abiotic material inputs with no abiotic depletion (ADP) characterization factor, average ADP characterization factor of abiotic inputs, total ADP of materials and data source for material inputs

| Material | Total mass of materials (t) | Abiotic material inputs per material ton (t/t) | Total abiotic material inputs (t) | Abiotic material inputs with no $ADP_{CF}$ (%) | $ADP_{AVG}$ of abiotic inputs (t Sb eq /t) | Total ADP of materials (kg Sb eq) | Data source for material inputs |
|---|---|---|---|---|---|---|---|
| Aluminium | 29 | 4.8 | 142 | 87.2% | $3.22 \times 10^{-6}$ | 0.46 | [75] |
| Concrete | 3549 | 1.4 | 5016 | 99.9% | $8.28 \times 10^{-9}$ | 0.04 | [76] |
| Copper | 4 | 6.0 | 26 | 99.2% | $1.90 \times 10^{-5}$ | 0.49 | [77] |
| Fossil materials | 90 | 2.8 | 256 | 99.9% | $7.40 \times 10^{-10}$ | 0.00 | [78] |
| Gravel | 629 | 1.9 | 1202 | 100.0% | – | – | [79] |
| Other minerals | 337 | 0.8 | 254 | 100.0% | $2.83 \times 10^{-10}$ | 0.00 | [80] |
| Steel | 83 | 3.5 | 291 | 91.6% | $1.86 \times 10^{-7}$ | 0.05 | [81] |
| Wood | 42 | 0.1 | 5 | 99.7% | $4.79 \times 10^{-9}$ | 0.00 | [82] |
| Wood boards | 200 | 0.6 | 129 | 99.7% | $4.84 \times 10^{-9}$ | 0.00 | [83] |
| Total | 4960 | – | 7319 | 99.3% | – | 1.05 | – |

**Table 2:** Total mass of materials, fossil energy inputs per ton material ton, total ADP Fossil of materials and data source for material inputs

| Material | Total mass of materials (t) | Fossil energy inputs per material ton (GJ/t) | Total ADP of materials (GJ) | Data source for material inputs |
|---|---|---|---|---|
| Aluminium | 29 | 37.0 | 1088 | [75] |
| Concrete | 3549 | 0.8 | 2720 | [76] |
| Copper | 4 | 17.5 | 75 | [77] |
| Fossil materials | 90 | 85.6 | 7696 | [78] |
| Gravel | 629 | 0.1 | 38 | [79] |
| Other minerals | 337 | 3.7 | 1259 | [80] |
| Steel | 83 | 15.7 | 1297 | [81] |
| Wood | 42 | 0.6 | 27 | [82] |
| Wood boards | 200 | 8.7 | 1728 | [83] |
| Total | 4960 | – | 15,900 | – |

**Table 3:** Total mass of materials, fossil energy inputs per ton material ton, total ADP Fossil of materials, abiotic material inputs per ton material ton, abiotic material inputs with no ADP characterization factor, average ADP characterization factor of abiotic inputs, total ADP of materials and data source for material inputs for soil stabilization of case-building

| Material | Total mass of materials (t) | Fossil energy inputs per material ton (GJ/t) | Total ADP Fossil of materials (GJ) | Abiotic material inputs per material ton (t/t) | Total abiotic material inputs (t) | Abiotic material inputs with no $ADP_{CF}$ (%) | $ADP_{AVG}$ of abiotic inputs (kg Sb eq /t) | Total ADP of materials (kg Sb eq) | Data source for material inputs |
|---|---|---|---|---|---|---|---|---|---|
| CEMII | 709 | 3.6 | 2558 | 1.7 | 1199 | 99.6% | 0.00045 | 0.53 | [84] |
| CaO | 709 | 5.4 | 3820 | 3.2 | 2303 | 100.0% | – | – | [85] |
| Total | 1 420 | 9 | 6380 | 5 | 3500 | 99.8% | – | 0.53 | – |

**Table 4:** Total weight of selected materials, ADP characterization factors for materials and total ADP for lighting, based on a single lamp type over a 50-year life-cycle with four replacements

| Material | Total weight of materials (kg) | $ADP_{CF}$ kg (Sb eq)/kg | Total ADP kg (Sb eq) | Data source for characterization factors |
|---|---|---|---|---|
| Mercury | 0.04 | 2.62 | 0.12 | [72] |
| Rare earth elements | 12.71 | 0.0006 | 0.007 | [86] |
| Total | 12.76 | – | 0.12 | – |

## Solar Panels

This section looks at the photovoltaic panels of solar panels and shows their contribution to the depletion of abiotic resources, in terms of ADP Elements. The selected panels are of two types: c-Si (Crystalline Silicone) and CIS/CIGS (Copper Indium Selenide/Copper Indium Gallium (di) Selenide). The main material is glass, which forms approximately 74% to 84% of the total mass. The remainder is aluminium (10% to 12% of the total mass) and other metals (4% to 16% of the totals), as summarized in a report to the European Commission [87]. Assuming that the lifetime of the panels is 15 years, the panels will need to be renewed three times over the lifetime.

The results in Table 5 show that the ADP for solar panels may vary from 180 to 174,000 kg of Antimony equivalents, or 60 g/m$^2$ to 60 kg/m$^2$.

**Table 5:** Total area of solar panels, ADP characterization factors per square meter of panel, and total ADP for solar panels over a 50-year life-cycle with three replacements

| Panel type | Total area of solar panels (m$^2$) | $ADP_{CF}$ Sb eq (kg/m$^2$) | Total ADP Sb eq (kg) |
|---|---|---|---|
| c-Si | 370 | 0.2 | 180 |
| CIS/CIGS | 370 | 157 | 174,380 |

## ADP of Material Transportations and Construction Work

This section presents the ADP fossil of material transportations and construction work. The total ADP Fossil from material transportations, and construction and demolition work is 2400 GJ, or 0.8 GJ/m$^2$. The results are as follows:

- Total mass (building) 4960 t;
- Fossil fuels (construction work) 0.249 GJ/t (Based on data presented in [49]);
- Fossil fuels (demolition work 0.137 GJ/t (Based on data presented in [49]);
- Fossil fuels (transportation) 0.10 t (Based on VTT LIPASTO traffic

emissions [88];

- Fossil fuels (total) 0.49 GJ/t;
- ADP fossil energy (building) 2413 GJ.

## ADP of the Lifetime Operational Energy Use and Comparison to the ADP of Materials

This shows the results for the ADP Fossil of the lifetime operational energy use of the case building. The calculation of the used conversion factors is also explained in this section. The case building uses a total of 3050 MWh of heating energy for spaces, 5350 MWh for hot water, and 7650 MWh of electricity over 50 years, in terms of end-use of energy (EUE). Heating and hot water is produced with district heat, whereas electricity is taken from the grid.

In order to relate the end-use of energy to the use non-renewable primary energy resources, primary energy conversion factors (PECFs) are needed. Here, these factors are based on the Finnish data [89] and those are 0.77 for district heating and 1.75 for electricity.

As the assessment period covers a 50-year timespan, these conversion factors will not remain constant due to developments in energy production. Therefore, another conversion factor is needed to translate the non-renewable energy consumption of today to match the expected average over the 50-year assessment period. The future use of non-renewable primary energy is expected to follow closely the estimated future development of Finnish GHG emissions (data prepared by the Finnish Ministry of Employment and the Economy, and presented, for example, in [28]). The conversion factors used here for translating contemporary PECFs to 50-year averages are 0.8 for district heat and 0.4 for electricity and they are named future conversion factors (FCFs).

The following table (Table 6) shows the ADP Fossil for the case building, in terms of total non-renewable primary energy use over the 50-year life cycle.

**Table 6:** End-use of energy, primary energy conversion factors, future conversion factors, and ADP Fossil for the case building from operational energy use over 50-year life cycle

| End-use energy of energy, purpose of use | End-use of energy (EUE) MWh | End-use of energy (EUE) GJ | Primary energy conversion factor (PECF) | Future conversion factor (FCF) | ADP Fossil/Total non-renewable primary energy use, 50a (GJ) |
|---|---|---|---|---|---|
| Heating energy | 3050 | 10,980 | 0.77 | 0.8 | 7063 |
| Hot water | 5350 | 19,260 | 0.77 | 0.8 | 12,389 |
| Electricity | 7650 | 27,540 | 1.75 | 0.4 | 19,278 |
| Total | 16,050 | 57,780 | – | – | 38,730 |

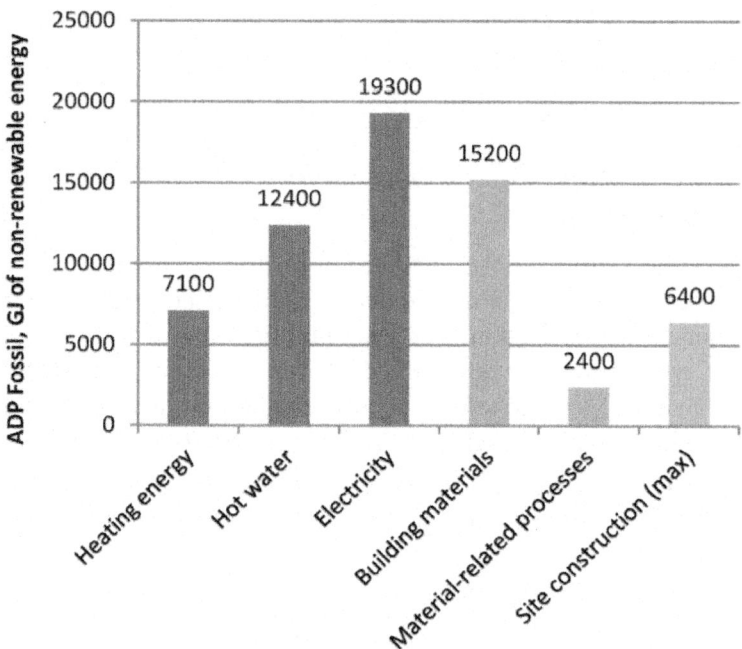

**Figure 1:** Fossil energy consumption (net calorific value) over the life cycle of the building.

Figure 1 combines the ADP Fossil values for operational energy use from Table 6, for building materials from Table 2, for soil stabilization from Table 3, and for material-related processes from Section 5.5. The respective ADP Fossil values for the different items are as follows: heating energy 7100 GJ (2.32 GJ/m$^2$), hot water 12,400 GJ (4.05 GJ/m$^2$), electricity 19,300 GJ (6.31 GJ/m$^2$), building materials 15,900 GJ (5.20 GJ/m$^2$), material-related processes 2400 GJ (0.78 GJ/m$^2$), and site construction (soil stabilization) 6400 GJ (2.09 GJ/m$^2$).

In summary, the APD Fossil due to operational energy totals 38,700 GJ (12.65 GJ/m$^2$) and material-related ADP Fossil is in total 17,600 GJ (5.75 GJ/m$^2$) or 24,000 GJ (7.84 GJ/m$^2$), depending on the stabilization needs. Therefore, the total lifetime ADP Fossil varies from 18.5 to 20.5 GJ/m$^2$. The result shows that the role of material-related non-renewable energy consumption for the case-building is at the level of 30% to 40% of lifetime total energy consumption.

## DISCUSSION

The case-study of this research aimed to fill in the gaps in the current knowledge, as identified in the literature review. It looked into the depletion of natural raw materials, through an assessment of lifetime abiotic depletion potential (ADP) of a residential multi-storey case-building with concrete structures, for both ADP elements and ADP fossil, as defined in current guidelines [24,25,63]. It should be highlighted that due to the case-study approach, the generalization of the results should be done with caution, especially considering the building type and location.

The material quantities were extracted from the building information model (BIM) of a real world building, so the data accuracy for initial material consumption can be considered high. The material losses, on the other hand, were estimated to be at the level of 5% of total material consumption. Commonly used values in literature vary from 0% to 10% [70]. Also, the lifetime material needs for replacements and refurbishments were assessed through simple estimates on service lives of different building components. An analysis on the impacts of estimation errors show that a change of 25% in these factors would increase/decrease the material amounts by some 10% for the case-building.

The case-study used the European reference life cycle database, ELCD [71] to derive the abiotic material inputs and energy requirements for each of the main materials of the building. The LCIs of the database are compiled mainly by process analysis. It can be argued that this method is associated with underestimation of the impacts, as the number of processes and the order of upstream processes are limited [68], and sufficient boundaries may be difficult to cover due to the complexity of upstream processes [69]. For basic building materials, for example, the incompleteness factor, often referred to as truncation error [66] is estimated to be at least 10% [69], some estimates being as high as 60% for residential buildings [67].

It should be noted that the data sources for the ELCD-database are drawn on a wider regional level and the energy inputs for the production processes use country-level statistics and national grid-mix information and they are not pure process based analyses. This enhancement of process-based information with IO-based data can be considered to make the profiles of ELCD profiles hybrid analyses in a sense [69].

The ADP characterization factors used for the calculation of ADP elements embody significant uncertainty in them. This research used the CLM database's base reserve figures [72], as recommended in European ILCD handbook [63]. However, the current standards [24,25] do not explicitly state which reserve estimates to use, and some LCAs and EPDs may still be assessed using the ultimate reserve figures, as this has been a past recommendation [64]. The ADP characterization factor for base reserves of copper, for example, is two times bigger than that for the ultimate reserves, for iron 30 times bigger and for aluminium, 23,000 times bigger. This makes it difficult to reliably compare the results of ADP studies between each other. However, the ADP of the case building, 1.05 kg of Antimony equivalents for almost five million kilograms (4960 t) of building materials can be compared to the production of some basic metals from virgin raw materials. The production of 420 kg of copper, 41,500 kg of aluminium, or 630,000 kg of iron from virgin raw materials would produce the same ADP of 1.05 kg [72]. These comparisons suggest that the result for ADP of the building is of very low level.

Only 0.7% of the abiotic material inputs of the case building have a characterization factor in the first place, making the ADP elements assessment practically worthless. The basic issue behind this is that such factors cannot be defined for any of the common building materials, such as gypsum, silica sand, construction sand, clays, limestone, and such, due to lack of data on material configurations, reserves, reserve bases and ultimate reserves for these materials [65]. Based on the results of the case-study, the benefits and purpose of calculating ADP elements for buildings is highly questionable in its current form. Methods, which would better account for local scarcity of resources [37] or land or social impacts [38], could fit the purpose better.

The assessment of advanced building systems resulted in ADP elements of 0.12 kg. For solar panels, the figures were 180 and 180,000 kg of Antimony equivalents. The results of advanced building systems show that such systems may be of relatively high importance, compared to the building itself.

The case-study of this research also assessed the APD Fossil for the materials of the case-building. The uncertainties related to these calculations, concerning the material quantities and the used LCI database are the same, which were discussed previously for ADP elements. As ADP fossil is defined in terms of non-renewable energy, the problem of characterization factors does not have an effect on the results. The assessment results showed that the material-related ADP Fossil totalled from 17,600 GJ (5.75 GJ/m²) to 24,000 GJ (7.84 GJ/m²). Research on similar buildings is limited but, for example, results of two residential buildings with concrete frame and floor area of some 1200 m² in Sweden, show embodied energy from 4.6 to 5.4 GJ/m² [90], as summarized in Ramesh *et al.* [44]. It should be pointed out that the embodied energy figures are not directly comparable to the ADP fossil figures, as the ADP fossil does not include the use of renewable energy.

The ADP fossil due to operational energy totalled to 38,700 GJ (12.65 GJ/m²) in the case-study. The results show that the material-related non-renewable energy consumption of the case building was at the level of 30% to 40% of lifetime total non-renewable energy consumption. These results are in line with a GHG assessment of the same building, done in a previous research, showing that material-related GHG emissions accounted for 40% to 50% of lifetime total emissions [28]. The comparable result is largely explained by the fact that GHG emissions are mainly due to consumption of fossil energy resources. As discussed above, ADP fossil does not contain renewable energy. In Finland, for example, the share of renewable energy sources in energy production was 27% in the year 2010 [91].

The operational energy consumption (end-use of energy) was assessed based on standard consumption figures, stated in Finnish regulations. The energy consumption of the case building was 105 kWh/m². The real consumption figures may vary from this significantly, due to user behavior, as shown in previous research [92]. However, assessment of user behavior was not the focus of this study and this variation was not considered in the assessment. In order to convert the end-use of energy into non-renewable primary energy use, Finnish national-level energy production information was used [89] and, in order to take the future development towards low-emission energy production, conversion factors based on [28] were used. Whereas the present-day ratio of non-renewable primary energy to end-use of energy can be thought to be a relatively reliable figure, the future conversion factors depend on political decisions in the future and cannot be predicted accurately. For example, a decrease of 25% in these factors would impact the results significantly, indicating higher than expected share of renewable energy in the future and lower than expected share of non-renewable energy. For the case-study, such change would decrease the ADP fossil from operational energy use from 12.65 GJ/m² to 9.5 GJ/m². This would increase the role of material-related energy consumption from the level of 35% to 45% of lifetime totals.

The study was founded on the premise that the importance of material efficiency is based on one or more of the following impacts:

- the depletion of raw materials and its long-term socio-economic impacts;
- land use change due to the extraction of raw materials and its environmental impacts and impacts on the landscape and future recreational use;
- the use of energy in production processes of materials and depletion of non-renewable energy;
- harmful emissions from production processes of materials and their local and/or global environmental impacts;
- material cost impacts due to the limited availability of raw materials or a higher need for energy and/or labor in the different phases of production processes.

This research did a comprehensive literature study to outline and draw conclusions about different aspects of the material efficiency of buildings.

Material efficiency is a complex issue to deal with in steering because there is no widely acknowledged way to make different materials commensurable. The impacts of material efficiency extend to all the aspects of resource efficiency, as shown with Equation (1) of this paper. The demand for new buildings is influenced by their durability, service life and flexibility. The use of lightweight structures impacts the average mass per product, and the yield ratio is affected by material losses on the building site. Finally, the use of secondary materials typically reduces the emissions from production. Due to the comprehensive nature of material efficiency, the focus of policy formulation should not be on its individual components, such as yield rates, average masses per products, and such, but on the impacts caused by material efficiency. Söderholm and Tilton [14] argue similarly, that it is better to avoid policies that directly encourage specific material efficiency options, and that policies should address particular environmental problems and information externalities to enhance material efficiency in instead.

The study was founded on the premise that the importance of material efficiency is based on some of its impacts. The importance of the different impacts (indicated with indicators) can be viewed from the perspective of sustainable development. An indicator can be validated as applicable to sustainable building if it fulfils two minimum requirements: it must be related to a subject of concern for sustainable development, and buildings must have a significant impact on that issue [93].

From the perspective of sustainable development, the greenhouse gas emissions from building sector are an example of an environmental problem, on which material efficiency has a significant impact on. Greenhouse gas emissions from building sector are a significant contributor on global warming, and material efficiency has a significant impact on the issue.

The consumption of non-renewable energy resources is near analogous to the greenhouse gas emissions, as the greenhouse gas emissions are mainly the result of consumption of non-renewable fossil energy in production processes of materials. This analogy was also partly illustrated by the results of the case-study.

However, the results on material-related land use showed that the importance of material-efficiency on land-use was practically negligible, as the footprint of the building was significantly more important than the land used for the extraction of non-renewable raw materials.

From the viewpoint of costs, the results showed that the role of materials is only small, some 10% to 40% of the construction costs. This means that both savings through improved material efficiency, and additional costs through future price increases in materials, have only a limited impact on total costs. The construction industry consumes significant amounts of raw materials globally. However, the most common building materials are also common in nature. The results suggest that the most common building materials have no significant impact on depletion natural raw materials globally, although locally this might be important. However, the case might be different for some scarcer resources, which are used in advanced building systems. The case of non-renewable energy resources is different, as discussed previously. The material efficiency has a significant impact on the consumption of non-renewable energy resources.

The impact indicators for material efficiency should be concrete and they should indicate problems, which have global significance. As such, the resource depletion indicators of the current guidelines for buildings do not fully support this. This research suggests that the material efficiency should focus on the significant global impacts of material efficiency, not on the individual factors of it. At the present-day, global warming and greenhouse gas emissions are among the biggest global problems, on which material efficiency has a direct and significant impact on. Therefore, this paper suggests that greenhouse gas emissions could be used as an indicator for material efficiency in building.

## CONCLUSION

Material efficiency is emphasized as an important aspect of sustainable building, as indicated by the inclusion of the ADP aspect in EN 15804 [24] and EN 15978 [25] and the inclusion of the new basic requirement for sustainable use of resources in the Construction Product Regulation [60]. The roadmap to a resource-efficient Europe [3] addresses buildings as one of the three key sectors. However, further research is still needed to clarify and draw conclusions about the correct indicators and methods to assess the material efficiency of buildings and construction.

This research studied the different aspects of material efficiency: scarcity, land use, and environmental impacts related to the manufacturing of materials.

The preliminary results received with the help of a comprehensive case study (which was aimed at all the materials used for the case building) revealed that basic building materials have only a minor effect on the results when assessed in terms of ADP elements (as recommended by ILCD [63]). Approximately 99% of building materials have no effect on the ADP value, and, thus, approximately 1% of the materials (by weight) determine the results. The basic building materials that affect the results are the metallic materials used in buildings (steel, aluminium and copper). The result also showed that very minor material flows (in terms of weight), such as lamps and solar panels, may have a significantly bigger effect than any of the basic building materials, including all the metal used. The result raises questions of whether the ADP elements assessment method is appropriate for the assessment of buildings and construction. On the other hand, the ADP fossil fuel calculations were able to capture the material impacts more effectively. When comparing the ADP fossil values from material-related sources with the values from operational energy use, the share of materials accounted for approximately 30% to 40% of the lifetime totals.

Despite the relatively low impact on the depletion of abiotic resources, the building materials still have local impacts on the landscape and natural environment. The impacts of the extraction of gravel on ground water may also be substantial on local level. The impact of land use of abiotic materials is small compared with the footprint of the building. The land use of the building itself dominates the results (unless the land used for wood used for heating energy production is taken into account). If the use of wood is taken into account, its impact dominates in terms of land use and but also with regard to biodiversity impacts.

The greenhouse gas emissions from building sector are examples of environmental problems, on which material efficiency has a significant impact on. Greenhouse gas emissions from buildings are affected by all the aspects of material efficiency, and improvements in material efficiency can have significant impacts on the amount of emissions. This paper suggests that greenhouse gas emissions could be used as an indicator for material efficiency in building.

## ACKNOWLEDGMENTS

This study was a part of the Sustainability and performance assessment and benchmarking of buildings (SuPerBuildings) project (FP7 EU Project 2010-2012) and the Ownership in sustainable building (OKRA) project (2011-2014) funded by TEKES – the Finnish Funding Agency for technology and Innovation.

## AUTHOR CONTRIBUTIONS

The article was done in collaboration of the authors. The case study was done by Antti Ruuska, the study of literature was done equally by both authors. The second author, Tarja Häkkinen was the supervisor while the main writing was done by Antti Ruuska.

# REFERENCES

1. The Federation of Finnish Technology Industries (Teknologiateollisuus) 2013. Kilpailukykyä ja uutta liiketoimintaa materiaalitehokkuudesta. (In Finnish). Available online: http://www.teknologiateollisuus.fi/file/15592/Materiaalitehokkuusjulkaisu2013.pdf.html (accessed on 13 February 2014).

2. European Commission. A resource-efficient Europe—Flagship initiative under the Europe 2020 Strategy. COM (2011) 21. Brussels, 26.1.2011. Available online: http://ec.europa.eu/resource-efficient-europe/pdf/resource_efficient_europe_en.pdf (accessed 23 June 2014).

3. European Commission. Roadmap to a Resource Efficient Europe. COM (2011) 571 final. Brussels, 20.9.2011. Available online: http://ec.europa.eu/environment/resource_efficiency/pdf/com2011_571.pdf (accessed 23 June 2014).

4. Berge, B. *The Ecology of Building Materials*, 2nd ed.; Elsevier: Italy, 2009; p. 427.

5. Azapagic, A. Developing a framework for sustainable development indicators for the mining and minerals industry. *J. Clean. Prod.* 2004, *12*, 639–662.

6. European Commission. Critical raw materials for the EU. Technical Report, June 2010. Available online: http://ec.europa.eu/enterprise/policies/raw-materials/files/docs/report-b_en.pdf (accessed on 21 January 2014).

7. Brent, A.C.; Hietkamp, S. The Impact of Mineral Resource Depletion. *Int. J. Life Cycle Assess.* 2006, *11*, 361–362.

8. Steen, B.A. Abiotic resource depletion. Different perceptions of the problem with mineral deposits. *Int. J. Life Cycle Assess.* 2006, *1*, 49–54.

9. Strauss, K.; Brent, A.; Hietkamp, S. Characterisation and normalisation factors for life cycle impact assessment mined abiotic resources categories in south Africa. The manufacturing of catalytic converter exhaust systems as a case study.*Int. J. Life Cycle Assess.* 2006, *11*, 162–171.

10. Meadows, D.H.; Randers, J.; Meadows, D.L. *Limits to Growth. The 30-Year Update*; Earthscan: Oxford, UK, 2005; p. 338.

11. Jolliet, O.; Müller-Wenk, R.; Bare, J.; Brent, A.; Goedkoop, M.; Heijnungs, R.; Itsubo, N.; Peña, C.; Pennington, D.; Potting, J.; *et al.* The LCIA Midpoint-damage Framework of the UNEP/SETAC Life Cycle Initiative. *Int. J. Life Cycle Manag.* 2004, *9*, 394–404.

12. Yellishetty, M.; Ranjith, P.G.; Tharumarajah, A.; Bhosale, S.A. Life cycle

assessment in the minerals and metals sector: A critical review of selected issues and challenges. *Int. J. Life Cycle Assess.* 2009, *14*, 257–267.

13. Finnveden, G. The Resource Debate Needs to Continue. *Int. J. Life Cycle Assess.* 2005, *10*, 372–372.

14. Söderholm, P.; Tilton, J.E. Material efficiency: An economic perspective. *Resour. Conserv. Recycl.* 2012, *61*, 75–82.

15. Vieira, M.D.M.; Ponsioen, T.C.; Goedkoop, M.J.; Huijbregts, M.A.J. Surplus cost as a life cycle impact indicator for mineral resource scarcity. Available online: http://www.lc-impact.eu/userfiles/D_1_4_mineral_ and_fossil_resource_use.pdf (accessed on 10 December 2013).

16. Ponsionen, T.C.; Vieira, M.D.M.; Goedkoop, M.J. Surplus cost as a life cycle impact indicator for fossil resource depletion. Available online: http://www.lc-impact.eu/userfiles/D_1_4_mineral_and_fossil_resource_ use.pdf (accessed on 10 December 2013).

17. Rees, W.; Wackernagel, M. Urban ecological footprints: Why cities cannot be sustainable—And why they are a key to sustainability. *Environ. Impact Assess. Rev.* 1996, *16*, 223–248.

18. Haberl, H.; Wackernagel, M.; Krausmann, F.; Erb, K.-H.; Monfreda, C. Ecological footprints and human appropriation of net primary production: A comparison. *Land Use Policy* 2004, *21*, 279–288.

19. European Environment Agency. Land accounts for Europe 1990–2000, towards integrated land and ecosystem accounting, EEA Report No 11/2006. Available online: http://www.eea.europa.eu/publications/eea_ report_2006_11(accessed 23 June 2014).

20. Eurostat. *Economy-wide Material Flow Accounts and Derived Indicators—A Methodological Guide*; Office for Official Publications of the European Communities: Luxembourg, Luxembourg, 2001. Available online: http://epp.eurostat.ec.europa.eu/portal/page/portal/ environmental_accounts/documents/3.pdf (accessed 23 June 2014).

21. Van der Voet, E.; van Oers, L.; Moll, S.; Schütz, H.; Bringezu, S.; de Bruyn, S.; Sevenster, M.; Warringa, G. *Policy Review on Decoupling: Development of Indicators to Assess Decoupling of Economic Development and Environmental Pressure in the EU-25 and AC-3 Countries*; Department Industrial Ecology, Leiden University: Leiden, The Netherlands, 2004.

22. Tukker, A.; Huppes, G.; van Oers, L.; Heijungs, R. *Environmentally Extended Input-Output Tables and Models for Europe*; Eder, P., Delgado, L., Neuwahl, F., Eds.; EC, JRC, IPTS Technical Report Series; EUR 22194 EN. Seville, Spain, 2006. Available online: http://ftp.jrc.es/

EURdoc/eur22194en.pdf (accessed on 23 June 2014).

23. Berger, M.; Finkbeiner, M. Correlation analysis of life cycle impact assessment indicators measuring resource use. *Int. J. Life Cycle Assess.* 2011, *16*, 75–81.

24. EN 15804:2012 Sustainability of construction works—Environmental product declarations—Core rules for the product category of construction products. 2012. Available online: http://standards.cen.eu/dyn/www/ f?p=204:110:0::::FSP_PROJECT:40703&cs=1C696AB3A6B08F09003 DC00E3E3B2DA17 (accessed 23 June 2014).

25. EN 15978:2011. Sustainability of construction works—Assessment of environmental performance of buildings—Calculation method. 2011. Available online: http://standards.cen.eu/dyn/www/ f?p=204:110:0::::FSP_PROJECT:31325&cs=16BA44316931 8FC086C 4652D797E50C47 (accessed 23 June 2014).

26. Stewart, M.; Weidema, B.P. A Consistent Framework for Assessing the Impacts from Resource Use—A focus on resource functionality. *Int. J. Life Cycle Assess.* 2005, *10*, 240–247.

27. Pennington, D.W.; Potting, J.; Finnveden, G.; Lindeijer, E.; Jolliet, O.; Rydberg, T.; Rebitzer, G. Life cycle assessment Part 2: Current impact assessment practice. *Environ. Int.* 2004, *30*, 721–739.

28. Ruuska, A.; Häkkinen, T.; Vares, S.; Korhonen, M.-R.; Myllymaa, T. Environmental Impacts of Building Materials (Rakennusmateriaalien ympäristövaikutukset), Finnish Ministry of Environment, 8/2013 (In Finnish). Available online: http://www. ym.fi/download/noname/%7B1FAF46B2-2649-41ED-B3AA-5EA789C9512F%7D/37571 (accessed on 31 January 2014).

29. Salmi, O.; Haapalehto, T.; Harlin, A.; Häkkinen, T.; Kangas, H.; Mroueh, U.-M.; Qvintus, P. *The Development of Material Efficiency in the Finnish Industries*; Technical Report for The Ministry of employment and the economy (In Finnish): Helsinki, Finland, 2013; p. 46.

30. Allwood, J.M.; Ashby, M.F.; Gutowski, T.G.; Worrell, E. Material efficiency: A white paper. *Resour. Conserv. Recycl.* 2011, *55*, 362–381.

31. Prior, T.; Giurco, D.; Mudd, D.; Mason, L.; Behrich, J. Resource depletion, peak minerals and the implications for sustainable resource management. *Glob. Environ. Chang.* 2012, *22*, 577–587.

32. Wouters, H.; Bol, D. *Material Scarcity*; Materials Innovation Institute: Delft, The Netherlands, 2009.

33. De Almeida, P.; Silva, P.D. The peak of oil production—Timings and

market recognition. *Energy Policy* 2009, *37*, 1267–1276.

34. European Commission. The raw materials initiative—Meeting our critical needs for growth and jobs in Europe. COM 699. Brussels, 4.11.2008. Available online: http://ec.europa.eu/enterprise/sectors/metals-minerals/files/com699_en.pdf(accessed 23 June 2014).

35. Ahmed, M.S.; Vidyadhara, H.S. Experimental study on strength behaviour of recycled aggregate concrete. *Int. J. Eng. Res. Technol.* 2013, *2*, 76–82.

36. Lohani, T.K.; Padhi, M.; Dash, K.P.; Jena, S. Optimum utilization of quarry dust as partial replacement of sand in concrete. *Int. J. Appl. Sci. Eng. Res.* 2012, *1*, 391–404.

37. Habert, G.; Bouzidi, Y.; Chen, C.; Jullien, A. Development of a depletion indicator for natural resources used in concrete. *Resour. Conserv. Recycl.* 2010, *54*, 364–376.

38. Yellishetty, M.; Mudd, G.M.; Ranjith, P.G. The steel industry, abiotic resource depletion and life cycle assessment: A real or perceived issue? *J. Clean. Prod.* 2011, *19*, 78–90.

39. U.S. Department of Energy, Critical Materials Strategy Summary. 2011. Available online: http://energy.gov/sites/prod/files/DOE_CMS2011_FINAL_Full.pdf (accessed on 7 January 2014).

40. Gleich, B.; Achzet, B.; Mayer, H.; Rathgeber, A. An empirical approach to determine specific weights of driving factors for the price of commodities—A contribution to the measurement of the economic scarcity of minerals and metals. *Resour. Policy* 2013, *38*, 350–362.

41. Sartori, I.; Hestnes, A.G. Energy use in the life cycle of conventional and low-energy buildings: A review article.*Energy Build.* 2007, *39*, 249–257.

42. Hernandez, P.; Kenny, P. Development of a methodology for life cycle building energy ratings. *Energy Policy* 2011, *39*, 2779–3788.

43. Stephan, A.; Crawford, R.H.; de Myttenaere, K. Towards a more holistic approach to reducing the energy demand of dwellings. In Proceedings of the 2011 International Conference on Green Buildings and Sustainable Cities, Procedia Engineering, Bologna, Italy, 15–16 September; 2011; pp. 1033–1041.

44. Ramesh, T.; Prakash, R.; Shukla, K.K. Life cycle energy analysis of buildings: An overview. *Energy Build.* 2010, *42*, 1592–1600.

45. Yung, P.; Lam, K.C.; Yu, C. An audit of life cycle energy analyses of buildings. *Habitat Int.* 2013, *39*, 43–54.

46. Aktas, C.B.; Bilec, M.M. Impact of lifetime on US residential building LCA results. *Int. J. Life Cycle Assess.* 2011, *17*, 337–349.

47. Rauf, A.; Crawford, R.H. The relationship between material service life and the life cycle energy of contemporary residential buildings in Australia. *Archit. Sci. Rev.* 2013, *56*, 252–261.

48. Stephan, A.; Crawford, R.H.; de Myttenaere, K. Multi-scale life cycle energy analysis of a low-density suburban neighbourhood in Melbourne, Australia. *Build. Environ.* 2013, *68*, 35–49.

49. Fuller, R.J.; Crawford, R.H. Impact of past and future residential housing development patterns on energy demand and related emissions. *J. Hous. Build. Environ.* 2011, *26*, 165–183.

50. Säynäjoki, A.; Heinonen, J.; Junnila, S. A scenario analysis of the life cycle greenhouse gas emissions of a new residential area. *Environ. Res. Lett.* 2012, *7*, 034037:1–034037:10.

51. Milà i Canals, L.; Bauer, C.; Depestele, J.; Dubreuil, A.; Freiermuth Knuchel, R.; Gaillard, G.; Michelsen, O.; Müller-Wenk, R.; Rydgren, B. Key elements in a framework for land use impact assessment within LCA. *Int. J. Cycle Assess.* 2007, *12*, 5–15.

52. Marshall, E.; Shortle, J. Urban Development Impacts on Ecosystems. In *Land Use Problems and Conflicts: Causes Consequences and Solutions*; Goetz, S., Shortle, J., Bergstrom, J., Eds.; Routledge Publishing: New York, NY, USA, 2005.

53. Scalenghe, R.; Marsan, F.A. The anthropogenic sealing of soils in urban areas. *Landsc. Urban Plan.* 2009, *90*, 1–10.

54. Wheater, H.; Evans, E. Land use, water management and future flood risk. *Land Use Policy* 2009, *26*, 251–264.

55. Häkkinen, T.; Helin, T.; Antuña, C.; Supper, S.; Schiopu, N.; Nibel, S. Land Use as an Aspect of Sustainable Building. *Int. J. Sustain. Land Use Urban Plan.* 2013, *1*, 21–41.

56. Finnish Ministry of Environment. Sustainable use of soil materials. Guidelines of the Ministry of environment. Available online: http://www.ymparisto.fi/download.asp?contentid=101195&lan=fi (accessed on 30 August 2013).

57. Goodrum, P.M.; Zhai, D.; Yasin, M. Relationship between Changes in Material Technology and Construction Productivity. *J. Constr. Eng. Manag.* 2009, *135*, 278–287.

58. European Parliament, European Council, Directive 2002/91/EC of the European Parliament and of the Council of 16 December 2002 on the energy performance of buildings, 2002. Available online: http://eur-lex.europa.eu/legal-content/EN/TXT/PDF/?uri=CELEX:32002L0091&fro

m=EN (accessed 23 June 2014).

59. European Parliament, European Council, Directive 2010/31/EU of the European Parliament and of the Council of 19 May 2010 on the energy performance of buildings, 2010. Available online: http://eur-lex.europa. eu/LexUriServ/LexUriServ.do?uri=OJ:L:2010:153:0013:0035:EN:PDF (accessed 23 June 2014).

60. Cansino, J.M.; Pablo-Romero, M.P.; Román, R.; Yñiguez, R. Promoting renewable energy sources for heating and cooling in EU-27 countries. Assessment of the sustainable building steering mechanisms in selected EU member states. *Energy Policy* 2011, *39*, 3803–3812.

61. Tuominen, P.; Klobut, K.; Tolman, A.; Adjei, A.; de Best-Waldhober, M. Energy savings potential in buildings and overcoming market barriers in member states of the European Union. *Energy Build.* 2012, *51*, 48–55.

62. European Union. The Construction Products Regulation (EU) No 305/2011 (CPR). 2011. Available online: http://eur-lex.europa.eu/ LexUriServ/LexUriServ.do?uri=OJ:L:2011:088:0005:0043:EN:PDF (accessed 23 June 2014).

63. European Commission Joint Research Centre (JRC). *ILCD Handbook: Recommendations for Life Cycle Impact Assessment in the European Context*; Publication Office of the European Union: Luxembourg, Luxembourg, 2011. Available online: http://publications.jrc.ec.europa. eu/repository/bitstream/111111111/26229/1/jrc61049_ilcd%20 handbook%20final.pdf(accessed on 4 February 2014).

64. Guinée, J.B.; Gorrée, M.; Heijungs, R.; Huppes, G.; Kleijn, R.; Koning, A.; de Oers, L.; van Wegener Sleeswijk, A.; Suh, S.; de Haes, H.A.; *et al. Handbook on Life Cycle Assessment. Operational Guide to the ISO Standards. I: LCA in Perspective. IIa: Guide. IIb: Operational annex. III: Scientific background*; Kluwer Academic Publishers: Dordrecht, The Netherlands, 2002; p. 692.

65. Van Oers, L.; de Koning, A.; Guinée, J.B.; Huppes, G. *Abiotic Resource Depletion in LCA, Improving Characterization Factors for Abiotic Resource Depletion as Recommended in the New Dutch LCA Handbook*; Road and hydraulic engineering institute: Amsterdam, The Netherlands, 2002.

66. EN ISO 14040:2006. In *Environmental Management. Life Cycle Assessment. Principles and Framework*; The International Organization for standardization: London, UK, 2006.

67. Crawford, R.H. Validation of a hybrid life-cycle inventory analysis method. *J. Environ. Manag.* 2008, *88*, 496–506.

68. Suh, S.; Lenzen, M.; Treloar, G.J.; Hondo, H.; Horvath, A.; Huppes, G.; Jolliet, O.; Klann, U.; Krewitt, W.; Moriguchi, Y.; Munksgaard, J.; Norris, G. Critical review: System Boundary Selection in Life-Cycle Inventories Using Hybrid Approaches. *Environ. Sci. Technol.* 2004, *38*, 657–664.

69. Treloar, G.J. Extracting Embodied Energy Paths from Input-Output Tables: Towards an Input-Output-based Hybrid Energy Analysis Method. *Econ. Syst. Res.* 1997, *9*, 375–391.

70. Dixit, M.K.; Culp, C.H.; Férnandez-Solís, J.L. System boundary for embodied energy in buildings: A conceptual model for definition. *Renew. Sustain. Energy Rev.* 2013, *21*, 153–164.

71. European Commission Joint Research Centre (JRC). European reference Life Cycle Database, ELCD, European Platform on Life Cycle Assessment. Available online: http://eplca.jrc.ec.europa.eu/?page_id=126 (accessed on 23 June 2014).

72. University of Leiden, CML-IA Characterisation Factors, CML, 2013. Available online: http://www.leidenuniv.nl/cml/ssp/databases/cmlia/cmlia.zip (accessed on 26 January 2014).

73. Soules, T.F.; Whitman, P.K.; Chirayath, D.R. Fluorescent lamp with phosphor coating of multiple layers. European Patent Specification EP 0807958B1, 30 October 2012.

74. U.S. Environmental Protection Agency. *Office of Solid Waste, Mercury Emissions from Disposal of Fluorescent Lamps*; Office of Solid Waste U.S. Environmental Protection Agency: Washington, DC, USA; 30; June; 1997.

75. European aluminium association (EAA). Aluminium extrusion profile, LCI data set, European reference Life Cycle Database, ELCD. 2013; Permanent dataset URI. Available online: http://lca.jrc.ec.europa.eu/lcainfohub/datasets/elcd/processes/09215eb0-5fc9-11dd-ad8b-0800200c9a66.xml (accessed on 23 June 2014).

76. PE International, Pre-cast concrete, LCI data set. European reference Life Cycle Database, ELCD. 2013; Permanent dataset URI. Available online: http://lca.jrc.ec.europa.eu/lcainfohub/datasets/elcd/processes/898618b0-3306-11dd-bd11-0800200c9a66.xml (accessed on 23 June 2014).

77. European Copper Institute, Copper tube, LCI data set. European reference Life Cycle Database, ELCD. 2013; Permanent dataset URI. Available online: http://lca.jrc.ec.europa.eu/lcainfohub/datasets/elcd/contacts/42a11490-573c-11dd-ae16-0800200c9a66.xml (accessed on 23 June 2014).

78. PE International, Polypropylene fibres (PP), LCI data set. European

reference Life Cycle Database, ELCD. 2013; Permanent dataset URI. Available online: http://lca.jrc.ec.europa.eu/lcainfohub/datasets/elcd/processes/db00901b-338f-11dd-bd11-0800200c9a66.xml (accessed on 23 June 2014).

79.  PE International, Gravel 2/32, LCI data set. European reference Life Cycle Database, ELCD. 2013; Permanent dataset URI. Available online: http://lca.jrc.ec.europa.eu/lcainfohub/datasets/elcd/processes/898618b2-3306-11dd-bd11-0800200c9a66.xml (accessed on 23 June 2014).

80.  Eurogypsum, Gypsum Plasterboard, LCI data set. European reference Life Cycle Database, ELCD. 2013. Available online: http://lca.jrc.ec.europa.eu/lcainfohub/datasets/elcd/processes/cc39e70e-4a40-42b6-89e3-7305f0b95dc4.xml(accessed on 23 June 2014).

81.  Steel sections, LCI data set. Worldsteel, European reference Life Cycle Database, ELCD. 2013; Permanent dataset URI. Available online: http://elcd.jrc.ec.europa.eu/ELCD3/resource/processes/09d61948-238a-40e7-8e1f-afdc0c98f902?format=html&version=03.00.000 (accessed on 23 June 2014).

82.  PE-International, Pine wood, European reference Life Cycle Database, ELCD. 2013; Permanent dataset URI. Available online: http://lca.jrc.ec.europa.eu/lcainfohub/datasets/elcd/processes/621e64d0-f471-4023-9ebc-a52cd8ee573f.xml(accessed on 23 June 2014).

83.  PE-International, Particle board, European reference Life Cycle Database, ELCD. 2013; Permanent dataset URI. Available online: http://lca.jrc.ec.europa.eu/lcainfohub/datasets/elcd/processes/bd7fdac9-40d5-4613-9374-6969803269d9.xml (accessed on 23 June 2014).

84.  Cembureau, Portland cement, European reference Life Cycle Database, ELCD. 2013; Permanent dataset URI. Available online: http://eplca.jrc.ec.europa.eu/ELCD3/resource/processes/600573dd-dfa5-44e5-b458-8727e793ffd7.xml(accessed on 23 June 2014).

85.  European Lime Association, Quicklime, European reference Life Cycle Database, ELCD. 2013; Permanent dataset URI. Available online: http://eplca.jrc.ec.europa.eu/ELCD3/resource/sources/7983f4c6-a355-4250-aaa8-5780a72cc1df.xml(accessed on 23 June 2014).

86.  European Commission Joint Research Centre (JRC). *Characterisation Factors of the ILCD, Recommended Life Cycle Impact Assessment Methods, Database and Supporting Information, JRC Technical Notes*; European Union, Publications office of the European Union: Luxembourg, Luxembourg, 2012.

87.  Bio Intelligence Service. *Study on Photovoltaic Panels Supplementing*

*the Impact Sssessment for a Recast of the WEEE Directive*, Final report to European Commission DG ENV. 14 April 2011. Available online: http://ec.europa.eu/environment/waste/weee/pdf/Study%20on%20PVs%20Bio%20final.pdf (accessed on 23 June 2014).

88. LIPASTO—A calculation system for traffic exhaust emissions and energy consumption in Finland, VTT Technical Research Centre of Finland. Available online: http://lipasto.vtt.fi/indexe.htm (accessed on 21 January 2014).

89. Keto, M. *Energy Factors, General Principles and Factors for Realized Production of Electricity and District Heating*; Technical Report for The Ministry of Environment; Aalto University: Espoo, Finland; November; 2010.

90. Adalberth, K.; Almgren, A.; Holleris, P.E. Life-Cycle assessment of four multi-family buildings. *Int. J. Low Energy Sustain. Build.* 2001, *2*, 1–21.

91. Statistics Finland, Energy Statistics Year book 2011. Available online: http://www.stat.fi/tup/julkaisut/tiedostot/julkaisuluettelo/yene_enev_201100_2012_6164_net.pdf (accessed on 23 June 2014).

92. Blom, I.; Itard, L.; Meijer, A. Environmental impact of building-related and user-related energy consumption in dwellings. *Build. Environ.* 2011, *46*, 1657–1669.

93. Häkkinen, T. *Sustainable Refurbishment of Exterior Walls and Building Facades*; VTT: Espoo, Finland, 2012. Available online: http://www.vtt.fi/inf/pdf/technology/2012/T30.pdf (accessed on 23 June 2014).

# Chapter 5

# GREEN TEMPLATE FOR LIFE CYCLE ASSESSMENT OF BUILDINGS BASED ON BUILDING INFORMATION MODELING: FOCUS ON EMBODIED ENVIRONMENTAL IMPACT

Sungwoo Lee[1], Sungho Tae[2], Seungjun Roh[1], and Taehyung Kim[1]

[1]Architectural Engineering, Hanyang University, Sa 3-dong, Sangrok-gu Ansan 426-791, Korea

[2]School of Architecture & Architectural Engineering, Hanyang University, Sa 3-dong, Sangrok-gu Ansan 426-791, Korea

## ABSTRACT

The increased popularity of building information modeling (BIM) for application in the construction of eco-friendly green buildings has given rise to techniques for evaluating green buildings constructed using BIM features. Existing BIM-based green building evaluation techniques mostly rely on externally provided evaluation tools, which pose problems associated with interoperability, including a lack of data compatibility and the amount of time required for format conversion. To overcome these problems, this study sets out to develop a template (the "green template") for evaluating the embodied environmental impact of using a BIM design tool as part of BIM-based building life-cycle assessment (LCA) technology development. Firstly, the BIM level of detail (LOD) was determined to evaluate the embodied environmental impact, and constructed a database of the impact factors of the embodied environmental impact of the major building materials, thereby adopting an LCA-based approach. The libraries of major building elements were developed by using the established databases and compiled evaluation table of the embodied environmental impact of the building materials. Finally, the green template was developed as an embodied environmental impact evaluation tool and a case study was performed to test its applicability. The results of the green template-based embodied environmental impact evaluation of a test building were validated against those of its actual quantity takeoff (2D takeoff), and its reliability was confirmed by an effective error rate of ≤5%. This study aims to develop a system for assessing the impact of the substances discharged from concrete production process on six environmental impact categories, *i.e.*, global warming (GWP), acidification (AP), eutrophication (EP), abiotic depletion (ADP), ozone depletion (ODP), and photochemical oxidant creation (POCP), using the life a cycle assessment (LCA) method. To achieve this, we proposed an LCA method specifically applicable to concrete and tailored to the Korean concrete industry by adapting the ISO standards to suit the Korean situations.

## INTRODUCTION

Since the mid-2000s, building information modeling (BIM) has increasingly been used for architectural design because it offers not only 3D building components for architectural drawing but also the detailed elements and attribute information necessary for quantity takeoff, cost reduction, and process management [1,2]. Along with the growing popularity of BIM, various technologies have been developed to support work processes using the salient features of BIM across the construction and building industry (e.g., architectural design, construction, and interior design), such that tools developed using these technologies are now used in related work processes. The number of BIM-based green building design cases is on the increase [3]. BIM-based energy simulation tools are used for predicting energy savings at the design phase of low-energy buildings [4,5,6]. In particular, Cofaigh *et al.* configured the shape and orientation of a low-energy building by using BIM, achieving a 40% reduction in the environmental load and financial burden relative to a conventional building shape and orientation [7]. Wang and Zmeureanu used a BIM-based energy analysis simulation tool to analyze the environmental impacts of different building materials and established parameters that were optimized for environmental impact evaluation [8].

BIM supports the individual input of attribute data for the major and auxiliary materials required for environmental impact evaluation. Using this salient feature of BIM, an increasing number of BIM-based studies are being undertaken to develop technologies for evaluating the embodied environmental impact of buildings, embracing the entire process of building materials production, construction, and demolition. IMPACT, developed by the UK Building Research Establishment (BRE) is representative of these technologies. IMPACT was designed to evaluate a range of environmental impacts by extracting BIM-implemented building structures in the standard Industry Foundation Classes(IFC) and gbXML BIM data formats. However, currently available embodied environmental impact evaluation tools, including IMPACT, lack the inter-data compatibility required during work sessions despite the use of the IFC and gbXML formats, while some of the standard data formats are entirely unsupported, which lowers the reliability of the results of an embodied environmental impact evaluation and increases the processing time, given the need for format conversion to ensure interoperability [9].

Against this background, this study was undertaken with the aim of developing a green template capable of evaluating embodied environmental impacts through the use of BIM tools, as part of life-cycle assessment (LCA) technology R&D.

## LITERATURE REVIEW

### BIM-Based Evaluation of Environmental Performance of a Building

BIM enables the highly efficient management of building material inventory by inputting a range of attribute data for each classified unit of the materials required for the environmental performance evaluation of a building. Simulations of building energy analysis were performed after implementing software interoperability, using international industrial standard formats such as IFC and gbXML [10].

The Korea Land and Housing Corporation and Public Procurement launched a design competition based on the outcomes of BIM-based building energy simulation [11]. The Korea Land and Housing Corporation called a design competition and evaluated eco-friendly design, insolation duration analysis, and energy analysis using IES/VE which is a building energy evaluation program and calculated an optimal layout from a range of building block layout plans. Additionally, using the Revit table feature, a quantity takeoff for each building block was derived, which was used to determine the $CO_2$ emission and energy efficiency grades [12,13].

The Norwegian government is promoting the application of BIM in the construction sector by providing a BIM manual. Architectural designers are selected via a two-step evaluation of their architectural designs. The first step involves a basic BIM model examination including a building concept and structural overview. Then, Step 2 requires the submission of $CO_2$ emission calculations along with a detailed BIM model [14]. BuildingSMART International implemented building modeling using ArchiCAD and performed building block layout and window planning using the results of Ecotect energy simulations such as insolation analysis based on the annual mean insolation and a daylight and shadow pattern analysis [15]. Project Chicago is a representative academia-industry collaboration initiated by the Autodesk Green Building Research Team in 2007 to deal with building modeling in the design process, an energy performance evaluation, and sustainable test methods, whereby a collaboration user interface was developed for building energy performance evaluation by introducing a multi-touch screen and using an instant feedback scenario [16]. Sutter Medical Center acquired LEED Silver for healthcare rating with a BIM energy simulation for the BIM-based reconstruction project commissioned by the State Government of California. The design team was granted financial support from the State Government upon the approval of the environmental impact report (EIR) on the building operational stage [17].

A review of the BIM-based building environmental performance evaluation methods revealed that most focus on energy simulation at the operational stage of a building with specific software tools.

## Use of Template in BIM

A BIM template is an input form with a predetermined structure and is intended for repeated use, being uniformly configured to obtain output values suitable for a specific evaluation scope and purpose, with the ultimate goal of acquiring the necessary BIM modeling information [18]. The user can acquire the desired outcomes by entering the data requested by the template. BIM templates are directly applied to work processes, and library-related data standardization helps designers to derive standardized information and outcomes in a consistent work environment. BIM templates can be classified according to the purpose of using BIM and their formats can vary greatly according to the purpose of the development.

The spatial BIM template developed by the U.S. General Services Administration (GSA) is utilized at the planning and design stage to check whether the design elements satisfy the space requirements. It uses IFC parameters to enhance the system coordination [19]. The Revit Start Kit developed by Japanese Autodesk also supports legal administrative affairs at an early stage of the design process by providing guidelines for checking the basic design elements against the graphic standards and construction-related laws and regulations, in addition to the template. The Korea Architecture BIM (KABIM) developed by a Korean architectural design company helps build up a coordination system within a company using Revit Architecture, which is a BIM program for small- and medium-size architectural companies. The Green BIM Template (GBT), also developed in Korea, helps satisfy the requirements of the Green Building Certification criteria by offering features for environmental analysis and simulation. Specifically, it offers a data input/

output environment for the template-based checking of a certificate program [20].

# RESEARCH METHODS

This chapter describes the process of generating the library and overview table for the main building elements for the evaluation of the embodied environmental impact of buildings within the scope of the BIM-based tools. Its application is limited to Autodesk Revit, the most frequently used BIM authoring tool in Korea. The major building materials for the embodied environmental impact evaluation include ready-mixed concrete, steel, glass, concrete block, insulation material, and gypsum board, based on the analysis results of a previous study [21].

## Generation of Major Building Element Library

### *Determination of LOD of BIM for the Embodied Environmental Impact Evaluation*

Figure 1 depicts the results of analyzing major building materials for which the environmental impacts can be derived according to the level of detail (LOD). The level of BIM was defined by the American Institute of Architects (AIA) to establish evaluation standards for the embodied environmental impact evaluation [22].

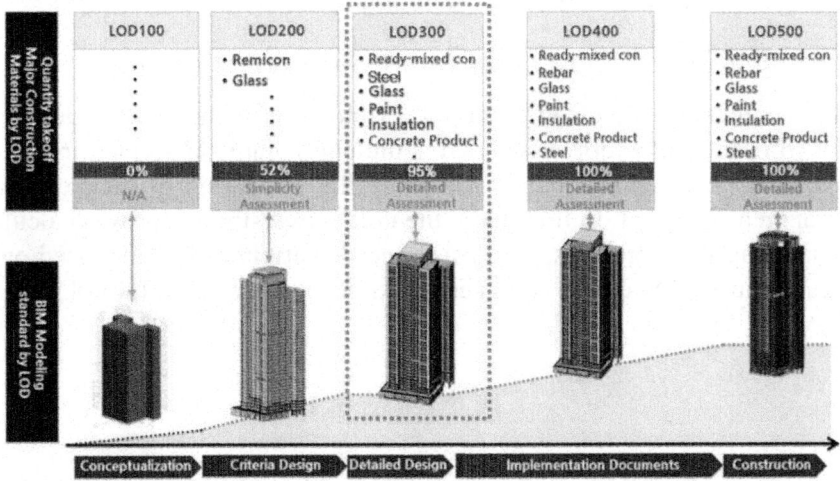

**Figure 1:** Material selection for library using AIA BIM guideline.

It has been shown therein that, of the main building materials for the embodied environmental impact evaluation (ready-mixed concrete, steel, glass, concrete block, insulation material, and gypsum board), the environmental impacts of five of those materials (except for steel) could be calculated at LOD 300 [23]. Therefore, this study set LOD 300 as the reference LOD for the evaluation of the embodied environmental impact of a building.

## Generation of a Database for the Embodied Environmental Impact Factor of Major Building Materials

Table 1 outlines the impact factors of major building materials by impact category (global warming potential (GWP), abiotic depletion potential (ADP), acidification potential (AP), eutrophication potential (EP), ozone depletion potential (ODP), and photochemical ozone creation potential (POCP)) as derived from analyses of the Korea life-cycle inventory (LCI) databases, compiled by the Ministry of Knowledge Economy and the Ministry of Environment and the national database for environmental assessment of building materials, compiled by the Ministry of Land, Infrastructure and Transport [24,25,26].

Additionally, the scenarios were applied for all stages of the life-cycles of the six major building materials and established by a calculation process for determining the environmental impact factor as a basis for evaluating the embodied environmental impact. This process involves the input of LCI databases, transportation information, and the fuel efficiency of the major building materials and the output of the embodied environmental impact factor databases of the major building materials according to preconfigured scenarios.

## Determination of Conversion Factor for Impact Factor

Since the quantity takeoff information for building materials provided by the Revit BIM authoring tool is obtained in units of volume or surface area, a unit conversion factor should be set to enable the evaluation of their embodied environmental impact. In this study, the unit conversion factor was calculated by analyzing the material-specific size and density information, based on the Korean Industrial Standard (KS) certification specifications (Table 1).

**Table 1:** Impact factors of major building materials by impact category

| Major Materials | Units | Environmental Impact Database | | | | | | Unit Conversion Factor | | |
|---|---|---|---|---|---|---|---|---|---|---|
| | | GWP (kg-CO$_{2eq}$) | ADP (kg) | AP (kg-SO$_{2eq}$) | EP (kg-PO$_4^{3-}$ eq) | ODP (kg-CFC-11$_{eq}$) | POCP (kg-Ethylene$_{eq}$) | Revit | 6EI | Factor |
| Concrete | m² | 419 | 1.56 | 0.6940 | 0.0818 | 0.0000461 | 1.13 | m³ | m³ | 1 |
| Rebar | ton | 352 | 2.79 | 2.3100 | 0.3480 | 0.0001104 | 0.3410 | m³ | ton | 7.85 |
| Steel | ton | 405 | 1.12 | 0.6450 | 0.1170 | 0.0000226 | 0.2930 | m³ | ton | 7.85 |
| Glass | ton | 788 | 6.97 | 3.6700 | 0.0523 | 0.0003040 | 0.8950 | m³ | ton | 3.45 |
| Concrete block | EA | 246 | 0.292 | 0.3140 | 0.0454 | 0.0000094 | 0.0262 | m² | 1000EA | 75 |
| Insulation | ton | 2060 | 174.000 | 40.50 | 2.75 | 0.0000289 | 6.390 | m³ | ton | 0.03 |
| Gypsum board | ton | 0.192 | 0.016 | 0.0313 | 0.00528 | 0.000000567 | 0.00761 | m³ | ton | 2.3 |

GWP: Global warming potential; POCP: Photochemical ozone creation potential; EP: Eutrophication potential; ODP: Ozone depletion potential; AP: Acidification potential; ADP: Abiotic depletion potential.

## Building Materials Classification System for Library Construction

A library of the major building materials was established in line with the construction information standard classification system that is internationally applied to construction, civil engineering, and plant classification [27].

Table 2 presents the classification system used in this study to systemize the building element library composition. The building materials were first divided into basic and secondary structural materials at Level 1. The Level 1 elements were then broken down into Level 2 elements (columns, beams, walls, slabs, and wall framings) depending on the component parts. Each Level 2 element was further broken down into Level 3 elements (concrete, steel, steel reinforced concrete, load-bearing walls, non-load-bearing walls, partitions, floors, and windows). By subdividing the Level 3 elements by individual input materials, a library of 34 building elements was constructed for standardized use. For example, by classifying walls as a Level 2 elements belonging to the basic structure and breaking these walls down into load-bearing and non-load-bearing walls at Level 3, a standard classification of building materials by element was implemented.

**Table 2:** Building materials classification system

| Building Materials Classification System | | | Library Items |
|---|---|---|---|
| Level 1 | Level 2 | Level 3 | Individual materials |
| Basic structure | Column | Concrete | Rectangular RC column |
| | | | Square RC column |
| | | | Round RC for column |
| | | Steel | H-shaped steel for column |
| | | | Square steel for column |
| | | | Round steel for column |
| | | SRC | Square SRC column |
| | | | Rectangular SRC column |
| | | | Round SRC column |
| | Beam | Concrete | H-shaped steel beam |
| | | | Square steel beam |
| | | | Round steel beam |
| | | | I-shaped steel beam |
| | | | L-shaped steel beam (equal legs) |
| | | | L-shaped steel beam (uneq. legs) |
| | | | L-shaped steel beam (thickness) |
| | | | C-shaped steel beam |
| | | | T-shaped steel beam |
| | | SRC | Rectangular SRC beam |
| | Wall | Load-bearing wall | Outer wall RC |
| | | | Inner wall RC |
| | | Non-load-bearing wall | Outer wall concrete block |
| | | | Inner wall concrete block |
| | | | Outer wall clay brick |
| | | | Inner wall clay brick |
| | | | Outer wall concrete block |
| | | | Inner wall concrete block |
| | | | Outer wall RC + concrete block |
| | | | Outer wall RC + clay brick |
| | | | Insulation |
| | Slab | Floor | Slab |
| Secondary structure | Wall framing | Partition | Secondary lightweight partition |
| | | Window | Single-hung window |
| | | | Double-hung window |

## Development of a Parametric Building Element Library

Parametric modeling is a BIM modeling technique for defining the inter-object relationships and operating data according to those relationships [28]. Accordingly, if the size of a structure changes, all of the information associated with the structure through relationship definition changes automatically [29]. This enables an instant output check for each aspect of the embodied environmental impact evaluation data whenever a design parameter changes, regardless of the complexity of the drawings.

For the construction of the building element library, the Revit content library and component library writing methods were used. A component library can be configured as a library file (*rfa), which can be reused in a different project and shared with other designers. A system library exists only in a template file as a library that is graphically associated with the predefined parametric set. Consequently, we applied the system library writing method to the wall and slab elements.

The component library writing method to the beam, column, and secondary wall-framing elements. Additionally, to match the information necessary for LCA with the library content, we utilized a shared parameter input method that is identical to that for the BIM data input method. Figure 2 illustrates an example of a parametric building element library developed as part of this study.

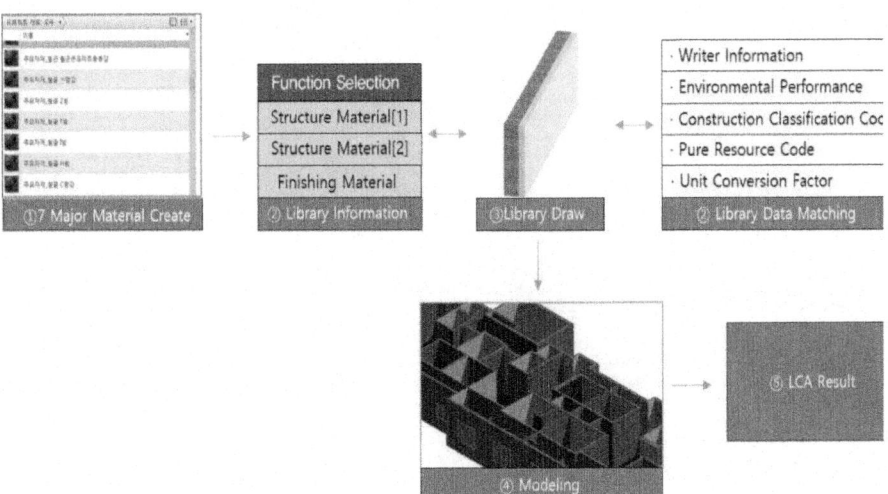

**Figure 2:** Parametric building element library Library production process(example : wall).

## Embodied Environmental Impact Evaluation Results Overview Table

Figure 3 shows the overview tables for the evaluated environmental impact of the major building materials modeled by the user on the basis of the major building element library. The overview table was integrated into the green template so that the embodied environmental impact evaluation results can be extracted in Revit without being linked to other programs used for the structure. This enables the user to directly check the evaluated values of the environmental impacts by the modeled building structure using the major building element library.

To sum up, the major building library evaluation result overview table provided in the green template presents the evaluated embodied environmental impacts of building elements (wall, floor, beam, column, and window). These two overview tables enable the user to directly check the embodied environmental impact of any given building element depending on its material composition, so that building contractors can use these two overview tables as the basis for efficient decision-making when selecting eco-friendly building materials [30]. First and foremost, along with the conventional building LCA technology, the evaluation result overview table of all the major building elements of a building structure can be efficiently used for quantity takeoff and material selection for building elements at the design stage, and it offers the great advantage of user convenience in terms of access.

Column Environmetal Impact Assessment Result

| Library Name | Type | Volume | Structure Material | U C F | GWP | GWP R | ADP | ADP R | AP | AP R | EP | EP R | ODP | ODP R | POCP | POCP R |
|---|---|---|---|---|---|---|---|---|---|---|---|---|---|---|---|---|
| Major element Lbrary Column RC | 600*60 | 1.44M3 | Major Material_Remicon_21MPa | 1 | 410 | 590.4 | 1.87 | 2.69 | 0.68 | 0.98 | 0.08 | 0.12 | 4.7E-05 | 6.8E-05 | 1.02 | 1.47 |
| Major element Lbrary Column RC | 600*60 | 1.44M3 | Major Material_Remicon_21MPa | 1 | 410 | 590.4 | 1.87 | 2.69 | 0.68 | 0.98 | 0.08 | 0.12 | 4.7E-05 | 6.8E-05 | 1.02 | 1.47 |
| Major element Lbrary Column RC | 600*60 | 0.72M3 | Major Material_Remicon_21MPa | 1 | 410 | 295.2 | 1.87 | 1.35 | 0.68 | 0.49 | 0.08 | 0.06 | 4.7E-05 | 3.4E-05 | 1.02 | 0.73 |
| Major element Lbrary Column RC | 600*60 | 0.72M3 | Major Material_Remicon_21MPa | 1 | 410 | 295.2 | 1.87 | 1.35 | 0.68 | 0.49 | 0.08 | 0.06 | 4.7E-05 | 3.4E-05 | 1.02 | 0.73 |
| Major element Lbrary Column RC | 600*60 | 0.72M3 | Major Material_Remicon_21MPa | 1 | 410 | 295.2 | 1.87 | 1.35 | 0.68 | 0.49 | 0.08 | 0.06 | 4.7E-05 | 3.4E-05 | 1.02 | 0.73 |
| Major element Lbrary Column RC | 600*60 | 0.72M3 | Major Material_Remicon_21MPa | 1 | 410 | 295.2 | 1.87 | 1.35 | 0.68 | 0.49 | 0.08 | 0.06 | 4.7E-05 | 3.4E-05 | 1.02 | 0.73 |
| Major element Lbrary Column RC | 600*60 | 0.72M3 | Major Material_Remicon_21MPa | 1 | 410 | 295.2 | 1.87 | 1.35 | 0.68 | 0.49 | 0.08 | 0.06 | 4.7E-05 | 3.4E-05 | 1.02 | 0.73 |
| Major element Lbrary Column RC | 600*60 | 0.72M3 | Major Material_Remicon_21MPa | 1 | 410 | 295.2 | 1.87 | 1.35 | 0.68 | 0.49 | 0.08 | 0.06 | 4.7E-05 | 3.4E-05 | 1.02 | 0.73 |
| Major element Lbrary Column RC | 600*60 | 0.72M3 | Major Material_Remicon_21MPa | 1 | 410 | 295.2 | 1.87 | 1.35 | 0.68 | 0.49 | 0.08 | 0.06 | 4.7E-05 | 3.4E-05 | 1.02 | 0.73 |
| Major element Lbrary Column RC | 600*60 | 0.72M3 | Major Material_Remicon_21MPa | 1 | 410 | 295.2 | 1.87 | 1.35 | 0.68 | 0.49 | 0.08 | 0.06 | 4.7E-05 | 3.4E-05 | 1.02 | 0.73 |
| Major element Lbrary Column RC | 600*60 | 0.72M3 | Major Material_Remicon_21MPa | 1 | 410 | 295.2 | 1.87 | 1.35 | 0.68 | 0.49 | 0.08 | 0.06 | 4.7E-05 | 3.4E-05 | 1.02 | 0.73 |
| Major element Lbrary Column RC | 600*60 | 0.72M3 | Major Material_Remicon_21MPa | 1 | 410 | 295.2 | 1.87 | 1.35 | 0.68 | 0.49 | 0.08 | 0.06 | 4.7E-05 | 3.4E-05 | 1.02 | 0.73 |
| Major element Lbrary Column RC | 600*60 | 0.72M3 | Major Material_Remicon_21MPa | 1 | 410 | 295.2 | 1.87 | 1.35 | 0.68 | 0.49 | 0.08 | 0.06 | 4.7E-05 | 3.4E-05 | 1.02 | 0.73 |
| Major element Lbrary Column RC | 600*60 | 0.72M3 | Major Material_Remicon_21MPa | 1 | 410 | 295.2 | 1.87 | 1.35 | 0.68 | 0.49 | 0.08 | 0.06 | 4.7E-05 | 3.4E-05 | 1.02 | 0.73 |
| Major element Lbrary Column RC | 600*60 | 0.72M3 | Major Material_Remicon_21MPa | 1 | 410 | 295.2 | 1.87 | 1.35 | 0.68 | 0.49 | 0.08 | 0.06 | 4.7E-05 | 3.4E-05 | 1.02 | 0.73 |
| Major element Lbrary Column RC | 600*60 | 0.72M3 | Major Material_Remicon_21MPa | 1 | 410 | 295.2 | 1.87 | 1.35 | 0.68 | 0.49 | 0.08 | 0.06 | 4.7E-05 | 3.4E-05 | 1.02 | 0.73 |
| Major element Lbrary Column RC | 600*60 | 0.72M3 | Major Material_Remicon_21MPa | 1 | 410 | 295.2 | 1.87 | 1.35 | 0.68 | 0.49 | 0.08 | 0.06 | 4.7E-05 | 3.4E-05 | 1.02 | 0.73 |
| Major element Lbrary Column RC | 600*60 | 0.72M3 | Major Material_Remicon_21MPa | 1 | 410 | 295.2 | 1.87 | 1.35 | 0.68 | 0.49 | 0.08 | 0.06 | 4.7E-05 | 3.4E-05 | 1.02 | 0.73 |
| TOTAL | | | | | | 7084.8 | | 32.31 | | 11.75 | | 1.38 | | 0.00081 | | 17.63 |

(a)

| Library Name | Type | Volume | Structure Material | UCF | GWP | GWP R | ADP | ADP R | AP | AP R |
|---|---|---|---|---|---|---|---|---|---|---|
| Major element Library Column RC | 600*60 | 1.44M3 | Major Material_Remicon_21MPa | 1 | 410 | 590.4 | 1.87 | 2.69 | 0.68 | 0.98 |
| Major element Library Beam Rebar H | 150*105 | 1.44M3 | Major Material_Remicon_21MPa | 1 | 410 | 590.4 | 1.87 | 2.69 | 0.68 | 0.98 |
| Major element Library Window | 1000*1000 | 0.72M3 | Major Material_Remicon_21MPa | 1 | 410 | 295.2 | 1.87 | 1.35 | 0.68 | 0.49 |
| Major element Library Window | 1000*1000 | 0.72M3 | Major Material_Remicon_21MPa | 1 | 410 | 295.2 | 1.87 | 1.35 | 0.68 | 0.49 |
| Major element Library Window | 1000*1000 | 0.72M3 | Major Material_Remicon_21MPa | 1 | 410 | 295.2 | 1.87 | 1.35 | 0.68 | 0.49 |
| Major element Library Column Rebar | 600*60 | 0.72M3 | Major Material_Remicon_21MPa | 1 | 410 | 295.2 | 1.87 | 1.35 | 0.68 | 0.49 |
| Major element Library Wall RC | 1500*1500*200 | 2.5M3 | Major Material_Remicon_21MPa | 1 | 410 | 295.2 | 1.87 | 1.35 | 0.68 | 0.49 |
| Major element Library Wall RC | 1500*1500*200 | 2.5M3 | Major Material_Remicon_21MPa | 1 | 410 | 295.2 | 1.87 | 1.35 | 0.68 | 0.49 |
| Major element Library Wall RC | 1500*1500*200 | 2.5M3 | Major Material_Remicon_21MPa | 1 | 410 | 295.2 | 1.87 | 1.35 | 0.68 | 0.49 |
| Major element Library Wall RC | 1500*1500*200 | 2.5M3 | Major Material_Remicon_21MPa | 1 | 410 | 295.2 | 1.87 | 1.35 | 0.68 | 0.49 |
| Major element Library Wall RC | 1500*1500*200 | 2.5M3 | Major Material_Remicon_21MPa | 1 | 410 | 295.2 | 1.87 | 1.35 | 0.68 | 0.49 |
| Major element Library Wall RC | 1500*1500*200 | 2.5M3 | Major Material_Remicon_21MPa | 1 | 410 | 295.2 | 1.87 | 1.35 | 0.68 | 0.49 |
| Major element Library Wall RC | 1500*1500*200 | 2.5M3 | Major Material_Remicon_21MPa | 1 | 410 | 295.2 | 1.87 | 1.35 | 0.68 | 0.49 |
| Major element Library Wall RC | 1500*1500*200 | 2.5M3 | Major Material_Remicon_21MPa | 1 | 410 | 295.2 | 1.87 | 1.35 | 0.68 | 0.49 |
| Major element Library Wall RC | 1500*1500*200 | 2.5M3 | Major Material_Remicon_21MPa | 1 | 410 | 295.2 | 1.87 | 1.35 | 0.68 | 0.49 |
| Major element Library Wall RC | 1500*1500*200 | 2.5M3 | Major Material_Remicon_21MPa | 1 | 410 | 295.2 | 1.87 | 1.35 | 0.68 | 0.49 |
| Major element Library Wall RC | 1500*1500*200 | 2.5M3 | Major Material_Remicon_21MPa | 1 | 410 | 295.2 | 1.87 | 1.35 | 0.68 | 0.49 |
| Major element Library Wall RC | 1500*1500*200 | 2.5M3 | Major Material_Remicon_21MPa | 1 | 410 | 295.2 | 1.87 | 1.35 | 0.68 | 0.49 |
| Major element Library Wall RC | 1500*1500*200 | 2.5M3 | Major Material_Remicon_21MPa | 1 | 410 | 295.2 | 1.87 | 1.35 | 0.68 | 0.49 |
| Major element Library Wall RC | 1500*1500*200 | 2.5M3 | Major Material_Remicon_21MPa | 1 | 410 | 295.2 | 1.87 | 1.35 | 0.68 | 0.49 |
| Major element Library Wall RC | 1500*1500*200 | 2.5M3 | Major Material_Remicon_21MPa | 1 | 410 | 295.2 | 1.87 | 1.35 | 0.68 | 0.49 |
| Major element Library Wall RC | 1500*1500*200 | 2.5M3 | Major Material_Remicon_21MPa | 1 | 410 | 295.2 | 1.87 | 1.35 | 0.68 | 0.49 |
| Major element Library Wall RC | 1500*1500*200 | 2.5M3 | Major Material_Remicon_21MPa | 1 | 410 | 295.2 | 1.87 | 1.35 | 0.68 | 0.49 |
| Major element Library Wall RC | 1500*1500*200 | 2.5M3 | Major Material_Remicon_21MPa | 1 | 410 | 295.2 | 1.87 | 1.35 | 0.68 | 0.49 |
| Major element Library Wall RC | 1500*1500*200 | 2.5M3 | Major Material_Remicon_21MPa | 1 | 410 | 295.2 | 1.87 | 1.35 | 0.68 | 0.49 |
| TOTAL | | | | | | 943.54 | | 5.77 | | 1.77 |

**(b)**

**Figure 3:** Overview tables of the LCA by building element. (**a**) Overview table of the building element library evaluation results; (**b**) Overview table of the embodied environmental impact evaluation results.

On the other hand, the embodied environmental impact evaluation result overview table was configured to allow the user to check the embodied environmental impacts of the individual elements of the building [31].

# DEVELOPMENT OF THE GREEN TEMPLATE

Figure 4 illustrates the composition of the green template developed in this study with a standard input form predefined in BIM software to enable the extraction of results suitable for the evaluation scope and purpose. In other words, the green template provides databases for the embodied environmental impact evaluation of a building at each modeling level, using Revit as the BIM authoring software. The designer can perform preliminary modeling for a building design using the green template guide, as shown in Figure 4. Furthermore, the green template guide contains instructions on how to construct a library and environmental information databases for major building elements. Using these features, the designer can develop a required building element library in addition to the existing libraries.

A designer can proceed to modeling using the building element library, thereby checking the evaluated embodied environmental impacts by impact category and LOD stage embedded in the green template. The green template-based modeling of a building structure should be implemented at the LOD 300 or a higher level, according to the BIM classification. The detail level here refers to the maturity of the information regarding the designations of the elements and materials used for the structure (ready-mixed concrete, glass, concrete block, insulation material, and gypsum board) as well as their respective volumes.

Furthermore, data on major building materials provided in the building element library of the green template can be modified or reconstructed using the shared parameter input method used in Revit Architecture.

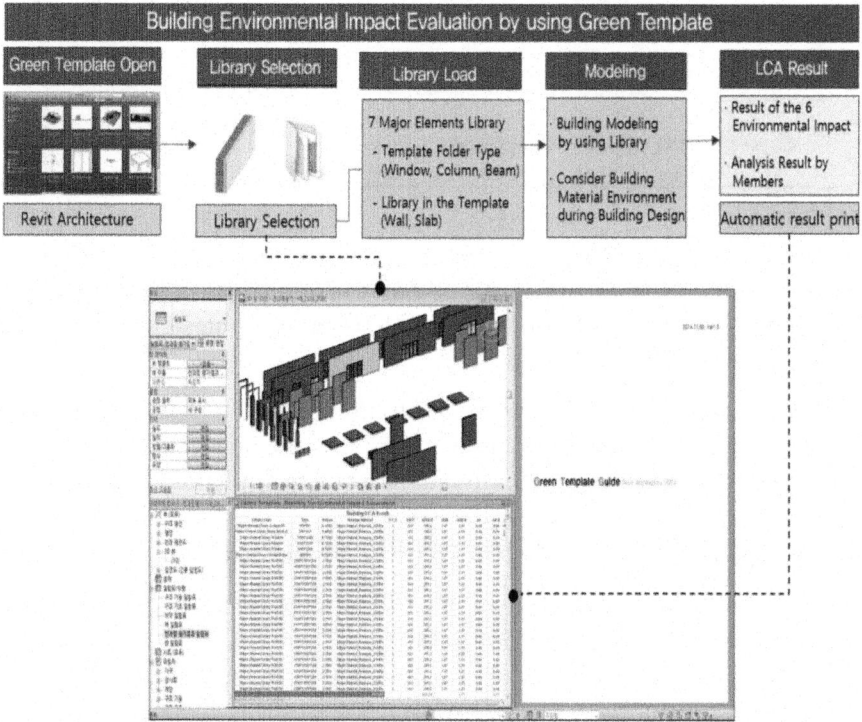

**Figure 4:** Modeling with the green template using the major element library and result output.

## CASE STUDY

A case study analysis was performed to test the applicability of the green template developed in this study. Table 3outlines the specifications of the test building (a standard Korean apartment building) that was used for the case study. We performed the modeling of this building using the green template at the LOD 300 level using the Revit Architecture 2014 BIM modeling tool.

**Table 3:** Specifications of test building used for case analysis

| | Project Name | Reconstruction of Busan Jugong Apartment Building | | | |
|---|---|---|---|---|---|
| | Structure | RC | | Above ground | 4581.94 m² |
| | Scale | 18 floors | Total floor area | Under ground | 732.06 m² |
| | Expected service life | 40 years | | Total | 5313.90 m² |
| Green template –based modeling | Heating method | Local heating method | Floor area ratio | | 282.95% |

## Evaluation Method

A case study analysis was performed to compare the values of the embodied environmental impacts evaluated using the green template with those calculated from the quantity takeoff of the major building materials used for the test building. First, we extracted the quantity takeoff for the building elements of the test building using the major element library embedded in the green template, and then performed an evaluation of the embodied environmental impacts of the building materials. The embodied environmental impact evaluation was performed for all of the major building materials except steel which is not considered at Revit architecture 2014 (ready-mixed concrete, glass, concrete block, insulation material, and gypsum board) for each of the six environmental impact categories.

## Evaluation Results

Table 4 compares the actual quantity takeoff (2D takeoff) of the building elements and the green template-based embodied environmental impact evaluation results. The values yielded by 2D takeoff and the green template and the error rates were 3055 m³ and 2946 m³, respectively, for ready-mixed concrete with an error of 3.49%, and 23.99 m³ and 23.02 m³ (3.98%) for glass. For concrete products, both the green template and 2D takeoff yielded 1832 m² ($n = 139961$). The values obtained after changing the number of concrete blocks to 130,432, using the unit conversion factor, revealed an error rate of 5.1%. A relatively higher error rate was yielded by the insulation material

(52.1901 m³ *vs.* 48.9390 m³; 6.17%), presumably because premiums were not considered in the BIM. The error rate for gypsum board between the 2D takeoff and green template was 4.66% (534.41 m³ *vs.* 509.1324 m³).

**Table 4:** 2D takeoff and the embodied environmental impacts evaluated using green template

| Division | Material Name | Material Quantity | Unit | GWP (kg-CO$_2$eq/Unit) | ADP (kg/Unit) | AP (kg-SO$_2$eq/Unit) | EP (kg-PO$_4^{3-}$eq/Unit) | ODP (kg-CFC-11eq/Unit) | POCP (kg-Ethyleneeq/Unit) |
|---|---|---|---|---|---|---|---|---|---|
| 2D Takeoff | Ready-mixed concrete | 3055.733 | m³ | 904 | 4.12 | 1.51 | 0.176 | 0.001103 | 2.25 |
| | Rebar | 389.01 | Ton | 98.80 | 0.78 | 0.65 | 0.098 | 0.000003 | 0.19570 |
| | Glass | 23.996 | m³ | 47.10 | 0.42 | 0.22 | 0.003 | 0.030018 | 0.05350 |
| | Concrete block | 139,961.52 | m³ | 24.80 | 0.03 | 0.03 | 0.005 | 0.000001 | 0.00265 |
| | Insulation material | 52.19051 | m³ | 2.33 | 0.28 | 0.05 | 0.003 | 0.0000000326 | 0.00722 |
| | Gypsum board | 534.41 | m³ | 0.17 | 0.01 | 0.03 | 0.005 | 0.000001 | 0.00675 |
| Green Template | Ready-mixed concrete | 2946.9489 | m³ | 872 | 5.15 | 1.46 | 0.170 | 0.000099 | 2.17 |
| | Glass | 23.026562 | m³ | 45.20 | 0.40 | 0.21 | 0.003 | 0.000017 | 0.05130 |
| | Concrete block | 130,542.11 | m² | 24.00 | 0.03 | 0.03 | 0.004 | 0.000001 | 0.00255 |
| | Insulation material | 48.939041 | m³ | 2.18 | 0.18 | 0.04 | 0.003 | 0.000000 | 0.00677 |
| | Gypsum board | 509.13241 | m³ | 0.16 | 0.01 | 0.05 | 0.004 | 0.000000 | 0.00643 |

A comparison was made between the 2D takeoff and green template results for the embodied environmental impact evaluation for the category of environmental impact, as shown in Figure 5, to analyze the contribution of major building materials to each of the environmental impact categories. The comparative analysis yielded an average error rate of about 5%, presumably ascribable to the quantity takeoff for steel that was excluded from the LOD 300 modeling with the green template.

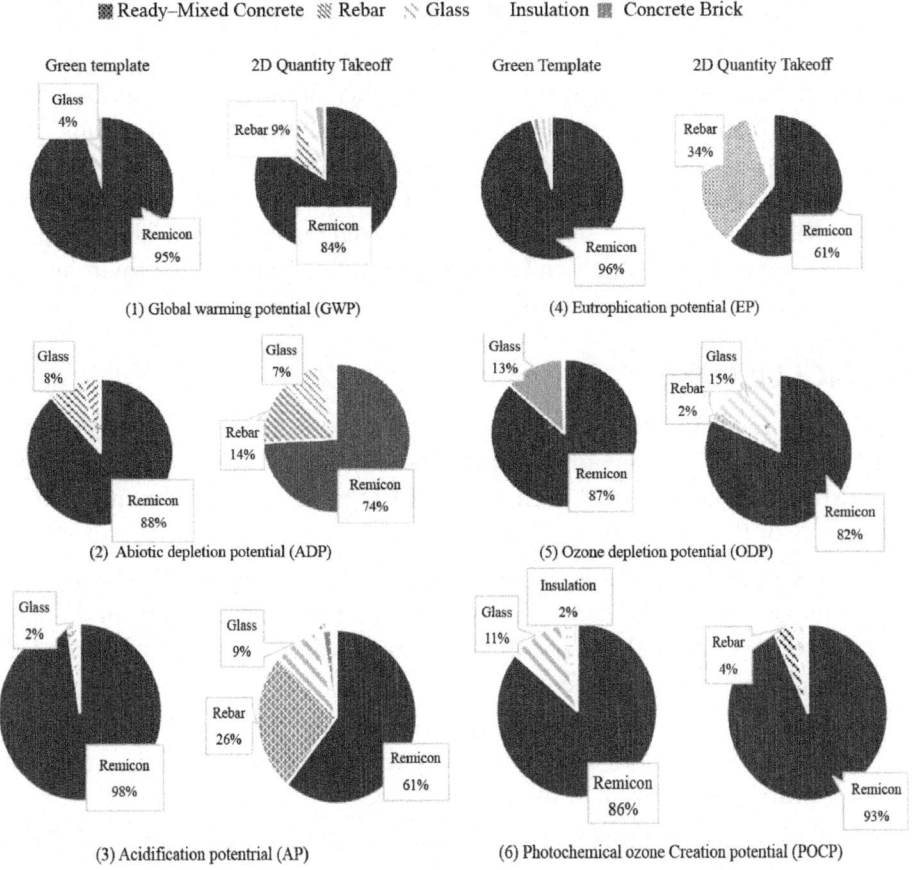

**Figure 5:** Comparison between green template and 2D takeoff for environmental impact category.

## DISCUSSION

The green template was developed to support users in the efficient production of an embodied environmental impact evaluation of a building based on BIM. In particular, six major building materials were proposed as evaluation targets, and six environmental impact categories, including greenhouse gases, were considered to enable consideration at a professional level. User convenience and the easy production of an embodied environmental impact evaluation were ensured by the building element library and evaluation result overview table. This study has two limitations. First, the lack of steel in the Revit Architecture environment may lower the reliability of the evaluation results, especially because steel accounts for over 20% of the overall embodied environmental impact of a building. Therefore, to improve the performance of the green template, steel and the premium rates for major building materials should be reflected in each LOD stage of the building modeling. Second, the system developed in this study is based on a standard Korean apartment building, without reflecting the current trend for using green technologies and materials to reduce the environmental impact of a building. Therefore, to enhance the applicability and reliability of the green template, future studies will have to include steel as a major building material, as well as premium rates for the building materials and also examine the comprehensive environmental impacts of buildings constructed using green technologies.

## CONCLUSIONS

This study resulted in the development of a green template, for use with a BIM authoring tool to evaluate the embodied environmental impacts, as part of BIM-based building LCA technology R&D project. The following outlines the results of this study.

- The green template developed in this study for embodied environmental impact evaluation using a BIM authoring tool consists of a major building material library, an evaluation result overview table, and the green template guideline.

- The impact factors of the six environmental impact categories, unit conversion factors, and the green template major building materials library reflecting them were established.

- An embodied environmental impact evaluation result overview table was constructed to allow a user to check the embodied environmental impacts in real time while designing a building using the building element library of the green template.

- The green template's applicability was verified by the small average error rate ($\leq 5\%$) between the results of the green template-based embodied environmental impact evaluation and the 2D takeoff.

## ACKNOWLEDGMENTS

This research was supported by Basic Science Research Program through the National Research Foundation of Korea (NRF) funded by the Ministry of Education, Science and Technology (NRF- 2015R1D1A1A01057925).

# REFERENCES

1.  Eastman, C.; Teicholz, P.; Sacks, R.; Liston, K. *BIM Handbook*; Wiley: New York, NY, USA, 2014.

2.  Ding, L.; Zhou, Y.; Akinci, B. Building information modeling (BIM) application framework: The process of expanding from 3D to computable nD. *Autom. Constr.* 2014, *46*, 82–93.

3.  Ibrahim, M.; Kate, C. Sustainable BIM-based Evaluation of Buildings. *Int. Proj. Manag. Assoc.* 2013, *74*, 419–428.

4.  Eddy, K.; Brad, N. *Green BIM: Successful Sustainable Design with Building Information Modeling*; Wiley: New York, NY, USA, 2008; pp. 70–71.

5.  Tracy, C.; Nashwan, D.; John, D. Energy profiling in the life cycle assessment of buildings. *Manag. Environ. Qual.* 2010,*21*, 20–31.

6.  Fan, S.L.; Skibniewski, M.J.; Hung, T.W. Effects of building information modeling during construction. *J. Appl. Sci. Eng.* 2014, *17*, 156–166.

7.  Cofaigh, E.O.; Fitzgerald, E.; Alcock, R.; McNicholl, A.; Peltonen, V.; Marucco, A. *A Green Vitruvius: Principles and Practice of Sustainable Architectural Design*; James & James (Science Publishers) Ltd.: London, UK, 1999.

8.  Wang, W.; Zmeureanu, R.; Rivard, H. Applying multi-objective genetic algorithms in green building design optimization. *Build. Environ.* 2005, *40*, 1512–1525.

9.  Du, J.; Qin, Z. Cloud and Open BIM-Based Building Information Interoperability Research. *J. Sci. Manag.* 2014, *7*, 47–56.

10. Korea Energy Management Corporation. *Building Energy Efficiency Rating Certification System Operating Rules*; Korea Energy Economics Institute: Seoul, Korea, 2011.

11. Kang, L.S.; Kwak, J.M. Integrated Code Classification System for work Sections in Standard Method of Measurement and Construction Standard Specification. *Korean J. Constr. Eng. Manag.* 2001, *2*, 80–91.

12. Roh, D.R. *BIM Handbook for Facilities Projects v1.2, PPS*; Building Smart Korea: Seoul, Korea, 2013.

13. Kim, S.W. *Environmental Life Cycle Assessment, Sigmapress*; Ministry of Environment: Seoul, Korea, 1998.

14. Roh, M.S.; Kim, I.S.; Kim, M.K.; Jun, H.J. Evaluation Environment for Evaluation Criteria of the Green Building Certification Criteria using a BIM-based Template. *Archit. Inst. Korea* 2013, *33*, 47–54.

15. Kang, T.W.; Hong, C.H. GIS-based BIM Object Visualization System Architecture Design using Open source BIM Server Cost-Effectively. *Korea Spat. Inf. Soc.* 2014, *22*, 45–53.

16. Virtual Construction System Development Agency. Available online: http://www.wckorea.org (accessed on 6 December 2015).

17. Jacobs, G. 3D Laser Scanning Reduces Risks. *Glob. Mag. Leica Geosystems* 2010, *62*, 818–825.

18. Kim, I.S.; Jun, H.J.; Kim, M.K. A Study on the Development of Evaluation System for G-SEED (Green Standard for Energy and Environmental Design) Using BIM. *Archit. Instit. Korea* 2013, *29*, 117–127.

19. Roh, S.J.; Tae, S.H.; Kim, T.H.; Kim, R.H. Characterization of Environmental Impact of Major Building Material for Building Life Cycle Assessment. *Archit. Inst. Korea* 2013, *29*, 93–100.

20. Lee, S.W.; Tae, S.H. The Analysis of BIM Modeling Standard(LOD) for Greenhouse Gases Assessment Point of View. In Proceedings of the SB13 Seoul International Conference, Seoul, Korea, 8–10 July 2013; pp. 428–431.

21. Tae, S.H.; Shin, S.W.; Woo, J.H.; Roh, S.J. The development of apartment house life cycle $CO_2$ simple assessment system using standard apartment houses of South Korea. *Renew. Sustain. Energy Rev.* 2011, *15*, 1454–1467.

22. Korea Environmental Industry & Technology Institute. Environmental Labeling Certification. Available online: http://www.edp.or.kr/edp (accessed on 6 December 2015).

23. Ministry of Land, Infrastructure and Transport. Building Materials Environmental Information DB Final Report. Available online: http://www.mlit.go.jp/en/index.html (accessed on 6 December 2015).

24. Korea Environmental Industry & Technology Institute. *Life Cycle Assessment Theory and Practice*; Ministry of Environment: Seoul, Korea, 2012. (In Korean).

25. AIA. *AIA Document E202*; American Institute of Architect: Washington, DC, USA, 2008.

26. Kim, J.A. A Study on the Realization of Object Library System Structure and the Utilization Scenarios for Open BIM-Based Building Energy Performance Assessment—Focused on the Implementation of Interperability of Building Object in OBES. Master's Thesis, Ulsan University, Ulsan, Korea, 2012.

27. Bang, J.S.; Tae, S.H.; Kim, T.H.; Roh, S.J. A Study on Developing BIM

Template based on Public Procurement Service Standard Construction Code to Improve the Efficiency in Carbon Dioxide Assessment of Buildings. *Archit. Inst. Korea*2013, *29*, 69–76.

28. Ministry of Land, Infrastructure and Transport. Available online: http://english.molit.go.kr/intro.do (accessed on 6 December 2015).

29. Shin, S.W. *Green Building Environmental Performance Assessment and Design*; Sustainable Building Research Center: Fairy Meadow NSW, Australia, 2007.

30. Ibem, E.O.; Laryea, S. Survey of digital technologies in procurement of construction projects. *Autom. Constr.* 2014, *46*, 11–21.

31. Cabeza, L.; Rincón, L.; Vilariño, V.; Pérez, G.; Castell, A. Life cycle assessment (LCA) and life cycle energy analysis (LCEA) of buildings and the building sector: A review. *Renew. Sustain. Energy Rev.* 2014, *29*, 394–416.

# Chapter 6

# A DFUZZY-DAHP DECISION-MAKING MODEL FOR EVALUATING ENERGY-SAVING DESIGN STRATEGIES FOR RESIDENTIAL BUILDINGS

Kuang-Sheng Liu[1], Sung-Lin Hsueh[2], Wen-Chen Wu[2], and Yu-Lung Chen[2]

[1]Department of Interior Design, Tung Fang Design University, No.110 Dongfang Road, Hunei Distract, Kaohsiung City 82941, Taiwan

[2]Graduate Institute of Cultural and Creative Design, Tung Fang Design University, No.110 Dongfang Road, Hunei Distract, Kaohsiung City 82941, Taiwan

## ABSTRACT

The construction industry is a high-pollution and high-energy-consumption industry. Energy-saving designs for residential buildings not only reduce the energy consumed during construction, but also reduce long-term energy consumption in completed residential buildings. Because building design affects investment costs, designs are often influenced by investors' decisions. A set of appropriate decision-support tools for residential buildings are required to examine how building design influences corporations externally and internally. From the perspective of energy savings and environmental protection, we combined three methods to develop a unique model for evaluating the energy-saving design of residential buildings. Among these methods, the Delphi group decision-making method provides a co-design feature, the analytical hierarchy process (AHP) includes multi-criteria decision-making techniques, and fuzzy logic theory can simplify complex internal and external factors into easy-to-understand numbers or ratios that facilitate decisions. The results of this study show that incorporating solar building materials, double-skin facades, and green roof designs can effectively provide high energy-saving building designs.

## INTRODUCTION

Building construction and operation contribute to more than one-third of the carbon emissions in the United States [1]. In addition, greenhouse gas emissions are caused by building construction. A total of 82%–87% of greenhouse gas emissions are from the embodied greenhouse gas emissions of building materials, 6%–8% are from the transportation of building materials, and 6%–9% are caused by the energy consumption of construction equipment [2]. Building construction contributes anthropogenic $CO_2$ emissions [3,4]. Although the construction industry is a significant indicator of economic development, it consumes a significant amount of energy and produces substantial pollution [5,6,7,8,9]. Construction has been blamed for causing various environmental problems, ranging from excessive consumption of global resources for construction and building operations to polluting the surrounding environment [10]. Based on estimations by the United Nations Environment Programme, the building sector accounts for 30% to 40% of global energy use [11]. Thus, improving construction practices to minimize their detrimental effects on the natural environment is an emerging issue [12,13]. The environmental impact of construction, green buildings, recycling, and eco-labeling of building materials has attracted the attention of building professionals worldwide [14,15,16].

The trend of developing green industries has resulted from policy pressure related to climate change, customer demand, and sustainable environmental protection. In addition, development of the green industry has become an effective method for promoting economic recovery and increasing employment levels [17]. Environmental protection and development of the green industry in the EU is particularly specific and active. In the E.U. Green Book, corporate social responsibility is a major tool for creating new jobs and sustaining economic development [18]. Thus, people are becoming increasingly convinced of the importance of corporate social responsibility, which is also an important factor for companies to achieve profits and stimulate socioeconomic development [19,20]. Many companies have recognized that corporate social responsibility is a source of future business opportunities and competitive advantages [21,22]. Therefore, green innovation for companies is not only a method of fulfilling corporate social responsibilities, but also a strategy to sustainably manage and create competitive advantages [23,24]. Recently, the energy-saving measures promoted by the construction industry include green supply chain management [25,26,27], green procurement [28], and green building design [29,30]. This issue is gaining importance in countries worldwide, and becoming a goal of policy objectives. For example, green building design projects in Taiwan are being given area ratio rewards, and additional green building-related guidelines and regulations are planned for the future. However, this is likely to cause problems in the management of traditional builders, building material suppliers, real estate developers, and architectural designers. Green transformation of the entire supply chain for both the upstream and downstream of the construction industry is an important development strategy for companies to create competitive advantages and sustainable operations.

Developing energy-saving design strategies for buildings requires a cross-disciplinary and cross-expertise design thinking model. Numerous factors are involved when considering energy-saving designs, particularly for large-scale carbon-neutral community building developments; examples include the use of renewable energies [31], eco-designs [32], solar energy [33,34,35], lighting [36], compressed shopper waste (CSW) blocks [37], waste disposal [8], air-conditioning facilities [38], ventilation designs [39,40], shading designs [41], heating systems [42,43], green roofs [44], building envelopes [45], and wall insulation for buildings and double-skin facades [46,47,48]. Therefore, comprehensive preparation in integration and design is required to demonstrate effectiveness. In addition, the energy-saving design of buildings may cause multiple issues in construction projects, including increased construction costs, increased complexity of contracting and procurement, greater difficulty integrating construction interfaces, influences on building quality, necessities of cross-disciplinary specialties, and influencing project progress. These also cause decision-makers to alter investment decisions regarding the energy-saving designs of buildings. However, because the green transformation of industries is a trend in future development, investment decision-makers should be prepared in advance to increase their company's competitive advantage. According to Pacheco, saving energy is a high-priority goal for developed countries. Therefore, energy-efficient measures are being increasingly implemented in all sectors [43]. To ensure the global environmental security of the future, green procurement is implemented when awarding international construction contracts in countries throughout the world, which significantly affects the operation of construction industries in developing and undeveloped countries.

The heat released from buildings is the largest contributor (89% to 96%) of global heat emissions [49]. People cannot ignore the high $CO_2$ emissions, which have already caused serious climate change and environmental damage, from various industries that consume energy. The environmental risks pose serious problems to individual and societal decision-making [50], and are serious issues that must be urgently addressed. The interactive relationships between people and society directly affect the level of concern individuals, families, and companies have for the environment. O'Neill stated that environmental protection is a national education project of personal and social responsibility that can be used in various specific institutional contexts and missions [51]. Education on social responsibility and policy promotions can help reduce the effects of anthropogenic $CO_2$ emissions. In addition, Hediger stated that corporate social responsibility is a strategy that acts as insurance against risks to corporate reputations, predicts damage to profits and corporate values, and satisfies external demands [52]. In the construction industry, a strategy of developing green supply chains, green procurement, green design, and green buildings is a manifestation of corporate social responsibility. Burtraw [53] noted that subsidies on green energy can aid in the return of the value of $CO_2$ emission allowances to households. Currently, Taiwan provides multiple subsidies for residential energy-saving measures. Therefore, the strategic approach of residential building developers for the energy-saving design and construction of residential buildings is a developmental strategy that matches the interior and exterior of a company. Investment decision-makers require objective assistive decision-making models to evaluate the influence of decision-making projects on the interior and exterior of their companies to provide recommendations that enable the best decision to be reached. For this study, we applied the group decision-making technique [54,55] to examine criteria suitable for this study. The multi-criteria decision-making technique [56,57] featured in the analytical hierarchy process (AHP), was used to verify the relative importance of each criterion. The quantifying ability of fuzzy logic theory was used to establish a model that evaluates energy-saving designs of residential buildings [1,58]. This enables effective resolution of energy-saving issues related to residential buildings during the early stages of design. In addition, the factors evaluated in this model consider corporate social responsibility, attitudes toward environmental protection, and long-term energy-saving factors after the completion of the residential buildings. This model shows the potential importance of each evaluating factor in the early design stages, which can provide professionals with decision-making references during the design stages of energy-saving residential buildings.

## MODEL OVERVIEW

The Delphi method is used to provide implicit expert assistance in research that is highly professional and objective. The Delphi method was developed by the U.S. RAND Corporation to assist management in predicting future events. However, its application scope is not restricted to predicting future events [1,58]. In this study, we explain the Delphi method as a group decision-making technique, including its uses, underlying assumptions, strengths and limitations, potential benefits to qualitative higher education research, and key considerations in its use. Use of the method was demonstrated in a recent national study to develop management audit assessment criteria that can benefit increases in research reliability [54].

The AHP method was first proposed by Saaty and has been extensively used to solve multi-criteria decision-making problems. AHP is also commonly applied to social, policy, and engineering decision-making issues [59,60]. AHP is employed in research for enhancing sustainable community development [58], the estimation and selection of building investments [61], maintenance selection problems [62], project management [63], maintenance strategy selections [64], evaluations of advanced construction technologies [65], decision-support systems in the housing sector [64], urban renewal proposals [66], and sustainable urban energy environment management [67].

Fuzzy set theory was developed by Professor Zadeh at the University of California, Berkeley, in 1965; it is the optimal quantitative tool for addressing fuzzy phenomena and fuzzy language. Fuzzy logic theory based on fuzzy sets is primarily used to express and quantify certain fuzzy concepts that cannot be clearly defined. This theory can provide excellent results when dealing with fuzzy language expressions. Fuzzy logic can manage vague information in natural human language, such as uncertainty, complexity, and tolerance for imprecision [68,69]. Fuzzy logic theory is extremely suitable for highly complex and difficult-to-quantify policy evaluations, especially group decision-making issues [70], such as the sustainable and efficient use of energy during corn production [71].

After determining the model evaluation factors using the Delphi method, we then applied fuzzy logic to build the model. During the model-building process, a rigorous inference system should first be completed to ensure effective and correct implementation and application of the evaluation model. The steps for building the fuzzy logic inference system are as follows:

1.  Define the fuzzy quantitative interval value and the high, moderate, and low quantitative values;

2.  Define the output score for fuzzy quantitative intervals and the quantitative high, moderate, and low values;

3.  Define the membership functions of various evaluation factors and output scoring values;

4.  Define the semantic logic of the inference system relevance (effect) to describe the inter-relationship logic of various scenarios based on various high, moderate, and low quantitative values;

5.  Establish a rule base and inference system according to the semantic

logic of various scenarios to use as the knowledge rule base for evaluating model inferences.

For this study, we developed a model that combines the Delphi method, AHP, and fuzzy logic theory. This model was highly rigorous and reliable because of the expert assistance we employed to examine the content, and form group decisions during the modeling process. Diagrams of these stages are shown in Figure 1. The 15 Delphi experts who assisted in this study had over 15 years of practical work experience in related fields; three experts were from the construction industry, three were from the real estate industry, three were architects, three were scholars, and three were from the public service sector. The group decision-making data collected from the Delphi experts provided the required information for a fuzzy logic model. In a Delphi Fuzzy- (DFuzzy) and Delphi AHP- (DAHP) model environment, appropriate criteria must first be selected from complex factors, and then each criterion hierarchy must be completed. After the quantitative natural language membership functions are selected, the fuzzy sets and fuzzy scale sets, the fuzzy logic inference systems (FLIS) of the "IF-THEN rule base" and the DFuzzy-DAHP models can function.

This model incorporates multiple characteristic benefits from DFuzzy, DAHP, Fuzzy-AHP, and DFuzzy-AHP, and its purpose is to address complex decision-making issues. Related studies have applied DFuzzy to human resources (HR) [72], maintenance strategy selections [73], and the reuse selection of historic buildings [74]. Researchers that have applied DAHP include Tavana and Liao [75,76]. Studies that have applied Fuzzy-AHP include those investigating optimum maintenance strategies [77], enhancing sustainable community developments [58], and evaluating the rankings of alternatives [78]. One study that applied DFuzzy-DAHP is mentioned in [79]. Models built by applying DFuzzy, DAHP, Fuzzy-AHP, and DFuzzy-DAHP all provide quantified group decision-making analysis, among which DFuzzy-DAHP is the most objective and rigorous.

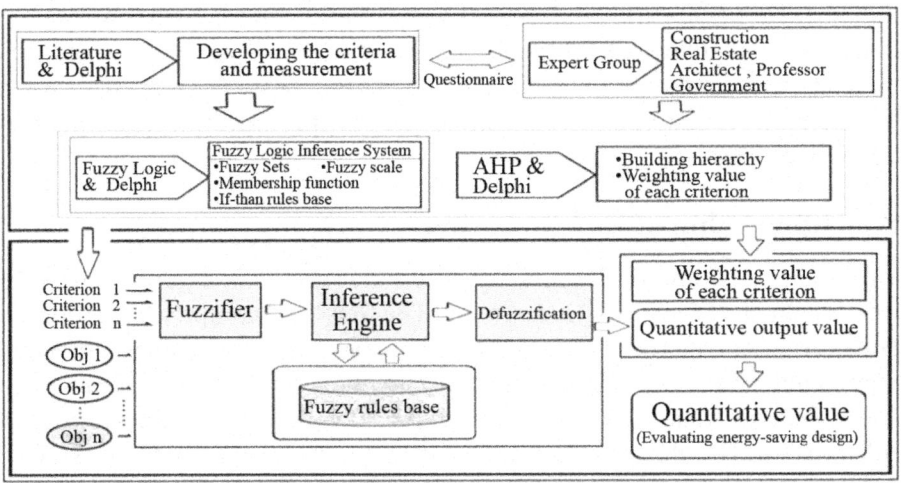

**Figure 1:** Framework of the evaluation model.

## DEVELOPING THE CRITERIA AND AHP HIERARCHICAL FRAMEWORK

This study completed the Delphi process in 6 months, and confirmed the criteria required for model building. Three main criteria for residential building energy-saving designs have received universal approval from experts; these criteria are interior design, building facades, and green attractions. These criteria contain nine sub-criteria, namely CSW blocks, shading designs, ventilation designs, green roofs, solar building materials, double-skin facades, cost differences, company images, and social responsibilities. Because building regulations in Taiwan do not enforce the implementation of green buildings and green procurement, they remain in the rewards and promotions stage. Most non-public sector building projects continue to follow traditional design and construction methods during construction. In this study, we obtained a consensus among Delphi experts on the energy-saving designs of residential buildings. Green transformation is believed to be an issue that companies must soon address. Therefore, investment decision-makers for residential building projects must reduce profits and increase green investments to adapt to changes in the investing environment. Faced with customer demand, green attractions are gradually having value-adding effects on company images with continuous policy and education promotions. This also highlights the social responsibility of green residential building projects, which benefits the sustainable operation of the company. The AHP framework of the criteria required for model building in this study, as determined by using the Delphi method, is shown in

Figure 1. Delphi experts unanimously agree that of the nine criteria, increased costs are more likely to be recognized by real estate developers and, thus, facilitate evaluations prior to the implementation of energy-saving designs. The relative weight of numerous evaluation factors can be determined using the AHP process.

## Weighting Value of Each Criterion

The hierarchical framework for each criterion is shown in Figure 2. Two levels were established for overall assessment. The first level comprises the following three criteria: interior designs, building facades, and green attractions. Each main criterion was then divided into three sub-criteria. Because AHP questionnaires frequently result in invalid responses, AHP is time-consuming. According to Hseuh, more than one year is required to complete the AHP [1].

We invited professionals to assist with the AHP questionnaires and obtained complete and valid questionnaire data. We adopted a strict attitude when completing the AHP. The experts who assisted with the 52 valid questionnaires during the AHP all had a minimum of 15 years of experience in their related fields. The scholars were a vice chancellor, a dean, and a senior professor at universities ranked among the top five universities in Taiwan. The industry experts included one construction expert, one real estate expert, one architect, a CEO, and a project manager with a master's degree. The research period for this study was from November 2011 to August 2012 to allow for the completion of the AHP. Table 1, Table 2, Table 3 and Table 4 show the relative weight calculations for each sub-criterion for each level. Table 5 shows the relative weight for each criterion in the overall assessment.

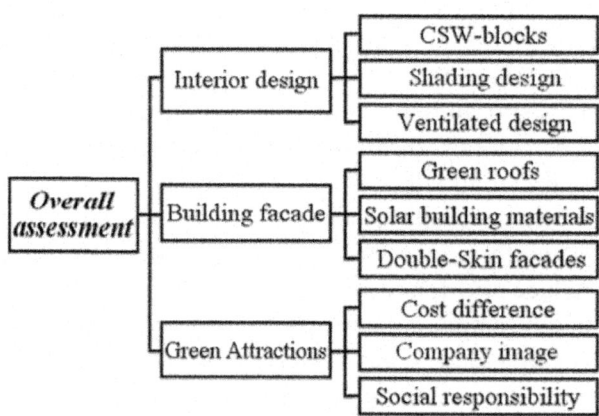

**Figure 2:** AHP hierarchical framework for each criterion.

**Table 1:** Weighting value of the main criteria: comparison of the interior design, building facades, and green attractions

| Attributes | Interior design | Building facade | Green Attractions |
|---|---|---|---|
| Interior design | 1 | 1/2 | 1/5 |
| Building façades | 2 | 1 | 1 |
| Green attractions | 5 | 1 | 1 |
| Eigenvector | 0.14 | 0.37 | 0.49 |

**Table 2:** Weighting value of interior design: comparison of CSW blocks, shading design, and ventilation design

| Attributes | CSW blocks | Shading design | Ventilation design |
|---|---|---|---|
| CSW blocks | 1 | 1/3 | 1/6 |
| Shading design | 3 | 1 | 1 |
| Ventilation design | 6 | 1 | 1 |
| Eigenvector | 0.11 | 0.4 | 0.49 |

**Table 3:** Weighting value of building facades: comparison of green roofs, solar building materials, and double-skin facades

| Attributes | Green roofs | Solar building materials | Double-skin facades |
|---|---|---|---|
| Green roofs | 1 | 1/3 | 1 |
| Solar building materials | 3 | 1 | 1 |
| Double-skin facades | 1 | 1 | 1 |
| Eigenvector | 0.2 | 0.6 | 0.2 |

**Table 4:** Weighting value of green attractions: comparison of cost difference, company image, and Social responsibility

| Attributes | Cost difference | Company image | Social responsibility |
|---|---|---|---|
| Cost difference | 1 | 1 | 1 |
| Company image | 3 | 1 | 1 |
| Social responsibility | 3 | 1 | 1 |
| Eigenvector | 0.14 | 0.43 | 0.43 |

**Table 5:** Weighting value of each criterion

| Main Criteria (wi) | Subcriteria (wi) | wi | Wi% |
|---|---|---|---|
| | CSW blocks (0.11) | 0.015 | 1.5% |
| Interior design (0.14) | Shading design (0.40) | 0.056 | 5.6% |
| | Ventilation design (0.49) | 0.069 | 6.9% |
| | Green roofs (0.20) | 0.074 | 7.4% |
| Building facades (0.37) | Solar building materials (0.60) | 0.222 | 22.2% |
| | Double-skin facades (0.20) | 0.074 | 7.4% |
| | Cost difference (0.14) | 0.069 | 6.9% |
| Green attractions (0.49) | Company image (0.43) | 0.211 | 21.1% |
| | Social responsibility (0.43) | 0.211 | 21.1% |
| Wi = wi × 100% | | 1.001 | 100.1% |

## Developing a Fuzzy Logic Inference System

The fuzzy logic inference method can be separated into two systems, that is, the Mamdani and the Sugeno systems. Typically, output from the Mamdani system is continuous, whereas that for the Sugeno system is discrete. To understand the change in continuous output, we adopted the Mamdani system. In addition, multiple types of membership functions existed; the membership functions commonly used include triangular functions and bell-shaped functions [80,81]. Therefore, triangular functions and bell-shaped functions were also adopted in this study for fuzzy set membership functions. The bottom section of Figure 1 shows the FLIS schematic diagram for the DFuzzy-DAHP decision-making model proposed in this study. The FLIS system was divided into the following four main parts: a fuzzifier, inference engine, fuzzy rule base, and defuzzification. The membership function and fuzzy range of the fuzzy set for each criterion must first be defined. The membership function and fuzzy range of the fuzzy set for the output value corresponding to each scenario must also be defined. After further completing the definition of the IF-THEN rule base in the FLIS system, the FLIS is then capable of inferences and quantified computations.

Fuzzy logic belongs to the field of artificial intelligence and can be used to process the complex and imprecise semantic meanings of people. For example, the expression "very good" does not have a "0" or "1" logical relationship. In a fuzzy decision environment [82], the membership function is used to define the degree of goodness. Fuzzy set theory expands traditional mathematical dichotomy theory (set value is 0 or 1) to an infinite number of continuous set values (between 0 and 1). This also renders fuzzy logic convenient for processing variables and inferences in language [83]. For this study, we applied fuzzy logic to define the evaluation content, which included risk factors, such as consumer-oriented future green energy requirements, corporate social responsibilities, corporate profitability, changes in operational environments, and policy changes. These influencing factors benefit decision-making for the overall project and investment risk considerations. Decision-making analysis of investments does not focus solely on profits. A company must understand the objective changes in demands in the external environment to form appropriate operating decisions, and to ensure the sustainable operation of the company. Once the FLIS system for the evaluation content examined in this study is complete, design evaluations of energy-saving residential building projects can be represented using easy-to-understand numbers or ratios. The potential importance of each criterion is also simultaneously presented, providing the decision-maker with additional information to facilitate decision-making.

## Defining the Fuzzy Set of the Input and Output Factors in Fuzzy Logic

For the three main criteria, the interior design criterion is composed of three sub-criteria, namely CSW blocks, shading designs, and ventilation designs. The building facades criterion is composed of three sub-criteria, namely green roofs, solar building materials, and double-skin facades. The last main criterion, green attraction, is also composed of three sub-criteria, namely cost differences, company images, and social responsibilities. Before the FLIS system can be constructed, the membership function, fuzzy set, and fuzzy range of each criterion must be defined. In addition, the membership function and fuzzy range for the fuzzy set containing the output value must also be defined. When the evaluation factors were quantitatively defined in the fuzzy set, the IF-THEN rule base of the FLIS system was used to perform the appropriate quantification process on the evaluation factors in various scenarios. Because the evaluation factor has various levels of influence on the evaluation of energy-saving designs of residential buildings, definitions of the membership function, fuzzy set, and fuzzy range are required to complete the variable computation and inferences from various input scenarios through the IF-THEN rule base [68,83], which presents the corresponding output evaluation, result.

The definition of the fuzzy set and fuzzy range for each sub-criterion of the three main criteria, and the fuzzy set and fuzzy range definition of the corresponding output values, are shown in Table 6, Table 7 and Table 8. The measurement scale defined in fuzzy logic is an arbitrarily defined fuzzy scale. For example, in the cost difference sub-criterion of green attractions, 60% profitability indicates a good outcome, 30% profitability indicates an ordinary outcome, and 10% profitability indicates a poor outcome. However, whether 40% is good or ordinary is defined by the membership function in the fuzzy logic scale according to its membership levels in good or ordinary. Defuzzification in FLIS then shows the results for the quantified output values. These problems are often difficult for traditional evaluation models to process.

**Table 6:** Definition of the input and output fuzzy set and fuzzy range for the interior design criteria

| Input Scenario | | | Fuzzy output value | |
|---|---|---|---|---|
| Criteria | Value range | Fuzzy sets | Description | Fuzzy sets |
| CSW blocks | 30% | Good | | |
| | 20% | Ordinary | Quantitative | Good (90%) |
| | 10% | Poor | value | Ordinary (75%) |
| Shading design | 90% | Good | | Poor (60%) |
| | 70% | Ordinary | | |
| | 50% | Poor | | |
| Ventilation design | 90% | Good | | |
| | 60% | Ordinary | | |
| | 40% | Poor | | |

**Table 7:** Definition of the input and output fuzzy set and fuzzy range for the building facades criteria

| Input Scenario | | | Fuzzy output value | |
|---|---|---|---|---|
| Criteria | Value range | Fuzzy sets | Description | Fuzzy sets |
| Green roofs | 90% | Very good | | |
| | 75% | Good | Quantitative | Very good (90% ↑) |
| | 60% | Ordinary | value | Good (80% ↑) |
| | 45% | Poor | | Ordinary (60%) |
| | 30% | Very poor | | Poor (45% ↓) |
| Solar building materials | 40% | Good | | Very poor (30% ↓) |
| | 25% | Ordinary | | |
| | 10% | Poor | | |
| Double-skin facades | 30% | Good | | |
| | 20% | Ordinary | | |
| | 10% | Poor | | |

**Table 8:** Definition of the input and output fuzzy set and fuzzy range for the green attractions criteria

| Input Scenario | | | Fuzzy output value | |
|---|---|---|---|---|
| Criteria | Value range | Fuzzy sets | Description | Fuzzy sets |
| Cost difference | 50% | Good | Quantitative value | Good (80% ↑) |
| | 35% | Ordinary | | Ordinary (55%) |
| | 20% | Poor | | Poor (30% ↓) |
| Company image | 80% | Good | | |
| | 60% | Ordinary | | |
| | 40% | Poor | | |
| Social responsibility | 90% | Good | | |
| | 70% | Ordinary | | |
| | 50% | Poor | | |

## Input Scenario and Output Mapping

The insertion of the input scenarios into FLIS first undergoes the IF-THEN rule base inference before the defuzzified result is quantified in the output value. This model has 99 input scenarios. The main criterion interior design comprises three sub-criteria, each of which has three fuzzy sets (three scenarios for each fuzzy evaluation: good, ordinary, and poor, or high, moderate, and low); thus, there are $3 \times 3 \times 3 = 27$ scenarios. The structure of the main criterion green attractions is the same as that for interior design; therefore, it also has $3 \times 3 \times 3 = 27$ scenarios. However, green roofs in three sub-criteria of the main criterion building facade comprise 5 fuzzy sets, and the other two criteria contain three fuzzy sets. Therefore, there are $5 \times 3 \times 3 = 45$ scenarios. The overall evaluation contains 99 scenarios, and each criterion in the scenario has a varying degree of influence on the energy-saving design strategy of residential buildings. In addition, the 99 evaluation scenarios comprise multiple properties and measurement units to handle the complex evaluation problem. The 3D mapping relationship chart for the input scenarios in the three main criteria and output is shown in Figure 3. This computation model is difficult to achieve manually. In addition, the fuzzy rule base is similar to the human brain in the overall FLIS. When the inference rules of FLIS are constructed, FLIS is then capable of inference computations. Providing the decision-maker allocates an input value to each evaluation factor, FLIS can automatically calculate a quantified performance evaluation value. Fuzzy logic is categorized as artificial intelligence; its scientific reasoning and computing mechanism is widely adopted for quantificational decision-making. Fuzzy logic possessed substantial objectivity and adaptability. The reasoning and computing of scenarios input into the FLIS are shown in Figure 4.

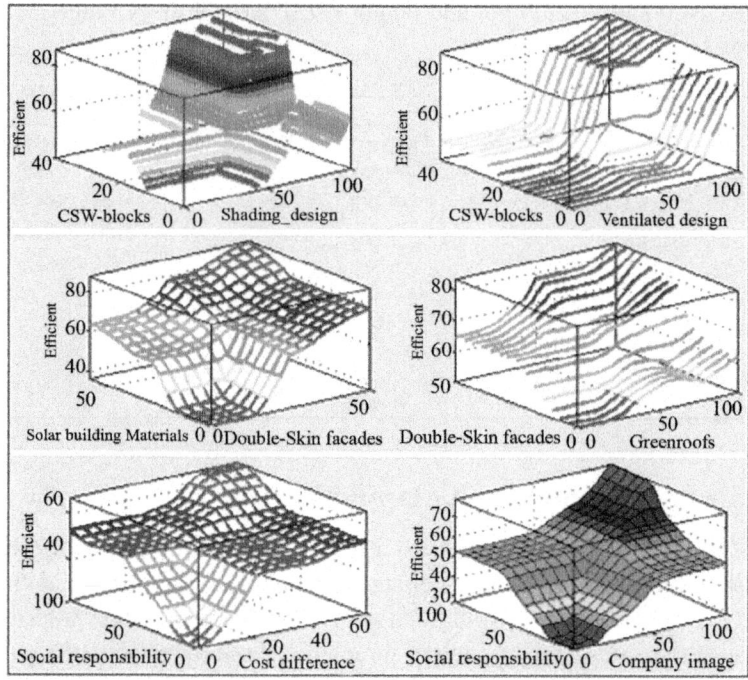

**Figure 3:** 3D mapping relationship diagram for input and output scenarios.

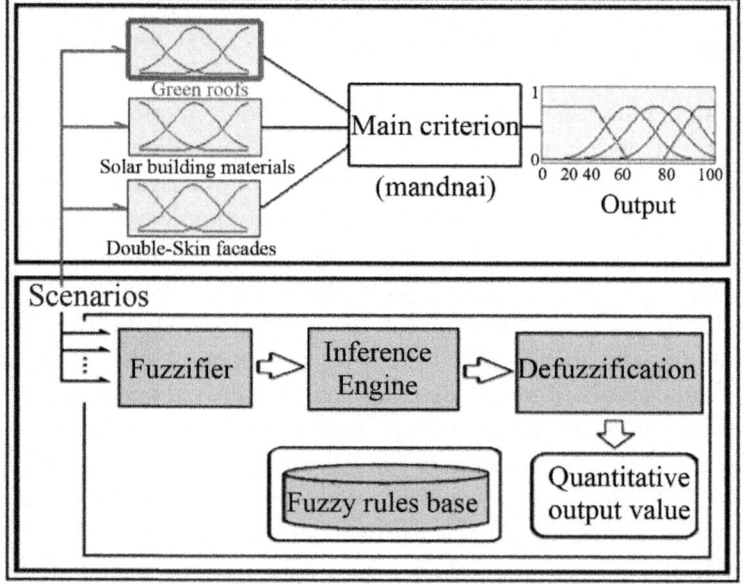

**Figure 4:** The reasoning and computing of scenarios input into the FLIS.

## Calculation of the Comprehensive Quantified Evaluation Value

The three main criteria function were first defined as follows: $y1$ represents interior design, and $f(y1)$ represents the fuzzy quantified output value of interior design; $y2$ represents building facades, and $f(y2)$ represents the fuzzy quantified output value of building facades; $y3$ represents green attractions, and $f(y3)$ represents the fuzzy quantified output value of green attractions. Once these definitions were completed, the weighting values derived from AHP, and the fuzzy quantified output values derived from FLIS ($yi$), could be used to calculate the quantified magnitude value of the level of energy-saving design compliance of the residential building project, which equals $\Sigma f(yi) \times (wi)$. The model also shows the potential importance of each criterion. Thus, the decision-making information is easy for the decision-maker to obtain. The optimum and worst quantified output values from FLIS computations are shown inTable 9. In addition, the input scenarios in Table 9 can be either quantified values or imprecise semantic meanings in natural language, such as good (high), ordinary (moderate), and poor (low). Because fuzzy logic can compute language variables and infer quantified language [68,83], this model can provide decision-makers the ability to further compare the magnitude of quantified values from scientific calculations prior to conducting project evaluations. This enhances the efficiency and effectiveness of decision-making, which reduces the risk of forming wrong decisions.

**Table 9:** The optimal and worst output value for each subcriterion

| Main Criteria | Subcriterion | Optimal | Worst | Case 1 | Case 2 |
|---|---|---|---|---|---|
| | CSW blocks | Good | Poor | Good | Good |
| Interior design ($y_1$) | Shading design | Good | Poor | Ordinary | Good |
| | Ventilation design | Good | Poor | Good | Good |
| Output value (%) | | 90.2 | 39.9 | 87.2 | 90.2 |
| | Green roofs | Very good | Very poor | Poor | Very good |
| Building façades ($y_2$) | Solar building materials | Good | Poor | Ordinary | Good |
| | Double-skin facades | Good | Poor | Poor | Good |
| Output value (%) | | 88.5 | 20.4 | 51.9 | 88.5 |
| | Cost difference | Good | Poor | Good | Poor |
| Green Attractions ($y_3$) | Company image | Good | Poor | Ordinary | Good |
| | Social responsibility | Good | Poor | Ordinary | Good |
| Output value (%) | | 82 | 31.4 | 57.7 | 62.7 |

# CASE STUDY

For this case study, a residential building project in Taiwan was used. The foundation was oriented to face south. Because the solar building materials are influential to residential building planning, the sunlight conditions must be modeled before planning to achieve design efficiency. Taiwan is located along 23.5° N latitude, as shown in Figure 5. The angle of elevation of sunlight in the four seasons in Taiwan and the changing relationship of the angle of azimuth are described.

- Each day, the elevation angle of sunlight increases from 0° at sunrise to the highest point at noon, returning to 0° at sunset. The angle of azimuth begins from the maximum value at sunrise, moving to 0° by noon before again reaching the maximum positive value at sunset.

- In one year, the range of variation for the angle of azimuth over one day changes from ±115.6° in the summer to ±64.4° in the winter. The value is 90° in the spring and autumn. The range of variation for the elevation angle changes from 90° in the summer to 43.6° in the winter. The buildings facing north in Taiwan receive minimal hours of sunlight in one year, whereas buildings facing south receive sunlight throughout the entire year.

- The horizontal face of the building receives numerous hours of sunlight in every season of the year.

- Excluding direct sunlight at noon in the summer (solar elevation angle at 90°), the angle of elevation changes in for the other seasons. Therefore, the angle of elevation and angle of azimuth are important aspects for building plans designed with solar building materials.

Case 1 is shown in Figure 5 and is an example of an ordinary traditional residential building design plan. Case 2 is shown in Figure 6. In this case, saving energy was considered before planning, and the company was required to sacrifice profit to improve green building construction. This action shows corporate transformation and the practice of social responsibility.

The results in Table 9 show not only the optimum and worst quantified output values of the three main criteria, but also the fuzzified quantified output values of the two residential building design cases after FLIS calculation, as shown in Figure 7. Because the evaluation content of energy-saving design in this evaluation model includes comprehensive evaluations of factors, such as corporate image and social responsibility, the quantified output values in Table 9 and the weighting values of each criterion in Table 5 can be used to calculate the value of $\Sigma f(yi) \times (wi)$ in Cases 1 and 2.

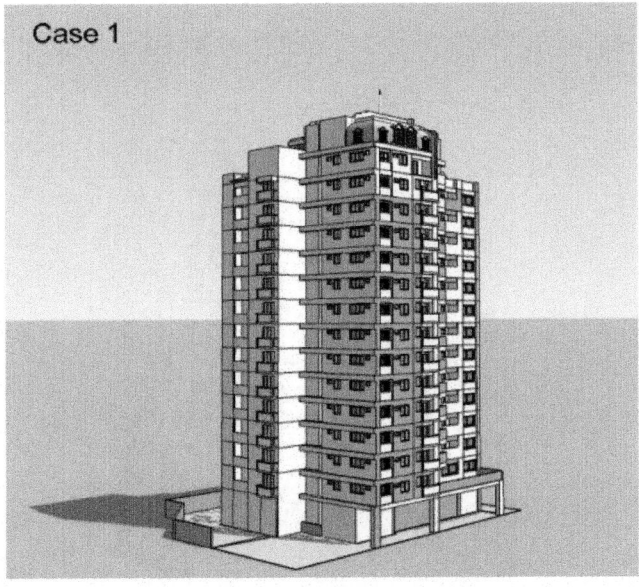

(a)

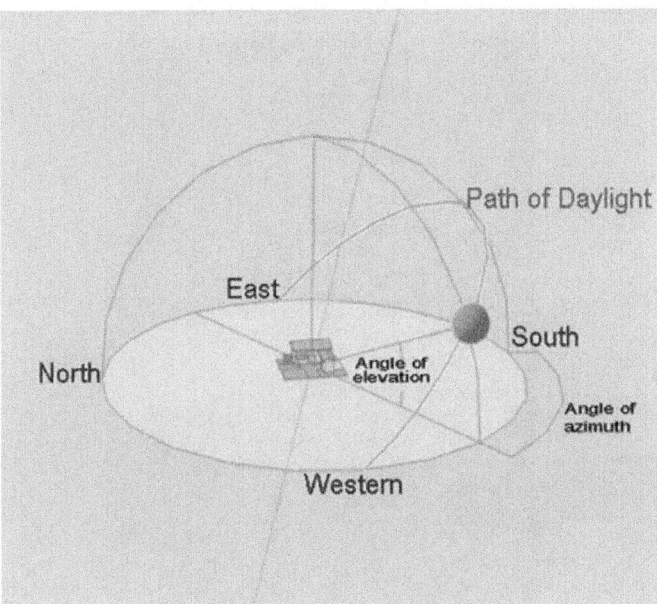

(b)

**Figure 5:** Traditional residential building planning designs for Case 1.

(a)

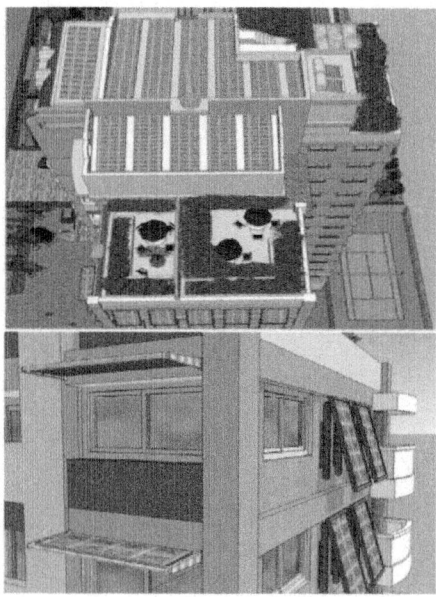

(b)

**Figure 6:** Design concepts for Case 2.

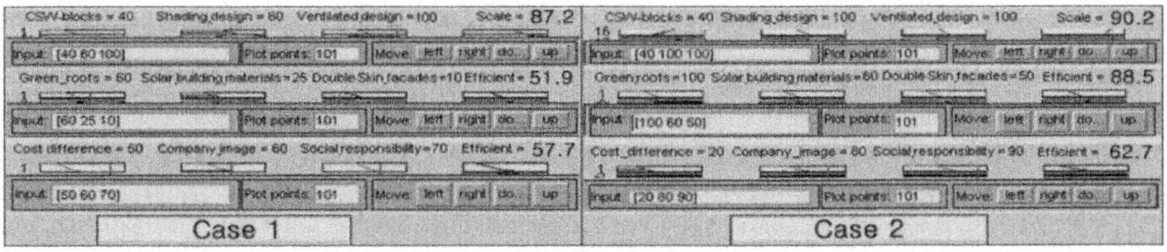

**Figure 7:** Quantified output values calculated by FLIS for Cases 1 and 2.

Table 10 shows that the $\Sigma f(yi) \times (wi)$ value of Case 2 was 76.16, which is higher than the value of Case 1, 59.75. The comprehensive evaluation of energy-saving design in Cases 1 and 2 shown in Table 10 indicates that Case 2 is the superior design proposal, which can serve as a decision-making reference in the early design stages. The design concept for Case 2 considered the various levels of influence of the evaluation factors in Table 10, and generated a comprehensive design decision consideration. This includes the use of solar shading panels, solar panels on the exterior walls, solar panels on the roof of the building, double-skin facades, and green roofs. The solar panel and plant cover area on the roof of the building approached 90%. In addition to enhancing the solar power generating efficiency, this effectively lowers the room temperature. The carbon sequestration effect of the green roof provides environmental benefits.

Residential buildings are an important indicator of economic progress. In addition, the residential sector is responsible for a significant portion of global energy consumption [43]. Reducing residential energy consumption is an issue that everyone should prioritize. Both newly built and older residential buildings can attain energy consumption savings through energy-saving design approaches. However, building projects that use energy-saving designs have increased construction costs, which directly impacts profits for investors.

**Table 10:** Comprehensive evaluation of energy-saving design for Cases 1 and 2

| Subcriteria (wi) | Case 1 | | | Case 2 | | |
|---|---|---|---|---|---|---|
| | Scenario | f (yi) | f (yi) × wi | Scenario | f (yi) | f (yi) × wi |
| CSW blocks (0.015) | Good | 87.2 | 1.31 | Good | 90.2 | 1.35 |
| Shading design (0.056) | Ordinary | | 4.88 | Good | | 5.05 |
| Ventilation design (0.069) | Good | | 6.02 | Good | | 6.22 |
| Green roofs (0.074) | Poor | 51.9 | 3.84 | Very good | 88.5 | 6.55 |
| Solar building materials (0.222) | Ordinary | | 11.52 | Good | | 19.65 |
| Double-skin façade (0.074) | Poor | | 3.84 | Good | | 6.55 |
| Cost difference (0.069) | Good | 57.7 | 3.98 | Poor | 62.7 | 4.33 |
| Company image (0.211) | Ordinary | | 12.18 | Good | | 13.23 |
| Social responsibility (0.211) | Ordinary | | 12.18 | Good | | 13.23 |
| $\Sigma f(yi) \times (wi)$ | | 59.75 | | | 76.16 | |

The use of green building materials increases construction complexity and affects building progress, which creates disinterest in general builders and land developers in energy-saving design planning for residential building projects. This not only affects the promotion of green building policies by the public sector, but also affects technological upgrading in the industry. Although companies are understandably oriented toward seeking profits, company decision-makers must be capable of judging the changing external environment and proposing recommendations to avoid risks. Haleblian [84] stated that the awareness-motivation capabilities of firms influence the timing of competitive action. Managers may occasionally sacrifice profits to improve their relative competitive standing [85]. The results for the main criterion of green attractions in this study show that sacrificing profit and assuming social responsibility can improve corporate image, which benefits sustainable operations and competitive advantages. According to Fernando [20], social responsibility enhances reputation, which improves the profitability of the firm.

## CONCLUSIONS

Whether working on public construction projects or various types of private construction projects, the construction industry is closely related to the lives of citizens. Although the construction industry provides economic development, it also causes environmental pollution. Various strategies of the construction industry, such as low-carbon construction, green buildings, zero-energy development, and carbon-neutral construction, can achieve the goals of project development through designs. Because the issue of carbon neutral design and construction is gaining attention, the construction industry should plan ahead for the green transformation. Company decision-makers should not consider risk analysis of decision-making as mere considerations of investment costs and profit; instead, decision-making analysis requires decision-making models that are objective and scientific to examine the future influence of project activities on both the interior and exterior of the company. Only in this manner can the effectiveness of a decision be enhanced. The model proposed in this study considered the lowest profit of the company, investing reduced profit into energy-saving design configurations. Considering multiple factors simultaneously, including corporate social responsibility, environmental protection, and reducing the energy consumed by residents, the model is highly adaptive. The model also benefits company image and improves intangible values, such as company reputation. Immediately adapting to carbon neutral construction methods is not easy for traditional residential building constructors. They are unwilling to sacrifice profits, which renders the effects of green building policy promotion in Taiwan insignificant. The model developed in this study enables decision-makers to understand multiple factors of corporate operation, including external environmental influences. From the various levels of influence from each criterion, decision-makers can develop strategies suitable for gradually improving the energy-saving decisions of projects during the early design stages.

## REFERENCES

1.  Lu, Y.; Zhu, X.; Cui, Q. Effectiveness and equity implications of carbon policies in the United States construction industry. *Build. Environ.* 2012, *49*, 259–269.

2.  Yan, H.; Shen, Q.P.; Fan, L.C.H.; Wang, Y.W.; Zhang, L. Greenhouse gas emissions in building construction: A case study of One Peking in Hong Kong. *Build. Environ.* 2010, *45*, 949–955.

3.  Narumi, D.; Kondo, A.; Shimoda, Y. Effects of anthropogenic heat release upon the urban climate in a Japanese megacity. *Environ. Res.* 2009, *109*, 421–431.

4.  Suga, M.; Almkvist, E.; Oda, R.; Kusaka, H.; Kanda, M. The impacts of anthropogenic energy and urban canopy model on urban atmosphere. *Annu. J. Hydraul. Eng.* 2009, *53*, 283–288.

5.  Zhang, L.; Tan, B.H. The pollution and measure in a building construction. *Sichuan Build. Sci.* 2005, *3*, 127–130.

6.  Cheng, W.; Chiang, Y.H.; Tang, B.S. Exploring the economic impact of construction pollution by disaggregating the construction sector of the input–output table. *Build. Environ.* 2006, *41*, 1940–51.

7.  Krisen, M.; Nigel, S.; Christopher, N.P. Stakeholder matrix for ethical relationships in the construction industry. *Constr. Manag. Econ.* 2008, *26*, 625–632.

8.  Begum, R.A.; Siwar, C.; Pereira, J.J.; Jaafar, A.H. Attitude and behavioral factors in waste management in the construction industry of Malaysia. *Resour. Conserv. Recycl.* 2009, *53*, 321–328.

9.  Sakr, D.A.; Sherif, A.; El-Haggar, S.M. Environmental management systems' awareness: An investigation of top 50 contractors in Egypt. *J. Clean Prod.* 2010, *18*, 210–218.

10. Ding, G.K.C. Sustainable construction the role of environmental assessment tools. *J. Environ. Manag.* 2008, *86*, 451–464.

11. Lam, T.I.; Chan, H.W.; Poon, C.S.; Chau, C.K.; Chun, K.P. Factors affecting the implementation of green specifications in construction. *J. Environ. Manag.* 2010, *91*, 654–661.

12.  Cole, R.J. Building environmental assessment methods: Clarifying intentions. *Build. Res. Inf.* 1999, *27*, 230–246.

13.  Holmes, J.; Hudson, G. An Evaluation of the Objectives of the BREEAM Scheme for Offices: A Local Case Study. In Proceedings of RICS Cutting Edge 2000 Conference, London, UK,, 6–8 September, 2000.

14.  Cole, R.J. Emerging trends in building environmental assessment methods. *Build. Res. Inf.* 1998, *26*, 3–16.

15.  Crawley, D.; Aho, I. Building environmental assessment methods: Application and development trends. *Build. Res. Inf.* 1999, *27*, 300–308.

16.  Rees, W. The built environment and the ecosphere: A global perspective. *Build. Res. Inf.* 1999, *27*, 206–220.

17.  Bauman, A. Salvage as a recession hedge: Green jobs and other economic stimuli. *Int. J. Environ. Tech. Manag.*2010, *13*, 84–95.

18.  Pop, O.; Dina, G.C.; Martin, C. Promoting the corporate social responsibility for a green economy and innovative jobs. *Procedia-Soc. Behav. Sci.* 2011, *15*, 1020–1023.

19.  Berkhout, T. Corporate gains. *Altern. J.* 2005, *31*, 15–18.

20.  Fernando, M. Corporate social responsibility in the wake of the Asian tsunami: Effect of time on the genuineness of CSR initiatives. *Eur. Manag. J.* 2010, *28*, 68–79.

21.  Jones, P.; Comfort, D.; Hillier, D. Corporate social responsibility and the UK construction industry. *J. Corp. Real Estate* 2006, *8*, 134–150.

22.  Sonja, P.L. The development of corporate social responsibility in the Australian construction industry. *Constr. Manag. Econ.* 2008, *26*, 93–101.

23.  Rao, P.; Holt, D. Do green supply chains lead to competitiveness and economic performance? *Int. J. Oper. Prod. Manag.* 2005, *25*, 898–916.

24.  Chen, Y.; Lai, S.; Wen, C. The influence of green innovation performance on corporate advantage in Taiwan. *J. Bus. Ethics* 2006, *67*, 331–339.

25.  Rao, P. The greening of suppliers—in the South East Asian context. *J. Clean Prod.* 2005, *13*, 935–945.

26.  Zhu, Q.; Sarkis, J. An inter-sectoral comparison of green supply chain management in China: Drivers and practices. *J. Clean. Prod.* 2006, *14*, 472–486.

27.  Shang, K.C.; Lu, C.S.; Li, S. A taxonomy of green supply chain management capability among electronics-related manufacturing firms in Taiwan. *J. Environ. Manag.* 2010, *91*, 1218–1226.

28. Varnäs, A.; Balfors, B.; Faith-Ell, C. Environmental consideration in procurement of construction contracts: Current practice, problems and opportunities in green procurement in the Swedish construction industry. *J. Clean Prod.* 2009, *17*, 1214–1222.

29. Pan, Y.; Yin, R.; Huang, Z. Energy modeling of two office buildings with data center for green building design.*Energy Build.* 2008, *40*, 1145–1152.

30. Castro-Lacouture, D.; Sefair, J.A.; Flórez, L.; Medaglia, A.L. Optimization model for the selection of materials using a LEED-based green building rating system in Colombia. *Build. Environ.* 2009, *44*, 1162–1170.

31. Airaksinen, M.; Matilainen, P. A carbon footprint of an office building. *Energies* 2011, *4*, 1197–1210.

32. Tsikaloudaki, K.; Laskos, K.; Bikas, D. On the establishment of climatic zones in Europe with regard to the energy performance of buildings. *Energies* 2012, *5*, 32–44.

33. Wright, C.; Baur, S.; Grantham, K.; Stone, R.B.; Grasman, S.E. Residential energy performance metrics. *Energies*2010, *3*, 1194–1211.

34. Yang, Z.; Wang, Y.; Zhu, L. Building space heating with a solar-assisted heat pump using roof-integrated solar collectors. *Energies* 2011, *4*, 504–516.

35. Liao, K.S.; Yambem, S.D.; Haldar, A.; Alley, N.J.; Curran, S.A. Designs and architectures for the next generation of organic solar cells. *Energies* 2010, *3*, 1212–1250.

36. Jacob, B. Lamps for improving the energy efficiency of domestic lighting. *Light. Res. Technol.* 2009, *41*, 219–228.

37. Shaukat, A.K.; Kamal, M.A. Study of visco-elastic properties of shoppers waste for its reuse as construction material. *Constr. Build. Mater.* 2010, *24*, 1340–1351.

38. Kikegawa, Y.; Genchi, Y.; Yoshikado, H.; Kondo, H. Development of a numerical simulation system toward comprehensive assessments of urban warming countermeasures including their impacts upon the urban buildings' energy demands. *Appl. Energy* 2003, *76*, 449–466.

39. Deru, M.; Pless, S.D.; Torcellini, P.A. BigHorn Home Improvement Center Energy Performance. In Proceedings of the 2006 ASHRAE Transactions Annual Meeting, Quebec City, Canada, 24–28 June 2006; pp. 349–366.

40. Jalalzadeh-Azar, A.A. Experimental evaluation of a downsized residential air distribution System: comfort and ventilation effectiveness. *ASHRAE J.* 2007, *113*, 313–322.

41. Tzempelikos, A.; Athienitis, A.K. The impact of shading design and control on building cooling and lighting demand. *Sol. Energy* 2007, *81*, 369–382.

42. Zago, M.; Casalegno, A.; Marchesi, R.; Rinaldi, F. Efficiency analysis of independent and centralized heating systems for residential buildings in Northern Italy. *Energies* 2011, *4*, 2115–2131.

43. Pacheco, R.; Ordóñez, J.; Martínez, G. Energy efficient design of building: A review. *Renew. Sustain. Energy Rev.* 2012, *16*, 3559–3573.

44. Sailor, D.J. A green roof model for building energy simulation programs. *Energy Build.* 2008, *40*, 1466–1478.

45. Lai, C.-M.; Wang, Y.-H. Energy-saving potential of building envelope designs in residential houses in Taiwan. *Energies* 2011, *4*, 2061–2076.

46. Shekarchian, M.; Moghavvemi, M.; Rismanchi, B.; Mahlia, T.M.I.; Olofsson, T. The cost benefit analysis and potential emission reduction evaluation of applying wall insulation for buildings in Malaysia. *Renew. Sustain. Energy Rev.* 2012, *16*, 4708–4718.

47. Roth, K.; Lawrence, T.; Brodrick, J. Double-skin façades. *ASHRAE J.* 2007, *49*, 70–73.

48. Shameri, M.A.; Alghoul, M.A.; Sopian, K.; Zain, M.F.M.; Elayeb, O. Perspectives of double skin façade systems in buildings and energy saving. *Renew. Sustain. Energy Rev.* 2011, *15*, 1468–1475.

49. Allen, L.; Lindberg, F.; Grimmond, C.S.B. Global to city scale urban anthropogenic heat flux: Model and variability. *Int. J. Climatol.* 2011, *31*, 1990–2005.

50. Böhm, G.; Pfister, H.-R. Consequences, morality, and time in environmental risk evaluation. *J. Risk Res.* 2005, *8*, 461–479.

51. O'Neill, N. Educating for Personal and social responsibility: Levers for building collective institutional commitment. *J. Coll. Character* 2011, *12*, 6.

52. Hediger, W. Welfare and capital-theoretic foundations of corporate social responsibility and corporate sustainability. *J. Socio-Econ.* 2010, *39*, 518–526.

53. Burtraw, D.; Parry, I.W.H. *Options for Returning the Value of CO$_2$ Emissions Allowances to Households*; RFF Discussion Paper 11-03; Resources for the Future: Washington, DC, USA, 2011.

54. Murry, J.W., Jr.; Hammons, J.O. Delphi: A versatile methodology for conducting qualitative research. *Rev. High. Educ.* 1995, *18*, 423–436.

55. Ziglio, E.; Adler, M. *Gazing into the Oracle: The Delphi Method and*

*Its Application to Social Policy and Public Health*; Jessica Kingsley: London, UK, 1996.

56.  Saaty, T.L. *The Analytical Hierarchy Process: Planning, Priority Setting, Resource Allocation*; McGraw-Hill: New York, NY, USA, 1980.

57.  Saaty, T.L.; Takizawa, M. Dependence and independence: From linear hierarchies to nonlinear networks. *Eur. J. Oper. Res.* 1986, *26*, 229–237.

58.  Hsueh, S.-L.; Yan, M.-R. Enhancing sustainable community development a multi-criteria Evaluation model for energy efficient project selection. *Energy Procedia* 2011, *5*, 135–144.

59.  Saaty, T.L. How to make a decision: The Analytic Hierarchy Process. *Eur. J. Oper. Res.* 1990, *48*, 9–26.

60.  Saaty, T.L. How to make a decision: The Analytic Hierarchy Process. *Interfaces* 1994, *24*, 19–43.

61.  Dziadosz, A. Estimation and selection of building investment using AHP. *Czas. Tech.* 2008, *1*, 41–51.

62.  Bertolini, M.; Bevilacqua, M. A combined goal programming—AHP approach to maintenance selection problem. *Reliab. Eng. Syst. Saf.* 2006, *91*, 839–848.

63.  Al-Harbi, K.A. Application of the AHP in project management. *Int. J. Proj. Manag.* 2001, *19*, 19–27.

64.  Bevilacqua, M.; Barglia, M. The analytic hierarchy process applied to maintenance strategy selection. *Reliab. Eng. Syst. Saf.* 2000, *70*, 71–83.

65.  Skibniewski, M.J.; Chao, L-C. Evaluation of advanced construction technology with AHP method. *J. Constr. Eng. Manag.* 1992, *118*, 577–593.

66.  Lee, G.K.L.; Chan, E.H.W. The analytic hierarchy process (AHP) approach for assessment of urban renewal proposals. *Soc. Indic. Res.* 2008, *85*, 155–168.

67.  Bose, R.K.; Anandalingam, G. Sustainable urban energy-environment management with multiple objectives.*Energy* 1996, *21*, 305–318.

68.  Zadeh, L.A. A fuzzy-algorithmic approach to the definition of complex or imprecise concepts. *Int. J. Man. Mach. Stud.* 1976, *8*, 249–291.

69.  Zadeh, L.A. Fuzzy logic = computing with words. *IEEE Trans. Fuzzy Syst.* 1996, *4*, 103–111.

70.  Hadi-Vencheh, A.; Mokhtarian, M.N. A new fuzzy MCDM approach based on centroid of fuzzy numbers.*Expert Syst. Appl.* 2011, *38*, 5226–5230.

71. Houshyar, E.; Azadi, H.; Almassi, M.; Sheikh Davoodi, M.J.; Witlox, F. Sustainable and efficient energy consumption of corn production in Southwest Iran: Combination of multi-fuzzy and DEA modeling. *Energy* 2012, *44*, 672–681.

72. Chang, P.T.; Huang, L.C.; Lin, H.J. The fuzzy Delphi via fuzzy statistics and membership function fitting and an application to human resources. *Fuzzy Set. Syst.* 2000, *112*, 511–520.

73. Jafari, A.; Jafarian, M.; Zareei, A.; Zaerpour, F. Using fuzzy Delphi method in maintenance strategy selection problem. *J. Uncertain Syst.* 2008, *2*, 289–298.

74. Wang, H.-J.; Zeng, Z.-T. A multi-objective decision-making process for reuse selection of historic buildings. *Expert Syst. Appl.* 2010, *37*, 1241–1249.

75. Tavana, M.; Kennedy, D.T.; Rappaport, J.; Ugras, Y.J. An AHP-Delphi group decision support system applied to conflict resolution in hiring decisions. *J. Manag. Syst.* 1993, *5*, 49–74.

76. Liao, C.-N. Supplier selection project using an integrated Delphi, AHP and Taguchi loss function. *ProbStat Forum* 2010, *3*, 118–134.

77. Wang, L.; Chu, J.; Wu, J. Selection of optimum maintenance strategies based on a fuzzy analytic hierarchy process. *Int. J. Prod. Econ.* 2007, *107*, 151–163.

78. Torfi, F.; Farahani, R.Z.; Rezapour, S. Fuzzy AHP to determine the relative weights of evaluation criteria and Fuzzy TOPSIS to rank the alternatives. *Appl. Soft. Comput.* 2010, *10*, 520–528.

79. Hsu, Y.-L.; Lee, C.-H.; Kreng, V.B. The application of Fuzzy Delphi Method and Fuzzy AHP in lubricant regenerative technology selection. *Expert Syst. Appl.* 2010, *37*, 419–425.

80. Yu, W.D.; Skibniewski, M.J. A neuro-fuzzy computational approach to constructability knowledge acquisition for construction technology evaluation. *Autom. Constr.* 1999, *8*, 539–552.

81. Perng, Y.H.; Hsueh, S.L.; Yan, M.R. Evaluation of housing construction strategies in China using Fuzzy-Logic system. *Int. J. Strat. Prop. Manag.* 2005, *9*, 215–232.

82. Bellman, R.E.; Zadeh, L.A. Decision making in a fuzzy environment. *Manag. Sci.* 1970, *17*, B141–B164.

83. Zadeh, L.A. The concept of a linguistic variable and its application to approximate reasoning—III. *Inf. Sci.* 1975, *9*, 43–80.

84. Haleblian, J.; McNamara, G.; Kolev, K.; Dykes, B.J. Exploring firm

characteristics that differentiate leaders from followers in industry merger waves: A competitive dynamics perspective. *Strat. Manag. J.* 2012, *33*, 1037–1052.

85.   Graf, L.; König, A.; Enders, A.; Hungenberg, H. Debiasing competitive irrationality: How managers can be prevented from trading off absolute for relative profit. *Eur. Manag. J.* 2012, *30*, 386–403.

# Chapter 7

# CFD SIMULATION OF A CONCRETE CUBICLE TO ANALYZE THE THERMAL EFFECT OF PHASE CHANGE MATERIALS IN BUILDINGS

Miguel A. Gómez, Miguel A. Álvarez Feijoo, Roberto Comesaña, Pablo Eguía, José L. Míguez, and Jacobo Porteiro

Industrial Engineering School, University of Vigo, Lagoas-Marcosende s/n 36310 Vigo, Spain

## ABSTRACT

In this work, a CFD-based model is proposed to analyse the effect of phase change materials (PCMs) on the thermal behaviour of the walls of a cubicle exposed to the environment and on the resistance of the walls to climate changes. The effect of several days of exposure to the environment was simulated using the proposed method. The results of the simulation are compared with experimental data to contrast the models. The effects of exposure on the same days were simulated for several walls of a cubicle made of a mixture of concrete and PCM. The results show that the PCM stabilizes temperatures within the cubicle and decreases energy consumption of refrigeration systems.

## INTRODUCTION

The study of energy efficiency is a highly topical subject, and the need to develop methods to improve the thermal performance of industrial systems is well recognized. Energy efficiency improvements are relevant in many fields. One of these fields is the construction industry, where energy efficiency can be improved by using appropriate materials. One type of material with interesting properties in this area is phase change materials (PCMs).

The interest in PCMs derives from their capacity to store energy as latent heat; these materials have been the focus of numerous studies [1,2,3]. This energy may be used in various applications according to the demand, whether in support of air-conditioning or free-cooling [4], walls that accumulate heat during peak hours at room temperature, then release heat in times of low temperature, heat exchanger modules [5], or simply to modify the properties of building materials by increasing thermal inertia through the addition of a certain amount of PCM [6,7,8]. Other works have researched PCMs encapsulated in different shapes in order to maximise the storage density [9,10], as well as improving the thermal conductivity of the PCM [11]. Most of these works study the thermal storage of heat. Studies of thermal storage for cooling applications [12,13,14] are more unusual.

Computational fluid dynamics (CFD) is a set of computer simulation techniques that help analyse and predict the performance of systems in which fluid motion plays an important role. This makes CFD an important tool that is frequently used to help design and improve products. These techniques can be applied to many fields of study, including the energy efficiency of buildings. To simulate air conditioning, works such as by Wang and Zhu [15] studied the movement of air through various rooms and analysed the performance of fans. Tanasic [16] studied the flow of air in an industrial building to determine the optimal placement of exhaust fans for ventilation. Transient heat transfer plays an important role in simulations in which thermal conditions are time dependent and affect the system. A Trombe wall was simulated in [17] with special attention to the natural advection air movement to study the convection heat transfer [18,19]. CFD simulations can be a useful tool to analyze the evolution of systems with PCM [20] in detail.

In this study, CFD techniques are used to simulate the thermal behaviour of a cubicle exposed to actual environmental conditions and compare it with the thermal behaviour under experimental conditions. The results of this simulation are compared with the data measured in the experimental cubicle. Thus, the ability of the simulation method to predict the thermal behaviour of the experimental construction is evaluated. Then, the cubicle is simulated under the condition of the cubicle walls containing PCM, and the results are compared with the simulation of the non-PCM-containing cubicle to study the effect of PCM on the cubicle's thermal behaviour.

## EXPERIMENTAL SETUP

A concrete cubicle which is located in Vigo, Spain and thermal data collection system were used to study the thermal behaviour of a building in response to environmental exposure. The cubicle is made of precast, self-compacting concrete panels, which are 2.4 m high, 2.6 m wide and 12 cm thick (Figure 1).

The cubicle rests on a base made of conventional concrete; the roof was also constructed with conventional concrete and contains expanded polystyrene insulation.

**Figure 1:** Experimental cubicle.

The enclosure also contains a heat pump, which functions to maintain a constant temperature inside the cubicle. The power of the heat pump is 2500 W for refrigeration and 3400 W for heating. The control of the heat pump is regulated in order to maintain a temperature set-point of 20 °C. When a temperature sensor detects a variation of 2.5 °C higher or lower than the set-point, the cooling or heating modes are activated. Once the cooling or the heating modes are working, the system operates with an on-off regulation when the sensor detects a deviation of 1.5 °C above or below the set-point.

The building is equipped with a data collection system consisting of internal thermometers, thermocouples embedded at a depth of a few mm from the inner and outer surfaces of each panel (located at the horizontal centre of each panel at 1 m and 2 m above ground level) and a weather station. The thermocouples are placed on the walls facing south, east and west, as these walls experience more thermal variation due to solar radiation. The heat flow meter is placed on the inner surface of the south wall. The cubicle also has a heat pump system to study the heating or cooling of each cubicle. Figure 2 shows a diagram comprising a front view and a top view to show the placement of these sensors and the heat pump in the cubicle.

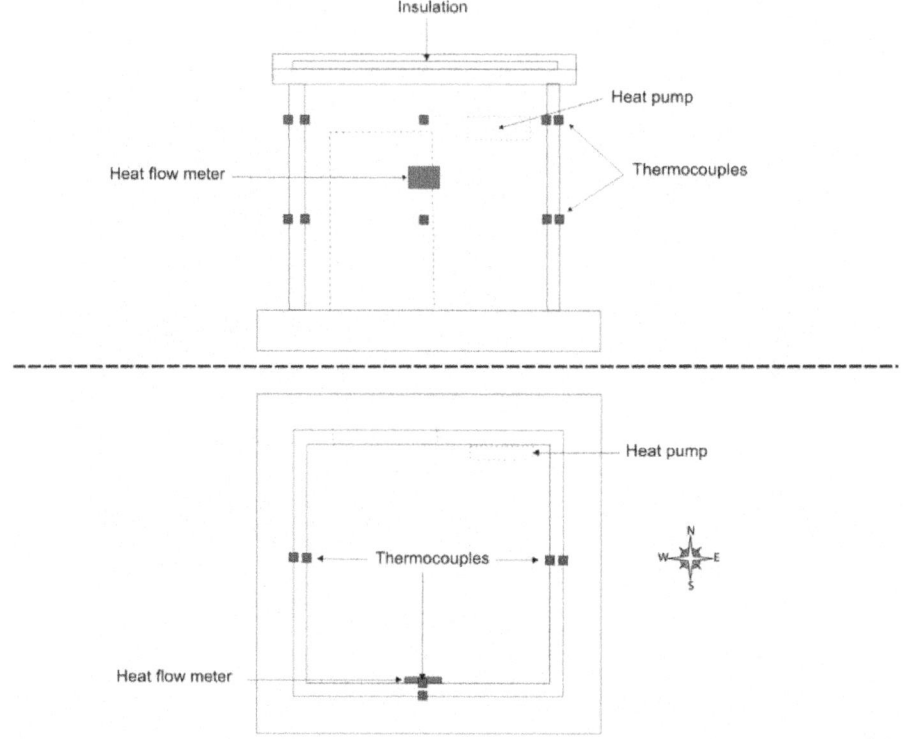

**Figure 2:** Schematic of the sensors location in the cubicle.

Regarding the electrical installation of the cubicle, an electrical breaker box is installed outside the cubicle to separate the electrical installation of the cubicle and the electric network. This breaker box consists of a three-phase thermal magnetic circuit breaker, an automatic differential circuit breaker and a three-phase electricity consumption meter to determine the total consumption of the cubicle installation. The consumption meter allows an evaluation of the variation in energy consumption when the heat pump operates in cooling or heating mode.

Next to the cubicle is a box with an industrial computer on which data are stored, a Wi-Fi router, a power supply unit, two 300 W sockets, a thermal magnetic circuit breaker for this portion of the installation and miscellaneous supplementary equipment.

The heat pump is installed in the door panel of the cubicle (left side entrance), along with a consumption meter, a circuit breaker, a 12 V power source, an 8-channel analog input module (model ADAM-4017 of B&B Electronics Ltd.), a humidity and temperature sensor, an electrical outlet and four temperature sensors in each of the east, south and west panels. The data acquisition program, developed for this study, allows recording of data (temperature, humidity, heat flow and consumption) at a user-selectable recording frequency. Figure 3 shows a flow chart of the complete setup of the data acquisition system.

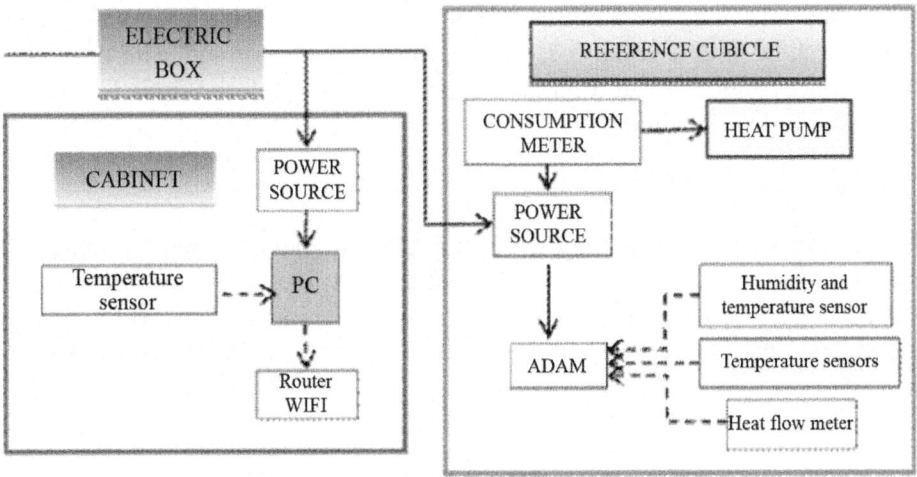

**Figure 3:** Flow chart of the data acquisition system.

## CFD SIMULATION

The aim of the CFD simulation presented in this paper is to reproduce the real behaviour of the system and to study other conditions without increasing the experimental investment. The methodology used for this simulation was based on modelling the geometry of the cubicle and then applying physical models that predict the system variables. Once the modelling is applied to the boundary conditions necessary for the material and thermal properties, it reproduces the real atmospheric conditions that the cubicles were subjected to during the days that were simulated. These boundary conditions include atmospheric temperature, wind speed and direction data, which were recorded by the weather station, and the calculated solar radiation during the simulated days. Ten consecutive days in October 2011 were simulated using the atmospheric data measured by the weather station, which are updated every five minutes. Solar radiation data were calculated as a function of the Sun's position with a resolution of one minute. The simulation was performed by calculating transient intervals of five minutes from October 5th to 15th, 2011, a period in which the weather was stable and sunny, which is suitable to avoid the difficulties of reproducing atmospheric variations. An unstructured, three-dimensional tetrahedral grid with approximately $1.5 \times 10^6$ elements was used. The size of the grid was selected to reach high accuracy of the solutions. The size of the mesh elements is very heterogeneous and dependent on the location of the element within the domain. Inside the cubicle, around the heat pump, the smaller cells have a 5 mm side length; outside the cubicle, it is not necessary to refine the cells. The mesh has an average angular skew of 0.3 and a maximum angular skew smaller than 0.7. Simulations were performed with eight 2.53 GHz processing cores in a parallel computation, which required 1.5 days to calculate each simulated day.

## Assumptions

The models used in this study were chosen to simulate the thermal behaviour of buildings under controlled conditions. However, some of the material properties and environmental conditions were simplified to avoid additional complications in the use of the models. The assumptions made for the calculations were as follows:

- Isotropic properties throughout the domain.
- The gas phase was assumed to be of an incompressible ideal gas with constant thermal properties.
- Air was assumed to be transparent to radiation, while concrete was considered opaque.
- The radiation model did not consider the effects of scattering.
- The solar calculator assumed fair weather conditions.
- The concrete properties (density, specific heat and thermal conductivity) were assumed to be constant.
- The PCM was assumed to be fully dispersed and homogenous within the concrete.
- The rated heating power and cooling power generated in the inner tubes of the heat pump were constant and homogeneous.

## CFD Models

To simulate the thermal behaviour of the cubicle, it is necessary to apply physical models to a geometry that represents the real system. The complete geometry of the cubicle and part of the environment was created using CAD software. Subsequently, all surfaces and volumes of the domain were meshed.

The full mesh was exported to ANSYS Fluent for the numerical solution of Navier-Stokes equations by finite volume methods. The equations (shown in theTable 1) were solved by an implicit formulation with the SIMPLE method and a temporary resolution to solve first-order temporal discretisation in a transient simulation with a time step of 5 minutes. Turbulence was modelled using the Realizable k-ε model with enhanced wall treatment. The radiation heat transfer was modelled by the DO (discrete ordinates) method and the ANSYS Fluent solar calculator was used to apply the solar radiation to the walls and the other surfaces. Continuity, momentum, turbulence and energy were modelled according to standard CFD procedures that are well documented [21].

**Table 1:** CFD models

| Model | Equation |
|-------|----------|
| *Mass conservation* | $$\frac{\partial \rho}{\partial t} + \nabla(\rho \vec{v}) = S_m$$ |
| *Turbulent flow (Navier-Stokes)* | $$\frac{\partial}{\partial t}(\rho \vec{v}) + \nabla(\rho \vec{v}\vec{v}) = -\nabla(p) + \nabla \cdot \left\{ \mu \left[ (\nabla \vec{v} + \nabla \vec{v}^T) - \frac{2}{3}\nabla \cdot \vec{v}I_t \right] \right\} + \rho \vec{g} + \vec{F}$$ |
| *k- ε Turbulence model* | $$\mu_t = \rho C_\mu \frac{k^2}{\epsilon}$$ $$\frac{\partial}{\partial t}(\rho k) + \frac{\partial}{\partial x_i}(\rho k v_i) = \frac{\partial}{\partial x_i}\left[ \left(\mu + \frac{\mu_t}{\sigma_k}\right)\frac{\partial k}{\partial x_i} \right] + G_k + G_b - \rho\epsilon - Y_M + S_k$$ $$\frac{\partial}{\partial t}(\rho \epsilon) + \frac{\partial}{\partial x_i}(\rho \epsilon v_i)$$ $$= \frac{\partial}{\partial x_i}\left[ \left(\mu + \frac{\mu_t}{\sigma_\epsilon}\right)\frac{\partial \epsilon}{\partial x_i} \right] + C_{1\epsilon}\frac{\epsilon}{k}(G_k + C_{3\epsilon}G_b) - C_{2\epsilon}\rho\frac{\epsilon^2}{k} + S_\epsilon$$ |
| *Incompressible ideal gas density* | $$\rho = \frac{P_{op} \cdot M}{T \cdot R}$$ |
| *Energy* | $$\frac{\partial}{\partial t}(\rho h) + \frac{\partial}{\partial x_i}(\rho v_i h) = \frac{\partial}{\partial x_i}(\lambda)\frac{\partial T}{\partial x_i} - \frac{\partial}{\partial x_i}\sum_j h_j \vec{J_j} + \frac{\partial p}{\partial t} + (\tau_{ij})\frac{\partial u_i}{\partial x_k} + S_h$$ |
| *Radiation DO* | $$\nabla(I(\vec{r},\vec{s})\vec{s}) = -(a_g + \sigma_S)I(\vec{r},\vec{s}) + \frac{a_g \cdot n^2}{\pi}\sigma T^4 + \frac{\sigma_S}{4\pi}\int_0^{4\pi} I(\vec{r},\vec{s'})\Phi(\vec{s}\cdot\vec{s'})d\Omega$$ |

There are three distinct regions in the control domain; the calculation requires different treatments of each region. First is the gas zone, which includes the air both inside and outside the cubicle; in this region, turbulence models are utilised, and the equations of mass and energy transport are solved. Second, there are the solid areas: the walls, ceiling and base of the cubicle. The calculations for the solid areas are simpler because the only thing which needs solving is the energy transport through the cells. Finally, the solid areas representing PCM walls require special treatment in calculating the effect of the PCM, which will be detailed below.

## Modelling the PCM Effect

The walls formed by self-compacting concrete mix and PCM have the ability to absorb or emit heat without experiencing an increase or a decrease in temperature because, at the phase change temperature, the heat is applied to melt or solidify the PCM and does not affect the concrete. To simulate this effect, it is necessary to define a new variable, $\phi$, which is programmed as a user defined scalar (UDS), representing the liquid fraction of PCM material, and to add to the energy transport equation a source term that represents the heat absorbed or released by the PCM when it is melting or solidifying. The liquid fraction of PCM in the cell is governed by a transport equation [Equation (1) in Table 2] that consists of the transient term and the source term (the convective and diffusive terms, which are the second term on the left side and first on the right side, respectively, do not apply in this case). The source term represents the mass of molten PCM in the cell at each time step. The melting process is governed by the total incoming energy in the cell and the latent heat of fusion of PCM material [Equation (2) in Table 2].

**Table 2:** Equations of the PCM model

| | |
|---|---|
| $$\int_V \frac{\partial \rho \varphi}{\partial t} dV + \oint_S \rho \varphi \vec{v} \cdot d\vec{A} = \oint_S \Gamma_\varphi \varphi_f \cdot d\vec{A} + \int_V S_\varphi \, dV$$ | Equation (1) |
| $$S_\varphi = \frac{\Delta H_S}{LH_{PCM} \cdot V} \Delta t$$ | Equation (2) |
| $$\frac{\partial}{\partial t} \rho e_s = \nabla \cdot (\lambda_S \nabla T_s) + S_s$$ | Equation (3) |
| $$S_S = -\frac{\Delta H_S}{V}$$ | Equation (4) |
| $$\Delta H_S = \nabla \cdot (\lambda_S \nabla T_s) + \dot{Q}_i + \rho \cdot V \cdot C_p \frac{T_s - T_{pc}}{\Delta t}$$ | Equation (5) |

The energy transport in the solid walls is governed by a transport equation [Equation (3) in Table 2] that is simplified for application to a non-moving solid material and includes a source term that is equal to the energy absorbed by the PCM with the opposite sign [Equation (4) in Table 2]. The total energy received by a solid cell [Equation (5) in Table 2] is calculated as the sum of the energy transmitted by solid diffusion between cells, the energy received from outside of the wall by convection and radiation and a term that takes into account the thermal inertia of the cell before the temperature change of the cell about the melting temperature of the PCM.

## Material Properties and Environmental Boundary Conditions

To simulate any system under conditions close to reality, it is necessary to choose material properties and boundary conditions that represent the experimental system. The materials used in the model were the concrete (walls, floor and ceiling of the cubicle), the steel (door and plates), the expanded polystyrene insulation boards in the ceiling, and the air of the environment. The properties of these materials are shown in Table 3.The specific heat of the air is considered to be a function of the temperature, calculated as a linear interpolation of values taken from Moran-Saphiro [22]. An important parameter to calculate the amount of solar radiation stored by the walls is the absorptivity. This parameter depends on the colour, brightness and surface finish, which are very heterogeneous, even within the same concrete panel, therefore, it is very difficult to estimate a representative value. Hence, this was used as an adjustable parameter to obtain a good agreement between simulation and experimental results. Different values were tested from 0.3 to 0.9, choosing 0.45 as the final value for all the simulations, which is considered a reasonable value as the concrete used has a highly reflective surface.

**Table 3:** Characteristics of the materials used in the simulation

| Properties | Air | Concrete | Steel | Polystyrene | PCM |
|---|---|---|---|---|---|
| Density (kg/m$^3$) | Incompressible ideal gas | 2400 | 8300 | 20 | 980 |
| Specific heat (J/kg·°C) | f(T) | 750 | 502.5 | 1500 | - |
| Thermal conductivity (W/m·°C) | 0.0242 | 1.5 | 16.27 | 0.03 | - |
| Viscosity (kg/m·s) | $1.789 \times 10^{-5}$ | - | - | - | - |
| Phase change temperature (°C) | - | - | - | - | 23 |
| Fusion latent heat (J/kg) | - | - | - | - | $110 \times 10^3$ |

Weather conditions were simulated using data from a weather station and entering the velocities and wind directions as well as the air temperatures outside the cubicle. The boundaries of the environment were surfaces defined as velocity-inlet and pressure-outlet, through which air entered and exited, respectively. The experimental outdoor temperature data from the weather station are shown in Figure 4. These data are crucial for the simulation of the environment behaviour. The wind velocity is another important parameter to take into account in the modelling of the environment. This velocity was measured and the data are shown in Figure 5. High and variable wind velocities can be observed around midday whereas the velocities remain stable the rest of the time. The wind direction angle was also measured in the weather station and it was introduced in the simulation as a boundary condition, considering the north direction as 0 degrees. The experimental data are shown in Figure 6. We see that the wind angle is significantly more chaotic than most of the boundary conditions and it is difficult to draw conclusions about its behaviour. The simulated atmospheric conditions were updated every hour, thereby making them coincide with the data obtained by the air station, which were averaged from the readings of that time interval.

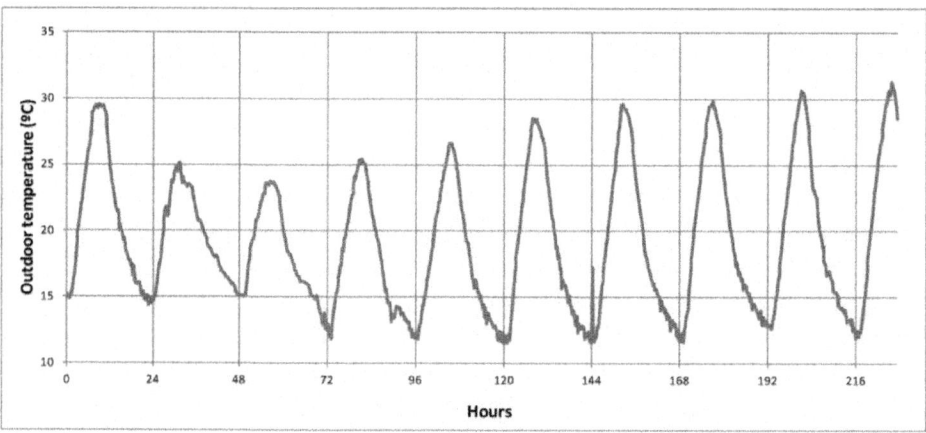

**Figure 4:** Outdoor temperature data measured in the weather station.

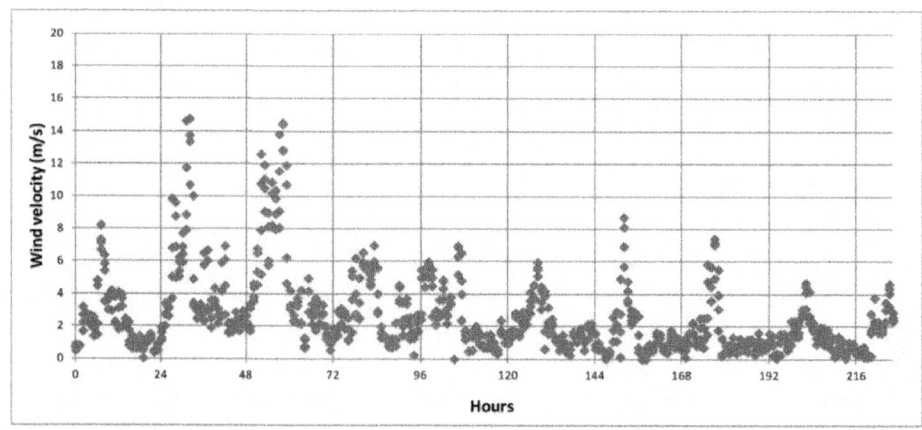

**Figure 5:** Wind velocities measured in the weather station.

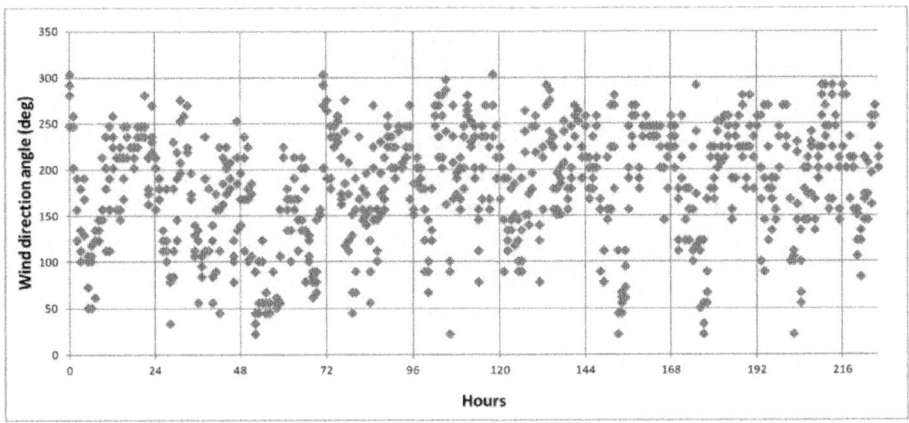

**Figure 6:** Wind direction angle measured in the weather station.

Solar radiation was introduced using the solar irradiation calculator of ANSYS Fluent, which imposed the incident radiation on the exposed surfaces according to the position of the sun, which varies as a function of time. The position of the sun depends on the day of the year and geographical coordinates, which were also entered into the solar irradiation calculator. The geographical coordinates where the cubicle was placed were latitude 42.13 and longitude 8.62.

## RESULTS AND DISCUSSION

The experimental system recorded data for several months during the year 2011. As it was difficult to take into account the variability of weather, a sunny weather period of ten days in the month of October was chosen to simulate climate conditions more accurately. Below, the proposed model is applied in a simulation of the experimental cubicle exposed to the measured weather conditions, and the results are compared with experimental data to contrast the proposed model and of the assumptions made. Subsequently, the parameters that affect the thermal behaviour of the system are presented and discussed in more depth. In addition, various figures are shown to analyse the variations of temperature fields and the PCM evolution throughout a typical day. Then, a similar simulation is performed in a cubicle in which panels exposed to solar radiation (south, east and west) consist of a mixture of concrete and 5% PCM. The results of this simulation are shown and compared with the results of the simulation without PCM.

## Comparison of Experimental and Simulation Results

Several measurements were taken in the experimental cubicle; the most representative measurements are the temperatures of the internal and external surfaces of the panels. As commented in Section 2, the heat pump is programmed to maintain a temperature of 20 °C inside the cubicle by working with on-off regulation. The temperature of the air inside the cubicle, which is measured by a set of thermometers, is shown in Figure 7 and compared with the predicted values (calculated as the average temperature inside the cubicle).

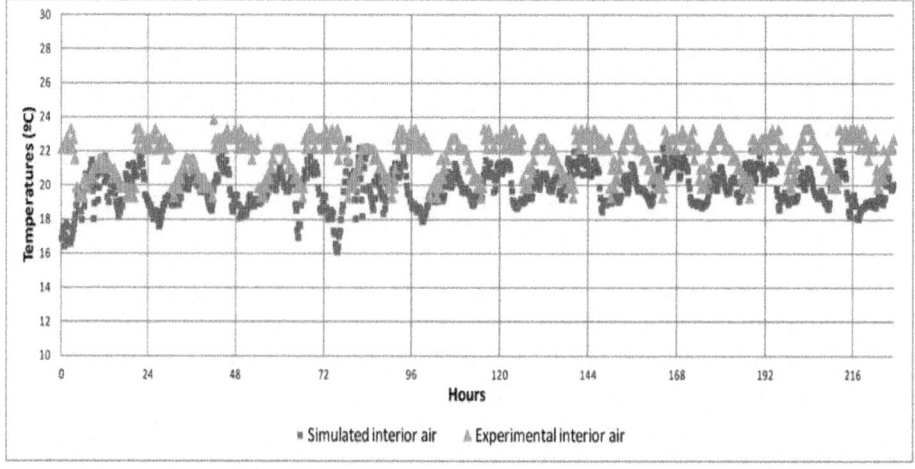

**Figure 7:** Measured and predicted temperatures of the air inside the cubicle.

Measurements and predictions show a reasonably similar pattern, however the predicted temperatures are generally 1–2 degrees lower. This could be because the actual efficiency of the heat pump is lower than the specifications. In some instants of the simulation the temperature drops to 16 degrees, this is caused by the time step size which causes excessive cooling time.

As the south wall receives the most solar radiation, the validation of the CFD simulation was performed by comparing the temperatures of this wall. Figure 8 shows the experimental south wall temperatures during ten consecutive days and the same temperatures predicted by the simulation for the same period, which begins at 8:00 AM on October 5th.

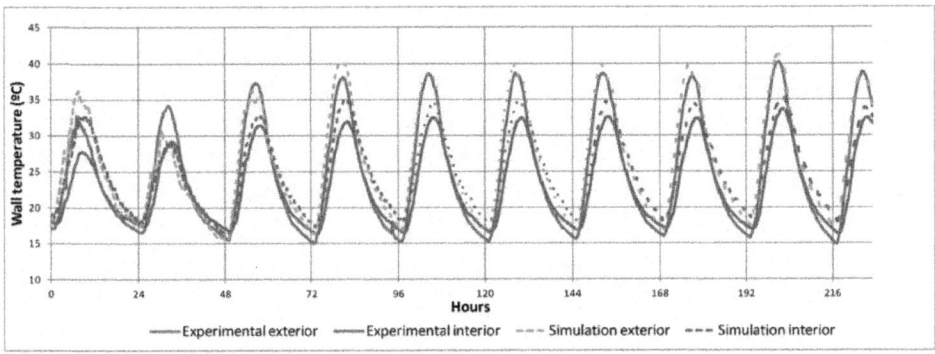

**Figure 8:** Measure and predicted temperatures of the south wall of the cubicle.

The overall behaviour of the two temperatures (internal and external) follows a reasonably similar pattern throughout the simulation period. The heating and cooling of the outer surface occur rapidly during daylight hours; however, the cooling is slower at night. The temperature of the inner surface responds to external changes following the pattern of the outer surface variations with a certain delay and slower temperature changes. As the simulation begins with initial values different from the real values, the predicted temperatures do not closely match the measured temperatures during the two first simulated days. In addition, the simulation predicts slightly higher maximum temperatures than are seen in the experimental data. This effect may be due to the assumed value of the concrete reflectivity, which may produce an excess of absorption of radiation. We should mention that in the experimental cubicle, the thermocouples are located at a depth a few millimeters from the surface, while the simulation shows the surface temperatures.

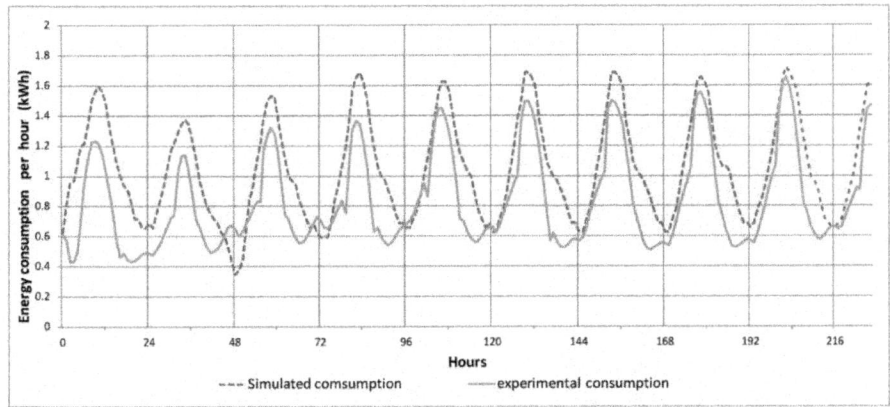

**Figure 9:** Measured *vs.* predicted thermal consumptions per hour in the heat pump.

The energy consumption of the heat pumps is a useful parameter to quantify the energy saving. Figure 9 compares the predicted values of thermal energy consumed per hour with the data registered by the consumption meter. The first conclusion we can draw from this comparison is the higher consumption predicted by the simulation than the experimental one in most days. This is consistent with the temperatures inside the cubicle shown in Figure 7 which may be caused by a lower efficiency of the real cooling system. Figure 10 shows the movement of air outside the cubicle due to the wind and inside the cubicle due to the air stream created by the heat pump. These air movements play an important role in enhancing the heat convection that cools the walls.

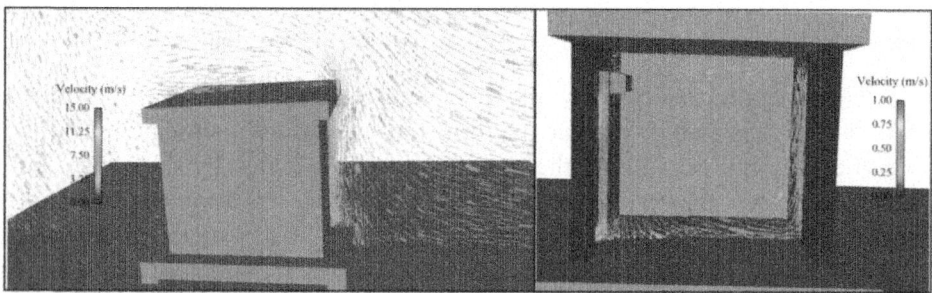

**Figure 10:** Vector fields of external and internal air velocities (m/s).

The surfaces exposed to direct sunlight experience significant heating as a result of the solar radiation.

This heat is transferred by conduction through the walls and by convection to the air near the wall surfaces. Figure 11 shows the variation of the temperature in the walls and floor of the cubicle due to the variation in sun position during the day and Figure 12 shows the variation between the air temperature and the wall temperature inside the cubicle.

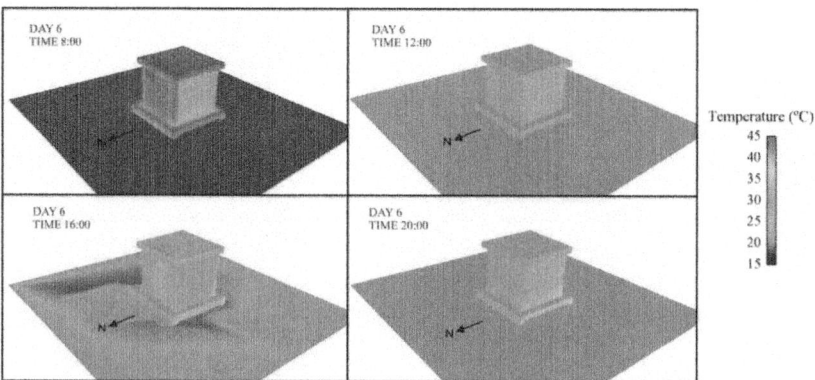

**Figure 11:** Evolution of the temperature field of the cubicle surfaces.

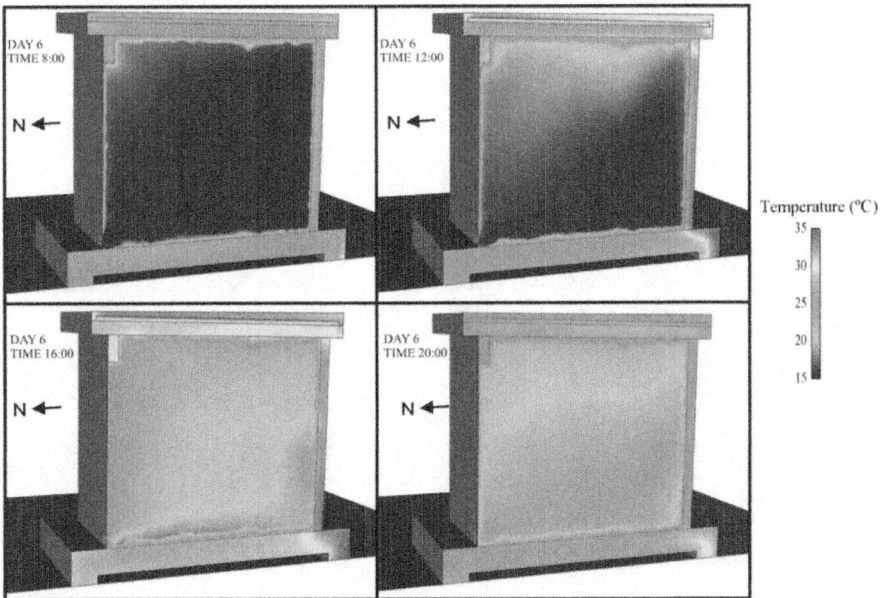

**Figure 12:** Evolution of the temperature field inside the cubicle and inside the walls.

## PCM Simulation

The same days shown in Figure 8 were simulated with the south, east and west panels (those exposed to solar radiation) consisting of a mixture of concrete and 5% PCM. In this way, the effect of the PCM on the thermal behaviour was evaluated; the results are analysed below.

In Figure 13, the daily cycles of the internal and external surface temperatures of the south wall are shown. Although some irregularities are observed during the simulated period, some general conclusions regarding the different thermal behaviour of the materials can be drawn. The most notable finding is that the maximum and minimum temperatures (internal and external) are clearly damped on the PCM concrete wall. The maximum and minimum values are reduced and increased, respectively, by approximately three degrees Celsius in the walls with PCM. Another clear difference observed in the inner surface temperature histories is that during cooling, the PCM acts when the phase change temperature is reached, reducing the slope of the interior temperature variation (cooling rate). This is because the PCM solidifies, and this process releases the energy stored as latent heat as the temperature remains constant. Therefore, the PCM slows the cooling rate of the wall. In the heating process, the PCM slows the heating rate of the wall, but this effect is less visible in the graph because the energy input due to solar radiation is relatively high.

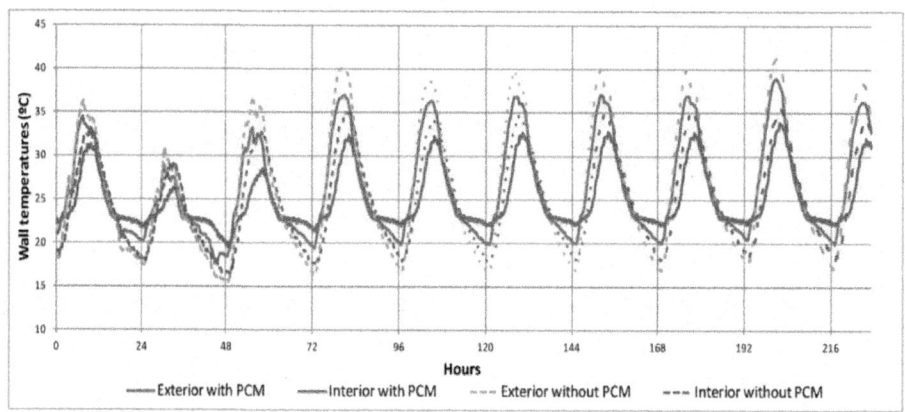

**Figure 13:** Predicted temperatures of the south wall of the cubicle in the simulation with and without PCM.

The comparison between the thermal consumption of the two simulations is shown in Figure 14 where the evolution of the consumptions per hour and the total consumption per day are observed.

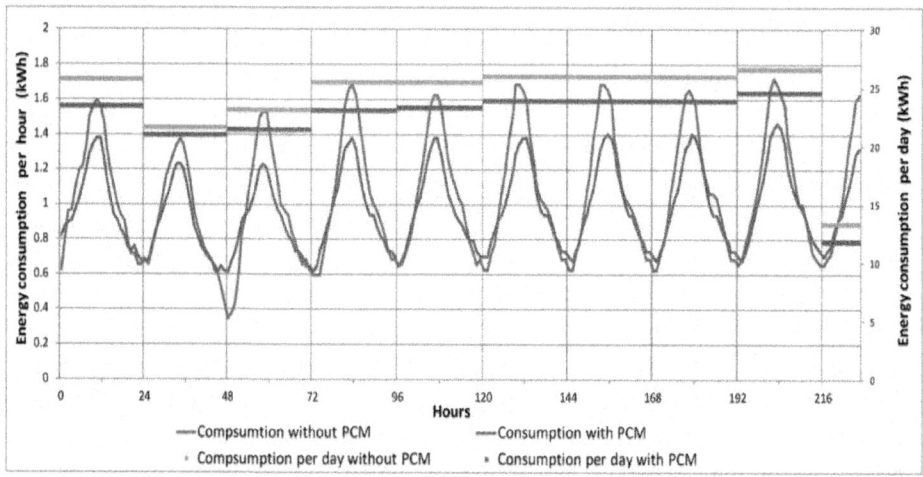

**Figure 14:** Predicted thermal consumption of the heat pump of the cubicle in the simulation with and without PCM.

The main conclusion is the lower energy consumed in the simulation with PCM during daylight hours, which shows that the PCM stores part of the energy received by the cubicle. The difference observed during the night when the PCM releases the energy stored, thereby producing a greater consumption of the heat pump, is less significant. The amount of the overall energy saved in the case of PCM was approximately 8%. This can be clearly seen through its daily consumption.

Figure 15 represents the evolution of the air temperature and the wall temperatures inside the cubicle with PCM. The temperature changes inside the walls are milder than in the cubicle without PCM (Figure 12).

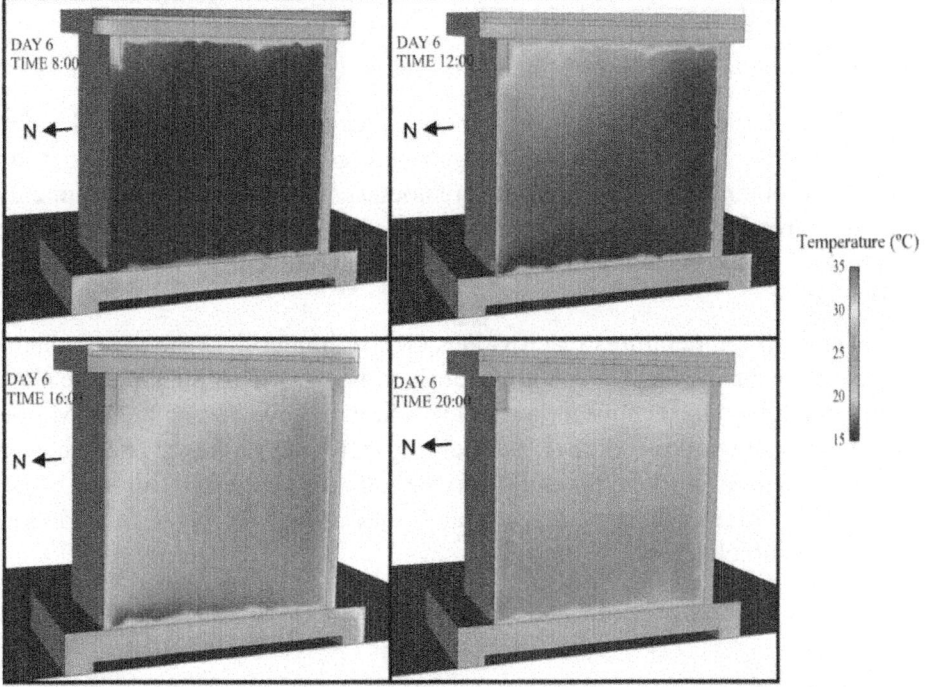

**Figure 15:** Evolution of the temperature field inside the cubicle and inside the walls in the PCM simulation.

Figure 16 shows the PCM liquid fraction [$\varphi$ in Equations (1) and (2)], which increases when the wall receives heat and its temperature reaches the phase change temperature while absorbing the energy required to melt the PCM. This process stops when the PCM totally liquefies ($\varphi = 1$). During the cooling process, the heat that has been absorbed by the PCM is released while the liquid fraction decreases (PCM solidification).

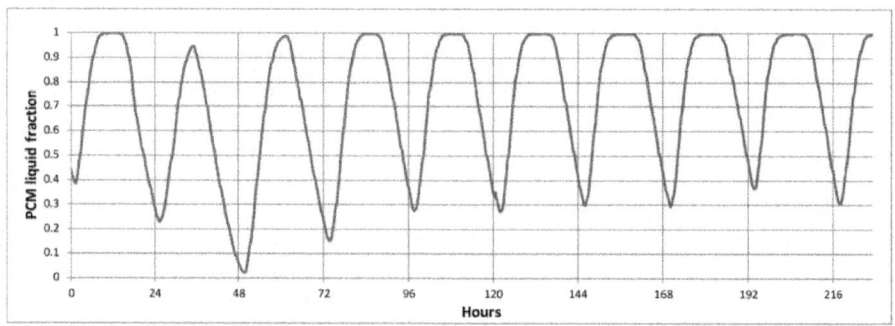

**Figure 16:** Evolution of the PCM liquid fraction.

Most of the days the PCM is totally charged ($\varphi = 1$) during the day; however, it does not fully discharge overnight ($\varphi > 0$). This means the amount of PCM is enough to hold a more severe nocturnal cooling. The environmental temperatures in the second and third day are moderate during daylight hours (Figure 4), so the PCM liquid fraction is less than one, due to the lower heating of the walls.

The wall temperatures reach values clearly higher than the phase change temperature during the day and lower during the night. This indicates the phase change temperature is appropriate to this climate, however, as the PCM does not completely solidify during the night, a PCM with a lower phase change temperature may work better and it may be fully discharged (solidified), which results in a greater heat storage capacity and lower temperature variations inside the cubicle.

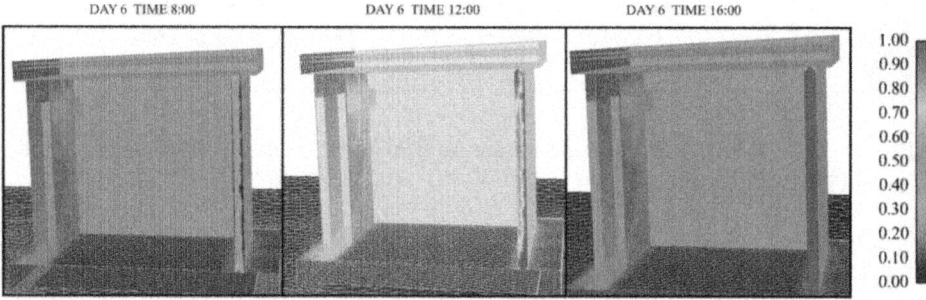

**Figure 17:** Evolution of the PCM liquid fraction in the south wall.

Figure 17 shows the variation of the liquid fraction of the PCM ($\varphi$ in Equations (1) and (2) in Table 2) in the south wall, which melts as the wall receives energy from solar radiation. The image of "Time 8:00", which represents one of the last moments before sunrise (when the night cooling ends), shows most of the wall section with a liquid fraction greater than zero (as shown in Figure 16). However, at "Time 16:00" (when the sun is still heating the wall), the PCM liquid fraction is equal to one in the whole wall section. This means the PCM effect disappears before the heating process ends.

## CONCLUSIONS

In this paper, we have presented a CFD model to simulate experimental buildings in order to study their behaviour under changing climatic conditions depending on their construction materials. The simulation allows an in-depth analysis of the main parameters that affect the behaviour of the building and helps predict the results of different experiments. The proposed model's simulation results have been compared with experimental data, and the results have been found to be reasonably similar.

The measured variations in the meteorological conditions of wind and sun over several consecutive days have been reproduced; the heat pump inside the cubicle and the thermal storage capacity of the walls have been simulated with the use of CFD simulation techniques and programming of user defined functions (UDF). The main difficulty in the modelling effort is to reproduce the climatic variation at every instant; there are limitations to correctly simulating the climatic conditions and intermittent rain clouds, which can lead to the implementation of approximate solutions. As a result, in this study only clear days have been simulated. Another limitation is the starting point of the simulation, as the initial values of the variables do not match the experimental values; the first day of the simulation is a transition period to allow for appropriate values on subsequent days.

The comparison of the simulation results of the cubicles with PCM and without PCM shows that the PCM provides temperature stability and resistance to the changes in external conditions, which makes these materials suitable for increasing the thermal comfort of buildings.

## ACKNOWLEDGMENTS

The work was funded in part the Ministry of Science and Innovation through a CDTI project.

## Nomenclature

| ag | Absorption coefficient |
|----|------------------------|
| Sφ | Source of the variable φ |

| $\vec{A}$ | Area vector |
|---|---|
| Ss | Energy source in the walls |
| Cp | Solid cell specific heat |
| t | Time |
| $C1\epsilon, C2\epsilon, C3\epsilon, C\mu$ | k-$\epsilon$ model constants |
| $\Delta t$ | Time step size |
| es | Solid cell energy |
| T | Temperature |
| $\vec{F}$ | External body forces |
| Tpc | Phase change temperature |
| $\vec{g}$ | Gravity vector |
| Ts | Solid cell temperature |
| Gk | Generation of kinetic energy due to velocity gradients |
| $\vec{v}$ | Velocity vector |
| Gb | Generation of kinetic energy due to velocity buoyancy |
| vi | Velocity component in direction $i$ |
| h | Enthalpy |
| V | Volume |
| $\Delta HS$ | Solid cell energy variation in a time step |
| xi | i coordinate |
| It | Unit tensor |
| xk | k coordinate |
| I | Radiation intensity |
| YM | fluctuating dilatation in compressible turbulence |
| $JJ\longrightarrow$ | Diffusion flux of species $j$ |
| $\Gamma\varphi$ | Diffusion coefficient of the variable $\square$ |
| k | Turbulent kinetic energy |
| $\epsilon$ | Turbulent dissipation rate |
| LHPCM | PCM Latent heat ($J \cdot kg_{-1}$) |
| $\lambda$ | Thermal conductivity |
| M | Molar mass |
| $\lambda s$ | Solid cell thermal conductivity |
| n | Refractive index |
| $\mu$ | Molecular viscosity |

| P | Pressure |
|---|---|
| ρ | Density |
| Pop | Operating pressure |
| σ | Stefan–Boltzmann constant |
| Q˙i | Total heat received by the solid cell from exterior |
| σS | Scattering coefficient |
| $\vec{r}$ | Radiation position vector |
| σk | Prandtl number for $k$ |
| R | Ideal gas constant |
| σϵ | Prandtl number for ϵ |
| $\vec{s}$ | Radiation direction vector |
| τij | Stress tensor |
| Sh | Volumetric heat sources |
| φ | PCM liquid fraction in a cell |
| Sm | Mass source |
| φf | PCM liquid fraction in a face |
| Sk | $k$ source term |
| Φ | Scattering function |
| Sϵ | ϵ source term |
| Ω | Solid angle |

# REFERENCES

1.    Sharma, A.; Tyagi, V.V.; Chen, C.R.; Buddhi, D. Review on thermal energy storage with phase change materials and applications. *Renew. Sustain. Energy Rev.* 2009, *13*, 318–345.

2.    Zalba, B.; Marín, J.M.; Cabeza, L.F.; Mehling, H. Review on thermal energy storage with phase change: Materials, heat transfer analysis and applications. *Appl. Therm. Eng.* 2003, *23*, 251–283.

3.    Butala, V.; Stritih, U. Experimental investigation of PCM cold storage. *Energy Build.* 2009, *41*, 354–359.

4.    Arkar, C.; Medved, S. Free cooling of a building using PCM heat storage integrated into the ventilation system.*Sol. Energy* 2007, *81*, 1078–1087.

5.    Anton Aroul Raj, V.; Velraj, R. Heat transfer and pressure drop studies on a PCM-heat exchanger module for free cooling applications. *Int. J. Therm. Sci.* 2011, *50*, 1573–1582.

6.  Bentz, D.P.; Turpin, R. Potential applications of phase change materials in concrete technology. *Cement Concrete Comp.* 2007, *29*, 527–532.

7.  Castellón, C.; Medrano, M.; Roca, J.; Cabeza, L.F.; Navarro, M.E.; Fernández, A.I.; Lázaro, A.; Zalba, B. Effect of microencapsulated phase change material in sandwich panels. *Renew. Energy* 2010, *35*, 2370–2374.

8.  Meshgin, P.; Xi, Y.; Li, Y. Utilization of phase change materials and rubber particles to improve thermal and mechanical properties of mortar. *Constr. Build. Mater.* 2012, *28*, 713–721.

9.  Lane, G.A. *Solar Heat Storage: Latent Heat Material. Volume I: Background and Scientific Principles*; CRC Press: Florida, FL, USA, 1983.

10. Wang, W.; Yang, X.; Fang, Y.; Ding, J.; Yan, J. Enhanced thermal conductivity and thermal performance of form-stable composite phase change materials by using beta-aluminum nitride. *Appl. Energy* 2009, *86*, 1196–1200.

11. Shukla, A.; Buddhi, D.; Sawhney, R.L. Solar water heaters with phase change material thermal energy storage medium: A review. *Renew. Sustain. Energy Rev.* 2009, *13*, 2119–2125.

12. Castella, A.; Beluskob, M.; Brunob, F.; Cabeza, L.F. Maximisation of heat transfer in a coil in tank PCM cold storage system. *Appl. Energy* 2011, *88*, 4120–4127.

13. Ismail, K.A.R.; Henriquez, J.R. Solidification of PCM inside a spherical capsule. *Energy Convers. Manag.* 2000, *41*, 173–187.

14. Eames, I.W.; Adref, K.T. Freezing and melting of water in spherical enclosures of the type used in thermal (ice) storage systems. *Appl. Therm. Eng.* 2002, *22*, 733–745.

15. Wang, S.; Zhu, D. Application of CFD in retrofitting air-conditioning systems in industrial buildings. *Energy Build.* 2003, *35*, 893–902.

16. Tanasic, N.; Jankes, G.; Skistad, H. CFD analysis and airflow measurements to approach large industrial halls energy efficiency: A case study of a cardboard mill hall. *Energy Build.* 2011, *43*, 1200–1206.

17. Gan, G. A parametric study of Trombe wall for passive cooling of buildings. *Energy Build.* 1998, *27*, 37–43.

18. Blocken, B.; Defraeye, T.; Derome, D.; Carmeliet, J. High-resolution CFD simulations for forced convective heat transfer coefficients at the facade of a low-rise building. *Build. Environ.* 2009, *44*, 2396–2412.

19. Coussirat, M.; Guardo, A.; Jou, E.; Egusquiza, E.; Cuerva, E.; Alavedra, P. Performance and influence of numerical sub-models on the CFD

simulation of free and forced convection in double-glazed ventilated façades. *Energy Build.* 2008, *40*, 1781–1789.

20. Onishi, J.; Soeda, H.; Mizuno, M. Numerical study on a low energy architecture based upon distributed heat storage system. *Renew. Energy* 2000, *22*, 61–66.

21. Ansys fluent 12.0 Theory Guide. 2009. Available online: https://www.sharcnet.ca/Software/Fluent12/html/th/main_pre.htm (accessed on 14 June 2012).

22. Moran, M.J.; Shapiro, H.N. *Fundamentals of Engineering Thermodynamics,*, 4th ed.; Wiley: New York, NY, USA, 2000.

# Chapter 8

# PERFORMANCE EVALUATION OF MODERN BUILDING THERMAL ENVELOPE DESIGNS IN THE SEMI-ARID CONTINENTAL CLIMATE OF TEHRAN

Shaghayegh Mohammad and Andrew Shea

The University of Bath, Department of Architecture and Civil Engineering, Bath BA2 7AY, UK

## ABSTRACT

In this paper we evaluate the thermal performance of a range of modern wall constructions used in the residential buildings of Tehran in order to find the most appropriate alternative to the traditional un-fired clay and brick materials, which are increasingly being replaced in favor of more slender wall constructions employing hollow clay, autoclaved aerated concrete or light expanded clay aggregate blocks. The importance of improving the building envelope through estimating the potential for energy saving due to the application of the most energy-efficient wall type is presented and the wall constructions currently erected in Tehran are introduced along with their dynamic and steady-state thermal properties. The application of a dynamic simulation tool is explained and the output of the thermal simulation model is compared with the dynamic thermal properties of the wall constructions to assess their performance in summer and in winter. Finally, the best and worst wall type in terms of their cyclic thermal performance and their ability to moderate outdoor conditions is identified through comparison of the predicted indoor temperature and a target comfort temperature.

## INTRODUCTION

In common with many other regions of the world, buildings in Iran account for approximately 36% of the energy consumption of the country [1], with the residential sector responsible for around 33% of total electricity consumption in the country [2]. The building envelope is the main interface between indoors and outdoors and has a significant role in moderating variations in the outdoor weather conditions, providing thermal comfort for occupants and consequently determining the heating/cooling loads of the building. In 2011, the 15–34 year age group accounted for 41% of the total population of Iran [3], which coupled with increasing migration to the cities has resulted in significant demand for new dwellings in the major cities of the country; a situation that appears to be replicated across the world. According to the Building and Housing Research Center [4] construction of 1.5 million residential units is required and statistics indicate an 18% increase in the construction of domestic buildings between March and October 2011 [5]. Energy consumption in the Iranian building sector is more than double the global average [6] and measures to reduce space heating and cooling energy use through intelligent modification of the building envelope materials should be promoted if the country is to play a part in mitigating the global problem of climate change and diminishing fossil fuel resources. The Iranian Ministry of Housing and Urbanism has introduced Code No. 19 [7], which requires the calculation of the whole building heat loss coefficient (W/K) using the steady-state thermal performance of the building envelope, *i.e.*, *U*-value, plus additional losses for thermal bridges. The code requires that the calculated whole building heat loss is less than the equivalent building constructed to *U*-values compliant with the Code [8]. In the vernacular architecture of this warm dry region of Iran, walls employing a high level of thermal mass and the use of local natural materials provide correspondingly high rates of thermal and moisture buffering effects and have delivered associated benefits in terms of thermal comfort for their occupants through the centuries. Common wall types combine a mixture of clay, mud, and straw or brick and are of the order of 50–60 cm thick. Today these materials have increasingly been superseded by predominantly lightweight materials such as hollow clay blocks, Light Expanded Clay Aggregate (LECA) and Autoclaved Aerated Concrete (AAC) blocks, with lower density and reduced thermal storage capacity. To improve thermal performance, some of these lightweight blocks are combined with insulating materials and form much more slender alternatives to the traditional wall constructions.

Thermal simulation software tools have been widely used by architects and engineers for many years and permit investigation and evaluation of various design alternatives influencing, for example, fabric performance and fenestration levels under varying casual gains and climatic conditions *etc*. Such tools provide insight to the dynamic behavior of whole buildings and enable building designers to estimate and optimize envelope thermal performance, occupant thermal comfort and, ultimately, the energy performance of the finished building. Crawley *et al.* [9] provide a thorough appraisal contrasting the capabilities of all the leading simulation tools. This study uses the Integrated Environmental Solutions Virtual Environment (IES-ve) dynamic thermal simulation package, which has been widely validated and its calculation methodology meets the requirements of a number of national and international standards such as the UK National Calculation Methodology (NCM) [10], ASHRAE 55 [11] and ISO 7730 [12]. The IES-ve software was used to model the dynamic thermal performance of a typical apartment building located in Tehran and simulated using hourly weather data over a complete year. The building was simulated for a range of different conventional wall construction systems and the resulting internal temperatures were compared using a simple comfort temperature criterion as presented by Heidari [13]. The results highlight that, whilst the national building regulation (Code No. 19) [7] and related guidelines focus on steady-state ($U$-value) performance and the application of insulation in wall construction to improve its thermal performance, the effect of thermal mass and cyclic behavior of materials in the overall thermal performance of buildings should not be neglected in thermal performance calculations.

## ENVELOPE THERMAL PROPERTIES

The value derived from the steady-state calculations ($U$-value) is not an appropriate indicator of the thermal performance of building elements by itself; as it is possible for two walls with the same $U$-value to absorb and release heat at different rates [14]. Steady-state analysis is concerned only with the thermal conductivity of the material; the influence of heat capacity is ignored. Intermittent occupancy and associated heating or cooling operation combined with external diurnal variations mean that the building is more often in a state of flux and, particularly in hot summer conditions, the dynamic behavior of the whole building should be assessed in order to optimize the selection of envelope materials for greatest combined thermal comfort and energy performance. The material bulk properties of heat capacity ($C$), density ($\rho$), and thermal conductivity ($\lambda$) play an important role in the cyclic performance of the construction, which is significant when the outdoor temperature is cycling below and above the desired indoor temperature. Materials with beneficial thermal properties are either insulating materials, or materials with thermal mass [15] and the effect of thermal mass and thermal insulation which are representatives of dynamic and steady state thermo-physical properties of materials must be taken into account simultaneously [14].

### The Role of Thermal Mass

High thermal mass materials can store more heat compared to other materials when exposed to a source of heat [16]. They also release their heat content more slowly when the source of heat is removed [16]. In winter days high thermal mass materials can store heat energy from incident solar radiation and then will release this heat into the indoor space later in the evening when the passive heat source is removed and more heat is demanded internally, thus reducing the mechanical heating load of the space [16]. In summer time, thermal mass sinks the heat caused by solar radiation in the internal space, preventing sudden peaks in indoor temperature and increased load on air-conditioning units. Having stored heat during the day, these building elements will release the heat content later in the evening, partly to indoor space, which with presence of sufficient time-lag can be dissipated by use of cooler outdoor air via natural ventilation, and partly to outdoors which can be accelerated by clear-sky radiation [14]. Additionally, the increased gradient between indoor warmer and the outside cooler environment will improve this purging process [17]. A thermal insulator decelerates the transfer of heat between different areas at different temperatures [14] and limits heat loss through the building fabric in winter and inward heat flow in summer.

Simple steady-state calculations ignore the (realistic) dynamic processes apparent in real buildings. Non-steady state, $i.e.$, dynamic, calculation methods permit evaluation of the thermal performance of buildings under real conditions. Such methods employ various parameters for including the mass effect in thermal performance analysis and methods vary in complexity. CIBSE presents the Admittance procedure which requires the calculation of three parameters: Admittance value ($Y$-value), decrement factor ($f$) and surface factor ($F$) in addition to thermal transmittance ($U$-value) [18]. Admittance relates to the storage of energy in the room surfaces following fluctuations in internal temperature. The Admittance value describes the ability of a material to exchange heat with a space, for each degree of deviation of the space temperature about its mean value [18]. The key variables involved in determining Admittance are heat capacity, density, thermal conductivity, surface resistance and the length of time available to get heat in and out of the material, which is typically assumed to be 24 h [19]. Accordingly, it is a function of the diffusivity and thickness of materials and can be considered as a cyclic $U$-value. The Admittance $Y$-value has the same units as the $U$-value (W/m$^2$ K). Greater Admittance confers lower amplitude indoor temperature fluctuations. Therefore, unlike $U$-values, high $Y$-values are desirable in a thermal mass perspective [18]. For a clearer differentiation between thermal admittance ($Y$-value) and thermal transmittance ($U$-value), it should be highlighted that it is possible to have different elements with the same insulation performance, indicated by the $U$-value, but different damping properties, indicated by the $Y$-value as presented in Table 1.

The decrement factor ($f$) represents the relationship between the indoor and outdoor daily temperature swings [20]. A low value indicates that the building fabric is capable of dampening the indoor temperature range, relative to outdoors. For a thin structure with low thermal capacity the value will be unity and will decrease with increasing thickness and/or thermal capacity [18]. The surface factor is the ratio of radiant heat flow (from shortwave sources, e.g., the sun) readmitted to the space from the surface, to the heat flow incident upon the surface [18]. The benefit of applying a surface factor is in quantifying the absorption and subsequent release of transmitted solar radiation. Higher heat capacity results in lower surface factors and greater time lag [21].

**Table 1:** Contrasting values of admittance and transmittance [14]

| Element | $Y$-value (W/m² K) | $U$-value (W/m² K) |
|---|---|---|
| Typical heavyweight wall (brick/blockwork with cavity insulation) | 4.0 | 0.6 |
| Typical lightweight wall (cladding, insulation, lining) | 1.0 | 0.6 |

## Local Climatic Context

Tehran, located at 35N and 51E, features a semi-arid continental climate, according to the Koppen climatic classification, with hot dry summers and cold winters. Figure 1 presents a 5-year average (2005–2009) of mean monthly maximum and minimum temperatures from the Geophysics station of Tehran and Figure 2 presents the 5-year average relative humidity and rainfall for the same location.

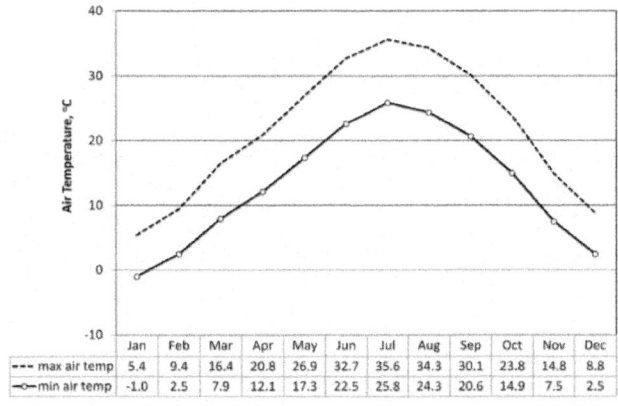

| | Jan | Feb | Mar | Apr | May | Jun | Jul | Aug | Sep | Oct | Nov | Dec |
|---|---|---|---|---|---|---|---|---|---|---|---|---|
| --- max air temp | 5.4 | 9.4 | 16.4 | 20.8 | 26.9 | 32.7 | 35.6 | 34.3 | 30.1 | 23.8 | 14.8 | 8.8 |
| —○— min air temp | -1.0 | 2.5 | 7.9 | 12.1 | 17.3 | 22.5 | 25.8 | 24.3 | 20.6 | 14.9 | 7.5 | 2.5 |

**Figure 1:** Five year average of mean monthly maximum and minimum temperatures for Tehran (Geophysics station, 2005–2009).

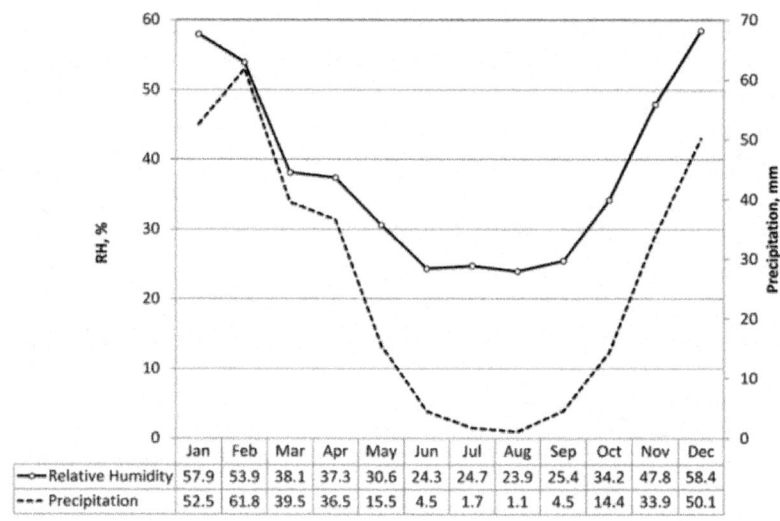

**Figure 2:** Five year average of mean monthly relative humidity and rainfall for Tehran (Geophysics station, 2005–2009).

**Table 2:** Categorization of conventional wall types

| Wall reference and type | Construction materials | Detail | $U$-value (W/m² K) | Thickness (cm) | Image of material |
|---|---|---|---|---|---|
| **HCB1** Hollow clay block | Plaster lining (3 cm) Clay block (15 cm) Sand & cement mortar (3 cm) Exterior stone finishing (2 cm) | 3  15  32 | 1.3 | 21 | |
| **HCB2** Hollow clay block | Plaster lining (3 cm) Clay block [15 cm with 2 cm expanded polystyrene (EPS)] Sand & cement mortar (3 cm) Exterior stone finishing (2 cm) | 3  15  32 | 1.08 | 21 | |
| **L1** LECA block | Plaster lining (3 cm) LECA block (20 cm) Sand & cement mortar (3 cm) Exterior stone finishing (2 cm) | 3  20  32 | 1.34 | 28 | |

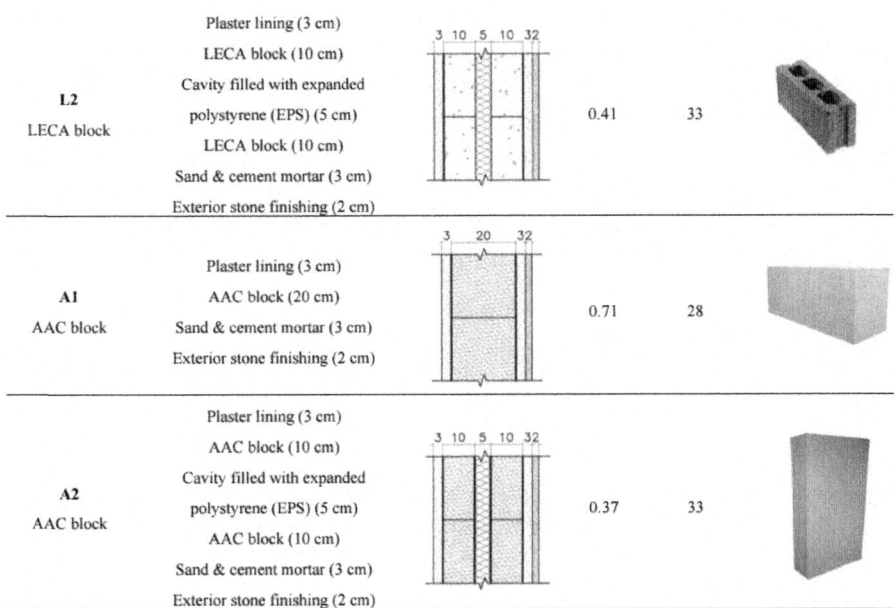

Through interviews with people involved in the construction industry and field observations of the author, conventional wall constructions which are used in residential buildings of the city of Tehran have been categorized in Table 2.

Walls are most commonly made of hollow clay blocks, LECA (Lightweight Expanded Clay Aggregates) and AAC (Autoclaved Aerated Concrete) blocks, with clay blocks being the most popular choice. Some practitioners favor the use of LECA over clay blocks due to a perceived thermal damping performance advantage. However, the results of this investigation indicate that LECA by itself is not the better choice, although LECA when combined with insulation does have significant benefits with regards to the attenuation of outdoor temperature swings. The application of AAC is less common due to some construction difficulties pertaining to a lack of adhesion between interior plaster and the AAC block. Based on the adaptive comfort standard, originally proposed by Humphreys and Nicol [22], Heidari [13] introduced an equation for the calculation of the comfort temperature for the people of Tehran. Comfort, or "neutral", temperature is the temperature at which people feel neither warm nor cold and is calculated using the monthly mean outdoor temperature as presented in Equation (1):

$$T_{comf.} = 0.555\, T_{out} + 12.8$$

(1)

A field-study methodology was employed by Heidari [13] and results showed good agreement between comfort temperature and mean outdoor temperature. The findings of Heidari's study [13] revealed that the people could achieve comfort at higher indoor air temperatures compared with the recommendations of international standard ISO 7730 [12]. According to adaptive comfort standards, the human body is capable of adapting to its environment. Therefore, as indicated by Equation (1), in different months of the year the human body feels comfortable at different temperatures.

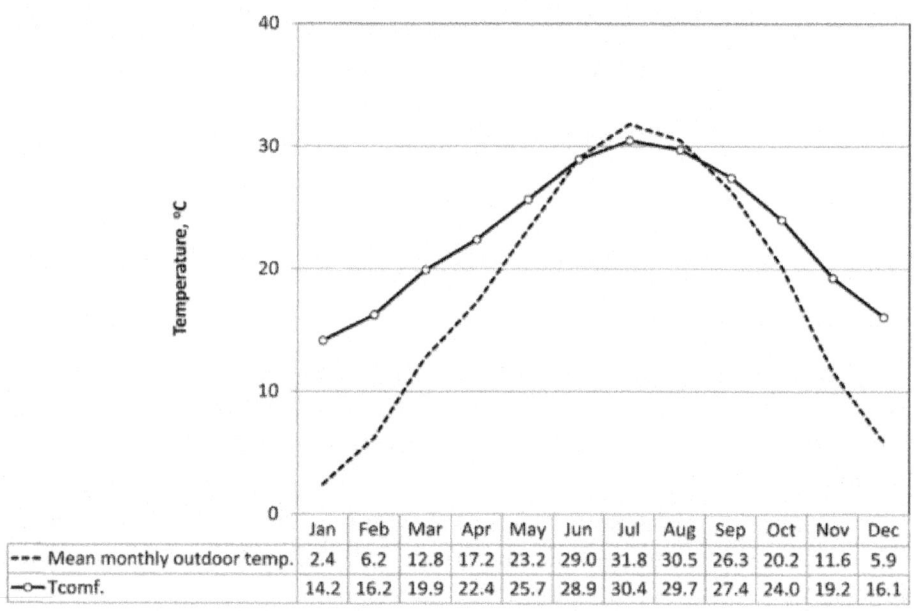

| | Jan | Feb | Mar | Apr | May | Jun | Jul | Aug | Sep | Oct | Nov | Dec |
|---|---|---|---|---|---|---|---|---|---|---|---|---|
| --- Mean monthly outdoor temp. | 2.4 | 6.2 | 12.8 | 17.2 | 23.2 | 29.0 | 31.8 | 30.5 | 26.3 | 20.2 | 11.6 | 5.9 |
| —○— Tcomf. | | 14.2 | 16.2 | 19.9 | 22.4 | 25.7 | 28.9 | 30.4 | 29.7 | 27.4 | 24.0 | 19.2 | 16.1 |

**Figure 3:** Mean monthly outdoor and occupant comfort temperature for Tehran.

$T_{comf.}$ for each month is presented in Figure 3 which has been calculated based on mean monthly temperature (5-year average) for the Geophysics station of Tehran. Whilst Heidari's field-based study encompassed both hot and cold seasons and presents findings that suggest the building occupants are accepting of both higher indoor air temperatures and a higher overall range of temperatures, the data were gathered from occupants in offices that ranged in indoor temperatures from approximately 21.8 °C to 33.2 °C and, therefore, the prediction of comfort temperatures outside of this range must be treated with some skepticism. However, our analysis seeks to use this simple metric as a proxy for space conditioning energy use; accepting possible shortcomings as a precise indicator of thermal comfort in mid-Winter months. For example, it would seem reasonable to suggest that a "free-running" building exhibiting a large deviation from a given comfort temperature will require more energy input in terms of mechanical heating or cooling than a building that is already close to the comfort temperature by virtue of its improved thermal envelope performance. Accordingly, employing Equation (1) to determine a base temperature, in a similar way to the application of Heating or Cooling Degree Days, provides a useful simple indicator of building energy performance that can be used to compare different thermal envelope solutions.

# SIMULATION

In order to study and analyze how different wall constructions perform under dynamic outdoor weather conditions, the thermal simulation package IES-ve was used. This tool is able to simulate the performance of a building under non-steady state conditions using real climate data, in this case for the city of Tehran. Some limitations exist in the representation of certain wall types in the simulation model, namely the effect of heat flow in parallel through the hollow core blocks. The input of the thickness of each layer of material in to the model does not permit the inclusion of different materials in the same plane, thus in walls that contain hollow core blocks with regular rectangular voids the material of the highest proportion in that plane was used, which is either air or expanded polystyrene insulation. This ignores the effect of the thermal bridge across the air or insulation layers, however, the dynamic performance is controlled mainly by the properties of the materials immediately adjacent to the interior, e.g., plaster, *etc.*, thus the presence of the bridge has little effect. In the wall types simulated in this study the inner layers comprise 30 mm of plaster followed by the first solid plane layer of the hollow block, providing at least 45 mm of material before any bridge is reached.

## Simulation Scenario

In this study, all sources of internal casual gains were omitted for running the IES simulation; ventilation, in the form of infiltration was included at a continuous background rate of 0.5 Air Changes per Hour (ACH) to represent expected typical air leakage through the fabric of the building. Accordingly, no heating/cooling system, occupant, lighting or appliance profiles (for internal heat gains) were defined or considered in the model. This approach was to make a clear evaluation of the performance of the building's fabric energy system when exposed to dynamic weather conditions. The geometric representation of the building in IES is presented in Figure 4.

The building is intended to represent a typical Tehran apartment block with a ground floor parking area plus four levels of accommodation above. This study is interested in the evaluation of different external wall constructions and, therefore, an intermediate (3rd Floor) level room was selected for detailed analysis. The room selected for analysis is outlined in Figure 4 and is south-facing and measures 3.45 m wide × 5.40 m deep and 2.80 m high. An argon-filled double-glazed window unit is located centrally in the external wall and measures 1.6 m × 1.7 m. Intermediate floors/ceiling are constructed from lightweight concrete beam and block units with ceiling level gypsum plaster finish and floor level ceramic tiles ($U = 1.4$ W/m$^2$ K). Internal walls are lightweight concrete block with a plaster finish. In all simulations only the external wall is modified and all other elements remain fixed. Materials and construction systems were analyzed and compared in terms of their basic thermal properties such as density ($\rho$), specific heat capacity ($C$), thermal conductivity ($\lambda$), diffusivity ($\alpha$) and their derived thermal properties including thermal admittance ($Y$), effective thermal mass ($C_m$), and decrement factor ($f$). Finally, indoor temperature was compared to the region-specific comfort temperature to demonstrate the ability of each fabric system to provide thermal comfort for occupants.

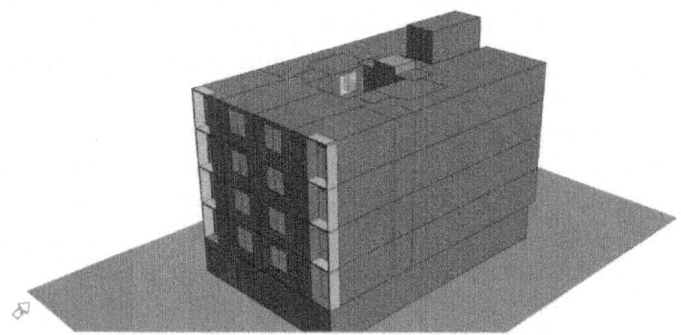

**Figure 4:** Integrated Environmental Solutions Virtual Environment (IES-ve) Pro 3-D model of a typical Tehran apartment building.

# RESULTS AND DISCUSSION

Dynamic simulation results provide valuable additional information in relation to wall type performance and result in a different appraisal of the optimum solution than would be found from $U$-value-based performance alone. The dynamic thermal performance of the building envelope is dependent upon three basic properties, namely, heat capacity, density, and thermal conductivity. Higher heat capacity and density maximize the amount of heat absorbed in every m$^3$ of the material and a moderate thermal conductivity is needed for a material to make its heat capacity advantageous. A moderate thermal conductivity enables the material to exchange heat with ambient air at an appropriate rate; some materials such as wood have high heat capacity but due to their relatively low thermal conductivity the heat exchange rate is so slow that their heat capacity would become ineffective [19]. The combination of these parameters is often expressed as thermal diffusivity ($\alpha$), as presented in Table 3.

The thermal diffusivity of a material is a measure of how fast the material temperature adapts to the surrounding temperature. The lower the diffusivity, the greater the time-lag of a material [23]. Therefore, for the same geometry and configuration, of the materials presented in Table 3, AAC and LECA would have less rapid response to temperature changes than clay or brick. It should be stated that the shape and configuration of the final construction product, e.g., hollow or solid blocks will clearly affect thermal performance of the whole construction. The thermal properties including derived parameters for each wall type are presented in Table 4.

**Table 3:** Material thermo-physical properties

| Material | $\rho$ (kg/m$^3$) | $\lambda$ (W/m K) | $C$ (J/kg K) | Diffusivity $\alpha$, (m$^2$/s) |
|---|---|---|---|---|
| Brick | 1700 | 1 | 840 | $7.00 \times 10^{-7}$ |
| Clay | 1300 | 0.5 | 837 | $4.60 \times 10^{-7}$ |
| LECA | 900 | 0.23 | 1000 | $2.56 \times 10^{-7}$ |
| AAC | 700 | 0.17 | 1000 | $2.43 \times 10^{-7}$ |
| EPS | 15 | 0.04 | 1340 | $1.99 \times 10^{-6}$ |

**Table 4:** Thermal properties including derived parameters for each wall type

| Wall ref. | $U$-value (W/m$^2$ K) | $C_m$ (kJ/m$^2$ K) | Admittance (W/m$^2$ K) | $f$ | $F$ | Thickness (m) |
|---|---|---|---|---|---|---|
| HCB1 | 1.30 | 71.5 | 3.52 | 0.81 | 0.72 | 0.21 |
| HCB2 | 1.09 | 71.7 | 3.64 | 0.80 | 0.73 | 0.21 |
| L1 | 1.34 | 72.5 | 3.57 | 0.81 | 0.71 | 0.26 |
| L2 | 0.41 | 72.5 | 4.01 | 0.44 | 0.70 | 0.31 |
| A1 | 0.71 | 94.5 | 3.77 | 0.39 | 0.69 | 0.26 |
| A2 | 0.38 | 94.5 | 3.86 | 0.29 | 0.68 | 0.31 |

$f$: Decrement factor
$F$: Surface factor
$C_m$: Effective thermal mass [a]

[a] Effective thermal mass is the product of density, thickness and heat capacity for each layer until one of the following three criteria is met: Working from inside to outside (i) 100 mm point is reached; (ii) Mid-point of the element is reached; (iii) Insulation layer is reached.

Based on a steady-state appraisal wall type A2, with the lowest $U$-value, has the best thermal performance and due to its lowest decrement factor the greatest capability with regards to attenuation of external temperature swings. However, wall type L2 has the highest thermal admittance ($Y$-value) and, as this value is calculated by considering many different thermal and physical properties, e.g., heat capacity, thermal conductivity, density, *etc.*, it can be concluded that this value can help to realize how effective a material will be in practice. Therefore, it is expected that L2 with $Y$-value of 4 is most capable of dampening weather conditions. In winter months the different wall types perform more according to their respective $U$-values than in summer months. This is because, as discussed by a number of authors, for example Givoni [24] and also De Saulles [19], the use of thermal mass is more advantageous when thermal conditions are fluctuating and the outdoor temperature is cycling below and above the indoor temperature. In other words, in cold weather conditions $T_{in}$ is always higher than $T_{out}$; therefore, heat flow occurs in one direction, from inside to outside. While in summer months building fabric is subject to both outward and inward heat flow and this is the phenomenon which challenges the primary assumptions inferred from steady-state calculations.

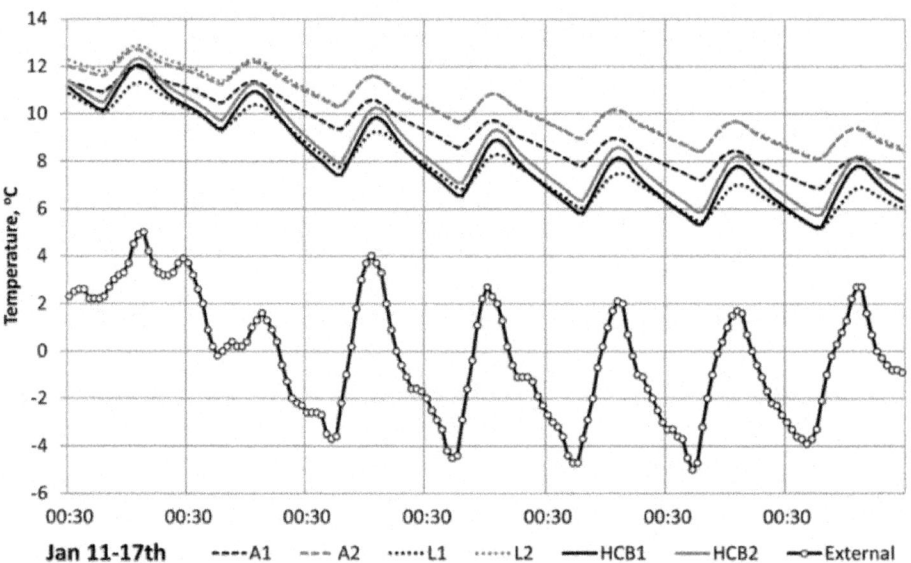

**Figure 5:** Outdoor temperature and indoor room temperatures for different wall constructions, for a seven day winter period.

Accordingly, in the summer months, instead of having distinct records of indoor temperature for each of the different wall types corresponding to their differing $U$-values, as is the case in the winter condition (Figure 5), internal temperatures begin to overlap one another during each 24-h period (Figure 6). This effect can be highlighted through comparison of A1, A2 and L2 performances in winter and in summer. A2 and L2 (with very similar *U*-values) have similar performances in winter (Figure 5); while in summer A1 and A2 (with very different *U*-values) perform similarly to one another, whilst L2 now exhibits very different response to A2 (Figure 6). This is due to thermal mass of these materials being effective in fabric interaction with its environment.

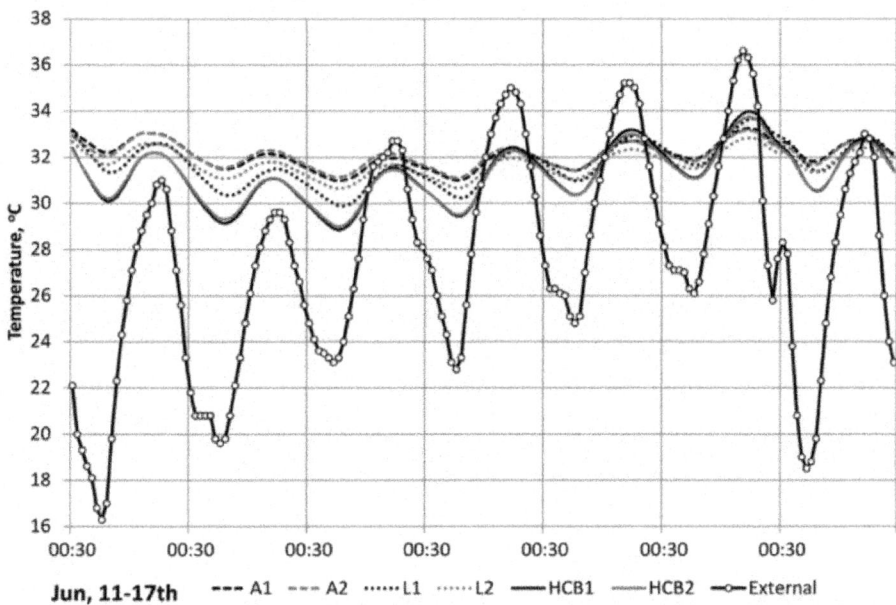

**Figure 6:** Outdoor temperature and indoor room temperatures for different wall constructions, for a seven day summer period.

Plotting the indoor temperature against the region-specific comfort temperature for each wall type (Figure 7) enables comparison of the degree of deviation from the comfort temperature. As indicated in Figure 7, in the colder months of January, February, March, April, November, and December, wall types L2 and A2, as was expected from their lower *U*-values, provide a higher degree of comfort for the building occupants whereas in the warmer months of the year, *i.e.*, May, June, July, August, September, October these wall types plus wall type A1 show a slightly greater deviation from the comfort temperature than the other wall types. For wall types L2 and A2 this is because, on summer nights when the outside temperature ($T_{out}$) falls below the indoor temperature ($T_{in}$), the insulation layer inside these wall types resists the outward heat flow and causes delay in discharge of the thermal mass of the building and consequently higher indoor temperatures. In wall type A1, however, this inability to maintain comfortable room temperatures is mostly due to the heat which is absorbed and stored in the depth of the available mass (solid block) which makes the transient performance worse than wall type L2 despite having a lower *U*-value and a layer of insulation. Furthermore, as the simulated building does not benefit from ventilation and convective cooling effects, beyond the moderate 0.5 ACH background rate, this trapped heat results in higher temperatures for the subsequent day.

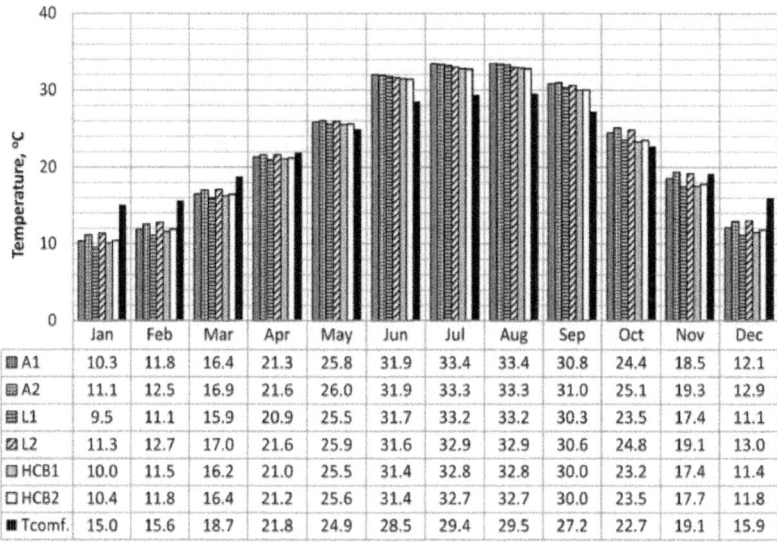

| | Jan | Feb | Mar | Apr | May | Jun | Jul | Aug | Sep | Oct | Nov | Dec |
|---|---|---|---|---|---|---|---|---|---|---|---|---|
| A1 | 10.3 | 11.8 | 16.4 | 21.3 | 25.8 | 31.9 | 33.4 | 33.4 | 30.8 | 24.4 | 18.5 | 12.1 |
| A2 | 11.1 | 12.5 | 16.9 | 21.6 | 26.0 | 31.9 | 33.3 | 33.3 | 31.0 | 25.1 | 19.3 | 12.9 |
| L1 | 9.5 | 11.1 | 15.9 | 20.9 | 25.5 | 31.7 | 33.2 | 33.2 | 30.3 | 23.5 | 17.4 | 11.1 |
| L2 | 11.3 | 12.7 | 17.0 | 21.6 | 25.9 | 31.6 | 32.9 | 32.9 | 30.6 | 24.8 | 19.1 | 13.0 |
| HCB1 | 10.0 | 11.5 | 16.2 | 21.0 | 25.5 | 31.4 | 32.8 | 32.8 | 30.0 | 23.2 | 17.4 | 11.4 |
| HCB2 | 10.4 | 11.8 | 16.4 | 21.2 | 25.6 | 31.4 | 32.7 | 32.7 | 30.0 | 23.5 | 17.7 | 11.8 |
| Tcomf. | 15.0 | 15.6 | 18.7 | 21.8 | 24.9 | 28.5 | 29.4 | 29.5 | 27.2 | 22.7 | 19.1 | 15.9 |

**Figure 7:** Monthly average indoor temperature for six different wall constructions compared with the comfort temperature.

Annual performance of wall types is presented in Figure 8, which indicates the deviation of $T_{in}$ from $T_{comf.}$ in summer (positive) and winter (negative) for an entire year. The summary of dynamic thermal properties for the six wall types (Table 4) shows that walls HCB1, HCB2 and L1 have higher decrement factors resulting in greater indoor temperature swings with relatively short time lags and lower effective thermal mass.

The summer-time performance, as indicated by the positive deviation in Figure 8, indicates that HCB1, HCB2, and L1 perform better than the other three wall types but that A1, A2 and L2 have better performance in winter. However, the sum for wall types L2, A2 and HCB2 reveals the lowest total deviation of $T_{in}$ from $T_{comf.}$, which supports the application of insulation in such walls. The application of wall type L2 in place of L1, results in 7.5 °C decrease in deviation of $T_{in}$ from $T_{comf.}$, leading to reduced energy demand for conditioning of the space. Heidari [13] estimates that each degree reduction in heating/cooling loads results in 7% energy saving.

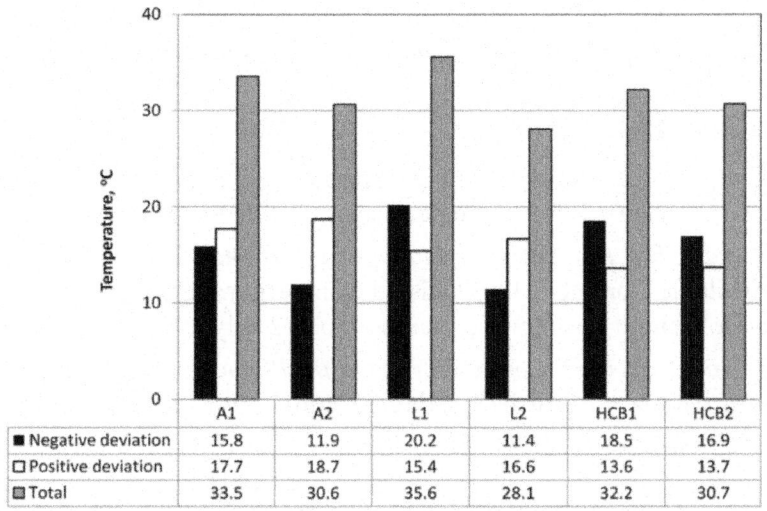

**Figure 8:** Annual average deviation of indoor temperature from comfort temperature for six different wall constructions.

## CONCLUSIONS

The aim of this research was to evaluate the thermal performance of modern wall constructions which are commonly used in place of their traditional counterparts in the new residential buildings of Tehran. The findings of this investigation can be summarized as follows: the steady-state calculation alone is not a true indication of thermal performance of building fabric under real climatic conditions. Additionally, the thermal insulation materials and thermal mass have different roles in the thermal performance of the building fabric energy system. Dynamic simulation using local weather data provides useful, cost-effective, insight in to these relative merits and the behavior of zones comprising different building envelope materials. In the absence of dynamic simulation, the appraisal of other performance indicators, such as the $Y$-value (Admittance), could provide useful guidance as part of simple standardized dwelling designs for use in building compliance codes such as Code No. 19. Whilst time-averaging such data removes some of the detail it does permit comparison with simpler methods including monthly average thermal comfort temperatures, which provides a useful parameter for appraisal of building performance when comparing the merits of many different wall types. This analysis shows that whilst wall type A2 has the lowest $U$-value amongst the introduced wall types, wall type L2 has the best overall performance in terms of moderating weather conditions and providing thermal comfort for occupants over a complete year. The intelligent use of natural ventilation, most likely employing night-time ventilation, could improve further thermal comfort and future work evaluating the optimization of such strategies is recommended.

## REFERENCES

1.    Energy & Electricity Planning Committee, *Energy Database for Iran for Year 2011–2012*; Department of Energy: Tehran, Iran, 2013.

2.    *Tehran: Electric Demand Management Bureau Online Reports*, 2013. Available online: http://edsm.tavanir.org.ir/motaleaate_modireyyate_ masraf/olgoo-khanegi.asp (accessed on 12 July 2013).

3.    *Tehran: Statistical Centre of Iran Online Reports*, 2011. Available online: http://www.sci.org.ir/SitePages/report_90/population_report.aspx (accessed on 12 July 2013).

4.    Building and Housing Research Centre (BHRC). New Construction Technologies. BHRC: Tehran, Iran, 2010. Available online: http://www. bhrc.ac.ir/portal/Default.aspx?tabid=668 (accessed on 12 July 2013).

5.    Khabar Online (Iranian News Website). Available online: http://www. khabaronline.ir/detail/218221/ (accessed on 12 July 2013).

6.    Nasrollahi, F. *Urban and Architectural Criteria for Reducing Building Energy Consumption*; National Energy Committee of Iran: Tehran, Iran, 2012.

7.    *Bureau for Compiling and Promoting National Regulations for Buildings*; Code No. 19: Energy Efficiency; Ministry of Housing and Urbanism IRI: Isfahan, Iran, 2010.

8.    Fayaz, R.; Mohammadkari, B. Comparison of energy conservation building codes of Iran, Turkey, Germany, China, ISO 9164 and EN 832. *Appl. Energy* 2009, *86*, 1949–1955.

9.    Crawley, D.B.; Jon, W.H.; Kummert, M.; Griffith, B.T. Contrasting the capabilities of building energy performance simulation programs. *Build. Environ.* 2008, *43*, 661–673.

10.   Building Research Establishment (BRE). *UK National Calculation Method*; BRE: Watford, UK. Available online: http://www.ncm.bre. co.uk/ (accessed on 12 July 2013).

11.   *ASHRAE Standard 55: Thermal Environmental Conditions for Human Occupancy*; ASHRAE: Atlanta, GA, USA, 2010.

12.   *ISO 7730: Ergonomics of the Thermal Environment—Analytical Determination and Interpretation of Thermal Comfort Using Calculation of the PMV and PPD Indices and Local Thermal Comfort Criteria*; ISO: Brussels, Belgium, 2005.

13. Heidari, Sh. Comfort temperature for Iranian people in the city of Tehran. *Honar-Ha-Ye-Ziba* 2009, *1*, 5–14.

14. McMullan, R. *Environmental Science in Buildings*, 6th ed.; Palgrave Macmillan: New York, NY, USA, 2007.

15. Clay Brick and Paver Institute, *The Role of Thermal Mass in Energy-Efficient House Design*; Austral Bricks: Langford, Australia, 2006.

16. Gregory, K.; Moghtaderi, B.; Sugo, H.; Page, A. Effect of thermal mass on the thermal performance of various Australian residential constructions systems. *Energy Build.* 2008, *40*, 459–465.

17. Balaras, C.A. The role of thermal mass on the cooling load of buildings: An overview of computational methods. *Energy Build.* 1996, *24*, 1–10.

18. Chartered Institution of Building Services Engineers (CIBSE), *Environmental Design: CIBSE Guide A*, 7th ed.; CIBSE: London, UK, 2006.

19. De Saulles, T. *Thermal Mass Explained*; CentER (The Concrete Centre): Camberley, UK, 2011.

20. Kruger, E.; Cruz, E.G.; Givoni, B. Effectiveness of indirect evaporative cooling and thermal mass in a hot arid climate. *Build. Environ.* 2010, *45*, 1422–1433.

21. Laughton, M.A.; Warne, D.F. *Eectrical Engineer's Reference Book*; Newnes: Oxford, UK, 2003.

22. Nicol, J.F.; Humphreys, M.A. Adaptive thermal comfort and sustainable thermal standards for buildings.*Energy Build.* 2002, *34*, 563–572.

23. Hegger, M.; Fuchs, M.; Stark, Th.; Zeumer, M. *Energy Manual: Sustainable Architecture*; Birkhauser: Berlin, Germany, 2008.

24. Givoni, B. *Climate Considerations in Building and Urban Design*; Van Nostrand Reinhold: New York, NY, USA, 1998.

# Chapter 9

## DETECTION AND CLASSIFICATION OF CHANGES IN BUILDINGS FROM AIRBORNE LASER SCANNING DATA

Sudan Xu, George Vosselman, and Sander Oude Elberink
Department of Earth Observation Science, Faculty ITC, University of Twente, 7500 AE Enschede, the Netherlands

## ABSTRACT

The difficulty associated with the Lidar data change detection method is lack of data, which is mainly caused by occlusion or pulse absorption by the surface material, e.g., water. To address this challenge, we present a new strategy for detecting buildings that are "changed", "unchanged", or "unknown", and quantifying the changes. The designation "unknown" is applied to locations where, due to lack of data in at least one of the epochs, it is not possible to reliably detect changes in the structure. The process starts with classified data sets in which buildings are extracted. Next, a point-to-plane surface difference map is generated by merging and comparing the two data sets. Context rules are applied to the difference map to distinguish between "changed", "unchanged", and "unknown". Rules are defined to solve problems caused by the lack of data. Further, points labelled as "changed" are re-classified into changes to roofs, walls, dormers, cars, constructions above the roof line, and undefined objects. Next, all the classified changes are organized as changed building objects, and the geometric indices are calculated from their 3D minimum bounding boxes. Performance analysis showed that 80%–90% of real changes are found, of which approximately 50% are considered relevant.

## INTRODUCTION

In urban areas, changes to buildings may be caused by natural disasters or geological deformation, but more often they are the result of human activities. These activities may lead to temporary or permanent changes, as, for example, discussed by Xiao *et al.* [1]. Detection of structural changes to urban objects, e.g., renovation of infrastructural objects and buildings, is important for municipalities, which need to keep their topographic object databases up-to-date.

Analysis of the geometric differences between point clouds from different epochs of data provides an insight into how areas change over time. The general approach in using point clouds for change detection is to focus on areas where points are present in one epoch of data and absent—at least in the close vicinity—in the other. As the need for detecting changes at higher levels of detail is growing, demand for more detailed interpretation of differences between epochs of data has also grown. There can be several reasons for absence of data points for a certain location in one epoch of data even if data points for that same location are present in another epoch, for example:

- the area around the location may have been occluded in one of the data epochs;
- the surface around the location may have absorbed the ALS laser pulses;
- outliers (either on the ground or inside buildings) may be present; or
- the object may have undergone change.

In the first three cases, differences in the data do not correspond with real changes. In the fourth (last) case, some of the changes may not be relevant for a municipality's databases, e.g., a parking place with a lot of changes due to the temporary presence or absence of cars, which results in different arrangements of point clouds. The main problem is that it is not known whether a difference between two data sets is caused by differences in scanning geometry, surface properties, or changes to an object—relevant or irrelevant. A better understanding of differences between epochs of data would allow us to precisely determine whether there has been a relevant change in an area and, if so, what kind of change it is.

Our study aims to develop a method for detecting relevant changes occurring in urban objects using point clouds from ALS data. Our focus is on changes to buildings, including changes to roof elements and those associated with car parking lots on top of buildings. The main challenge is how to separate irrelevant differences from relevant object changes. We consider changes on objects with an area of <4 m$^2$ to be irrelevant changes (although we can detect them), because we assume that 4 m$^2$ is the minimum area that can accommodate a person for the purpose of checking the building or property.

With the above aims in mind, we set up the following change detection procedure: first, in each point cloud, objects are classified as "ground", "water", "vegetation", "building roof", "roof element", "building wall", or "undefined object" using a rule based classifier with different entities as seen in Xu *et al.* [2]. This classification step is referred to as "scene classification" in the remainder of this paper. Next, a surface difference map is generated by calculating the point-to-plane distances between the points in one epoch of data to their nearest planes in the other epoch. By combining scene classification information with the surface difference map, changes to building objects can be detected. Points on these changed objects are grouped and further analyzed in a rule-based context to classify them as changes related to: a roof, a wall, a dormer, cars on flat roofs, a construction on top of a roof, or undefined objects. Reference data was collected manually to determine the accuracy of this approach.

To begin, we describe the state-of-the-art in change detection techniques in Section 2. The explanation of the method in Section 3 looks at our use of surface difference maps (Section 3.1), our change detection algorithm (Section 3.2), the classification of changes (Section 3.3), and our analysis of object-based changes (Section 3.4). The data sets used are described in Section 4, followed by presentation of results and their analysis in Section 5. We offer our conclusions in Section 6.

# RELATED RESEARCH

Research related to change detection is discussed in two parts. First, an overview is given of previous research done on change detection from imagery. In the second part, we discuss techniques using lidar point clouds for detecting changes in buildings.

## Change Detection from Imagery

According to Mas [3], change detection from imagery can be grouped into three categories divided by the data transformation procedures and the analysis techniques: (1) image enhancement; (2) multi-date data classification; and (3) comparison of two independent land cover classifications. The first group of image enhancement includes image differencing, image regression, image rationing, vegetation index differencing, change vector analysis (CVA), IRMAD, new kernel based methods, and transformation methods such as Principle Component Analysis (PCA) [4] and RANSAC [5]. The main problem of all the algorithms is to determine the threshold for the change areas. Among all these methods, CVA performs best. CVA is an extension of image differencing and it computes the change direction and magnitude of all bands for all dates that the images were taken.

The multi-date data classification approach is based on pixels being directly classified as changed or unchanged according to differences distinguished between features of the pixels and/or those of their neighborhoods [6]. The third approach is based on comparing two independent classified images and is also called a post-classification comparison [3]. Both images are classified independently and differences in pixel labels are assumed to be caused by changes in, for example, land cover [7]. Obviously, the quality of the change detection depends on the quality of the classification. Classification methods can be supervised and unsupervised. Yang and Zhang [6] and Di *et al.* [7] both used the Support Vector Machine (SVM) as a supervised classifier. Unsupervised methods have been described by Tanathong *et al.* [8], who defined a classifier agent and an object agent, and Kasetkasem and Varshney [9], who introduced Markov Random Field Models for change detection.

## Change Detection in Lidar Point Clouds

Similar to image differencing, the first approach used for detecting change in multi-temporal lidar point clouds involved the subtraction of two DSMs [10] from each other. Classification is also used to directly distinguish changed objects from unchanged ones; this is usually applied in disaster assessment. For an example of this method of classification, see Khoshelham *et al.* [11], who, using an ALS data set, performed several supervised classifiers on a small number of training samples to distinguish damaged roofs from undamaged roofs.

In cases of 3D change detection in buildings, depending on the nature of the application and the availability of data sets, comparisons can be made between data sets of multi-temporal images, between multi-epoch lidar point clouds, and between image and lidar point clouds. With the improvements in the accuracy of obtained images as well as improvements in processing skills, 3D point clouds can also be derived from images by, for example, dense matching algorithms [12].

There are two basic approaches to the problem of change detection, each determined by the availability of original data: (1) original data is available from both epochs to be used for detecting change in cases of disasters or geological deformation; (2) original data is only available for one epoch, while the other epoch of data is an existing map or database [13]. The literature of these two approaches is reviewed below:

### Approach 1

In 1999, Murakami *et al.* [10] used two ALS data sets, one acquired in 1998 and the other in 1996, to detect changes to buildings after an earthquake in a dense urban area in Japan. A difference map was obtained by subtracting one DSM from the other. This difference map was laid over an ortho-image and an existing GIS database to identify changes to buildings. Vögtle and Steinle [14] presented their research as a part of a project using DSMs from two ALS data sets to detect changes after strong earthquakes. Segmentation was run to find all the buildings in the area. The changes were identified by the overlay rate of all the buildings in both epochs of data. Rutzinger *et al.* [15] extracted buildings from DSMs of two data epochs using a classification tree. Shape indices and the mean height difference of the building segments were compared. Differences between classified DSMs derived from two epochs of ALS data have also been used by Choi *et al.* [16] to detect changes.

## *Approach 2*

Vosselman *et al.* [13] compared ALS data to an existing medium-scale map. Change detection was done by segmentation, classification, and the implementation of mapping rules. Pixel overlay rates on a classified DSM and a raster map were used to finally identify changes. This method was then improved by using aerial images to refine the classification results [17]. Rottensteiner [18] employed data fusion with the Dempster-Shafer theory for building detection. He then improved his method by adding one more feature to the data fusion, which makes the classification suitable for building-change detection. Champion *et al.* [19] detected building changes by comparing a DSM with a vector map. The similarity measure between building outlines in the vector map and the DSM contours was used to identify the demolished, modified, and unchanged buildings. New buildings were detected separately in the DSM. Chen *et al.* [20] used lidar data and an aerial image to update old building models. After registration of the data, the area of change was detected and height differences were calculated between the roof planes in the lidar data and planes in the old building models. A double-thresholding strategy was used to identify the main-structures in changed and unchanged areas, while uncertain parts were identified from the line comparisons between building boundaries extracted from an aerial image and the projection of the old building models. The double-thresholding method was reported to improve overall accuracy from 93.1% to 95.9%. Awrangjeb *et al.* [21] detect building changes from lidar point cloud by determining the extended or demolished part of a building with a connected component analysis and they conclude that no omission errors are found.

## Problems of Existing Methods and Our Contribution

Approaches described above enable changes to be detected in 2D (maps) or 2.5D (DSMs), which may result in information loss such as changes under tall trees. To avoid these problems, methods that can compare the lidar data directly are required. In 2011, Hebel *et al.* [22] introduced an occupancy grid to track changes explicitly in multi-temporal 3D ALS data. In order to analyze all the laser beams passing through the same grid, they defined belief masses "empty", "occupied", and "unknown" for each voxel in the object space. For the data sets they compared, they computed belief masses resulting from all the laser beams, with conflicts in belief masses denoting a change. By adding an extra attribute to indicate the smoothness and continuity of a surface, they were able to achieve reliable change detection results, even when occlusion had occurred in either of the data epochs. Subsequently, Xiao *et al.* [1] introduced an occupancy grid to the mobile laser scanning system to detect permanent objects in street scenes, where occlusions frequently occur. They too reported that occupancy grids are resistant to occlusion. However, when point density is too low, for example on walls, and the size of the occupancy grid is small, parts of walls were identified as occluded because no occupied points were included in the grid due to the low point density. The size of the occupancy grid may, therefore, influence the detection result when the point density of the lidar data varies from place to place. Besides the occlusion problem, pulse absorption may cause geometric differences in the data, although no real change occurred, as explained in the introduction (Section 1). Compared to the existing methods, our contribution consists of: (1) avoiding information loss by directly comparing the lidar data (not DSM) using the surface difference map when changes occur under vegetation; (2) solving problems of occlusion and pulse absorption on building tops without using occupancy grids; (3) evaluating the accuracy per changed object, be it just a small dormer or a large building (or part of a building).

## METHODOLOGY

Our approach can be labelled as a post-classification comparison on multi-epoch 3D point clouds. To interpret changes in a scene, we start by calculating the differences between two data sets. A surface difference map is generated by calculating the point-to-plane distance between the points from one data epoch to their nearest planes in the other (see Section 3.1 for more details). As the points on building roofs, walls, and roof elements are considered to be extracted in the scene classification step, it is possible to combine the scene classification result with surface difference information. This combination enables us to not only detect changes on building objects but also detect occluded areas, where it is not known whether there has been a change or not (see Section 3.2). Points on the changed objects are grouped and analyzed according to a rule-based context. In addition to changes to roofs and walls (seeSection 3.3), changes to roof elements are further classified as cars, construction, dormers, or undefined objects. The classified changes are finally grouped into building objects and analyzed (see Section 3.4). Reference data were collected manually at random locations to determine the accuracy of our method.

## Generating a 3D Surface Difference Map

The 3D surface difference map records the disparities of points between two epochs of ALS data. The disparity per point is computed as the distance from a point to its nearest fitted plane from another epoch. Surface difference was employed by Vosselman [23], who evaluated the quality of data using overlapping strips. In our paper, the surface difference map is used to indicate 3D differences between two lidar data sets.

For every point in one epoch of data, we search within its 1.0 m (radius) 3D neighborhood to check whether there is a point from the other epoch. If there are no points from the other epoch, this implies a difference greater than 1.0 m. In these cases we record a difference value >1.0 m. If there are points from the other epoch of data, we define the surface difference as the distance from the selected point to the nearest fitted plane in the point cloud of the other epoch.

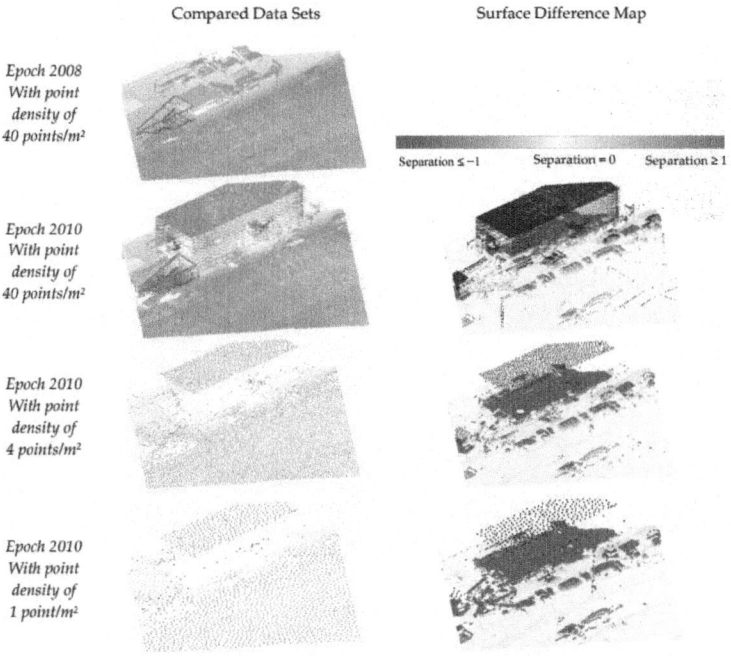

**Figure 1:** Examples of the surface difference map for various point densities. Comparison of data sets with different densities from the year 2010 with the data from the year

2008 showed that the locations of the changes are the same in all the difference maps. This indicates that the surface difference map generated with a neighborhood of 1.0 m points is suitable for all the data sets that have a point density larger than 1 point/m$^2$.

A 3D neighborhood of 1.0 m radius points is chosen because the point density is greater than 1 point/m$^2$ in most locations of our dataset. This ensures that the comparison is not affected by the point density. Although we have two data sets that have similar point densities, in order to make sure that a 1.0 m neighborhood is suitable for comparisons between data sets with different point densities. We simulated several data sets with our test data by reducing the point density of one of the epochs (see Figure 1).

The difference map contains the geometric indication of whether there is a change or not. However, not every difference is a change (Figure 2), and sometimes part of a change is not represented as a large value in the difference map (Figure 3). Figure 2c shows the difference map containing several points with a surface difference greater than 10 cm. However, these differences are caused by lack of data in the other epoch and therefore cannot be considered as a change in an object. Conversely, points with a difference value less than 10 cm may belong to a changed object.Figure 3 is an example that displays points with a surface difference of less than 10 cm in the connections between a dormer and a roof; the change was detected because the dormer was newly built.

(a)

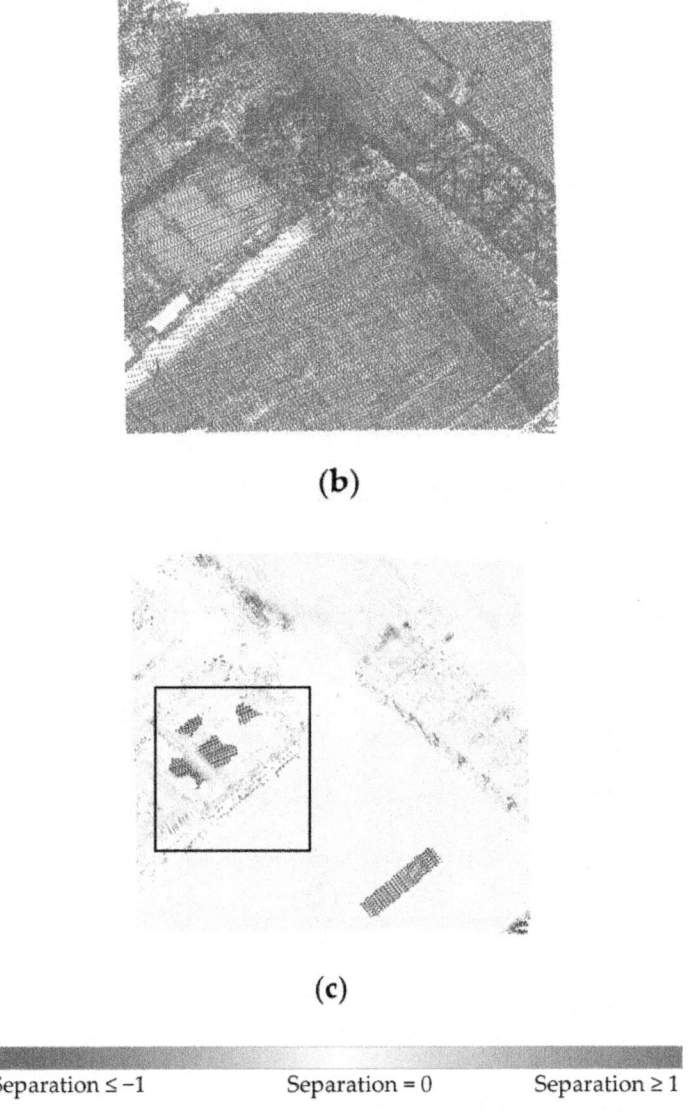

**(b)**

**(c)**

Separation ≤ −1              Separation = 0              Separation ≥ 1

**Figure 2:** (**a,b**): Different colors represent different epochs of data, (**a**) 2008, (**b**) 2010; (**c**) Surface difference map derived from the two epochs. Points with a great difference do not always denote a change. For example, in (**c**) blue points in the rectangle are unknown points because there is no data for the roof in thr 2008 epoch data (perhaps caused by water on the roof), although they have high difference values.

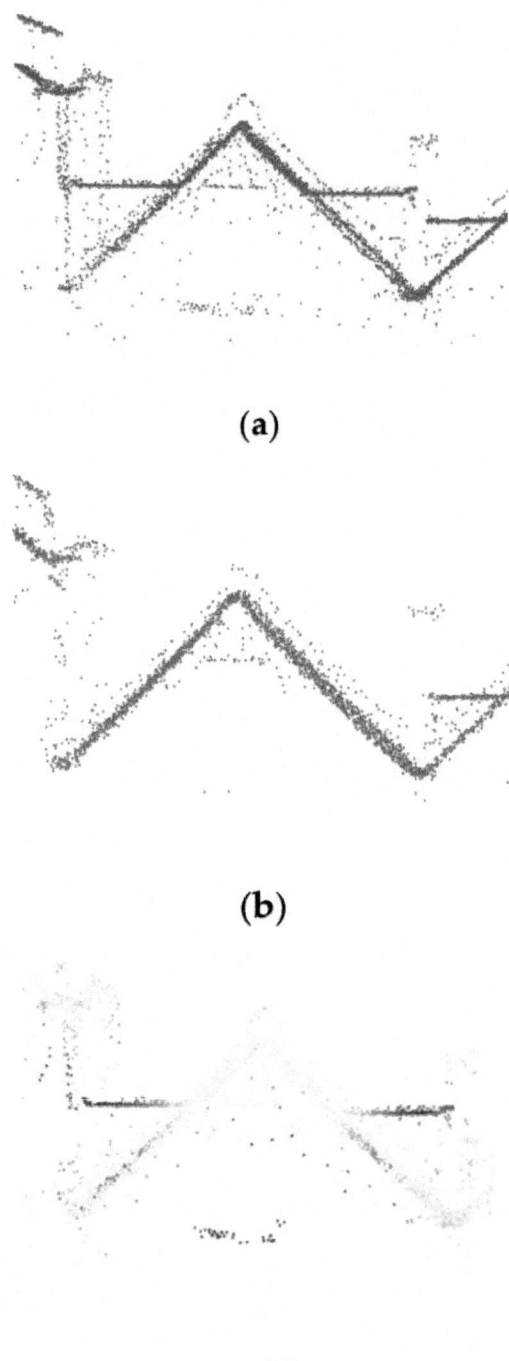

**(a)**

**(b)**

**(c)**

Separation ≤ −1              Separation = 0              Separation ≥ 1

**Figure 3:** (**a,b**) Different colors represent different epochs of data, (**a**) 2012, (**b**) 2010; (**c**) Surface difference map—both epochs merged. The points in the rectangle have a distance difference ≤0.08 m, although there is indeed a change. The difference value is small because the points on the roof from the other epoch are in the neighborhood of the points on the dormer and they are quite close to each other. The problem is to identify the areas in the rectangle as being changed when the difference value is low.

Our task is to assign the labels of "changed", "unchanged", and "unknown" to each point according to the surface difference map.

## Detecting a Change

The strategy for interpretation of the surface difference map is as follows: points classified as building (wall, roof, or roof element) in the scene classification are selected. All building points are assumed to be unchanged except for those distinguished as "unknown" or "changed".

Points with a difference value greater than 1 m in one epoch for which, even in a two dimensional neighborhood, no nearby points can be found in the other epoch are considered "unknown" (see Figure 4) . This occurs in areas where there is a lack of data in one epoch because of occlusions or pulse absorption by the surface. For walls, large surface difference values may occur. However, our algorithm will label the points for the wall as "unknown" instead of "changed" if there has been no change to the roof of that same building. Due to a lack of evidence as to what happened with the wall, the points are labelled as "unknown".

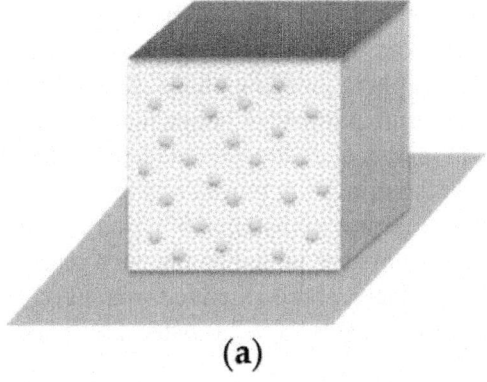

(a)

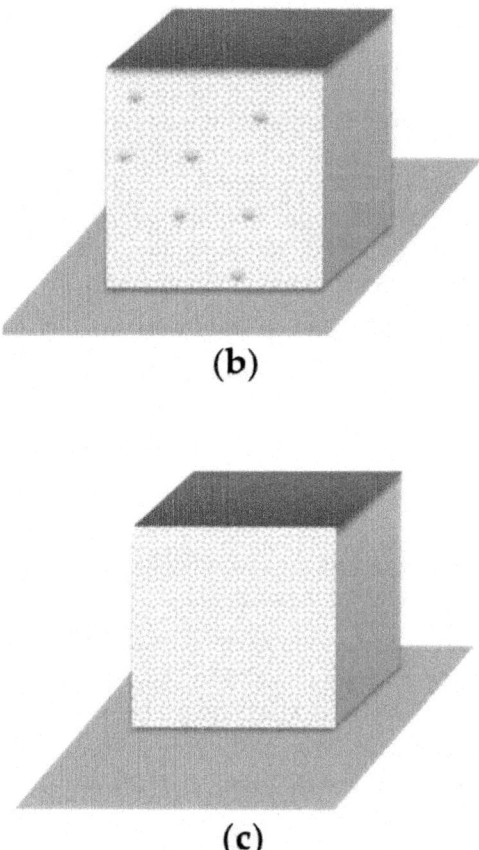

**(b)**

**(c)**

**Figure 4:** Points labelled as "unknown" for walls. (**a**) a wall scanned with dense points; (**b**) the same wall scanned with sparse or (**c**) no points in another epoch.

In Table 1, the "seed neighborhood radius" and the "growing radius" indicate the maximum radius for allowing a point participating in the seed plane extraction and the plane growing phase. The "minimum number of seed points" means the minimum number of points that are required for initialization of a plane. The plane parameters are obtained from the Hough space and 10 points are required as the minimum number of seed points to ensure a reliable plane extraction. The "maximum distance of a point to the plane" ensures that only nearly co-planar points are added to the seed plane. The 1.0 m radius is set to ensure that the surface can be extended with neighbouring points on the same surface. This setting needs to be adjusted for processing datasets with a different point density.

**Table 1:** Parameters for the surface growing and connected component algorithm.

| Surface Growing | | | | Connected Component |
|---|---|---|---|---|
| Parameters for seed selection | | Parameters for growing | | Maximum distance of component |
| Seed neighborhood radius | 1.0 m | Growing radius | 1.0 m | 1.0 m |
| Minimum number of seed points | 10 pts | Maximum distance of a point to the plane | 0.1 m | |

Generally, "changed" points have a high difference value, although they cannot be selected using a simple threshold in the difference map. Therefore, we derived a rule-based decision-tree to detect "changed" points in surface difference maps. The decision-tree is shown in Figure 5. The "unknown" points are first excluded from the data sets. The remaining points are grouped into planar segments using the surface growing method [24]. Within each segment, points are further separated into connected components with a smaller radius than during the planar surface growing step. This separates two nearby objects that are located in same plane. For example, Figure 6a shows two dormers belonging to the same planar segment, and after deriving connected components, they are separated into two components. These components are used as the basic units for identifying changed points. Parameters used in the surface growing and connected component algorithms are listed in Table 1.

If, for each connected component, the vast majority of points have surface differences greater than the maximum strip difference, the whole component is very likely to be a change and will be labelled as "changed". Otherwise, the component may be unchanged or only partly changed. If partly changed, the changed part is only labeled as "changed" by checking its area. The parameters for the vast majority, the maximum strip difference, and the area are listed in Table 2 and the reason for the parameter setting is given.

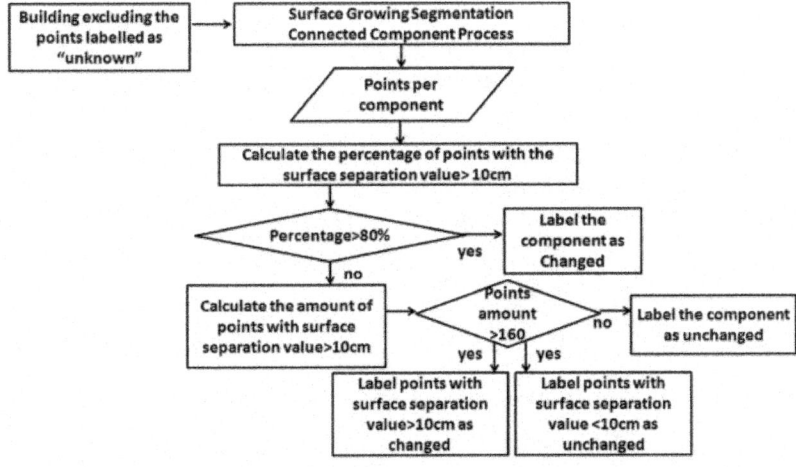

**Figure 5:** Identification of changed and unchanged points using a decision-tree.

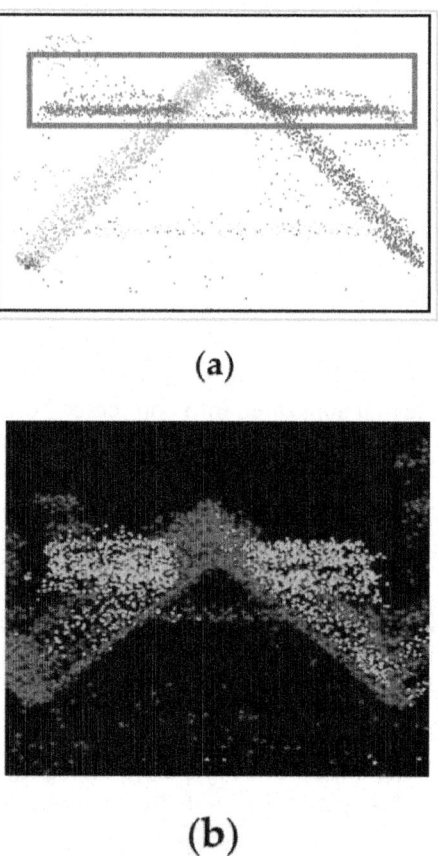

**(a)**

**(b)**

**Figure 6. (a)** The two dormers in the red rectangle belong to the same planar segment and therefore need connected components analysis to separate them (different colors indicate different segments); **(b)** Both changes to roofs and dormers are detected (two data epochs are merged).

**Table 2:** Parameters for deciding on change

| Name | Thresholds | Reason for the Choice of Thresholds | Suitability to Transfer to Other Datasets |
|---|---|---|---|
| Vast majority | 80% | Tested | To be adjusted according to the training data. |
| Maximum strip height difference between two epochs of data | 10 cm | Registration error in each epoch was shown to be below 5 cm. | To be adjusted to the maximum registration error. |
| Area | 160 points (4 m²) | The average point density is 40 pts/m², and all buildings larger than 4 m² should be mapped. | To be adjusted to the point density and the definition of relevant change. |

## Classification of Building Change

Points detected as changes in a building are extracted and a second classification step is performed on these changes. This second classification step is required to understand the activities that could have possibly led to the changes detected for a building. For example, by identifying changes to roof elements on top of a building (Figure 7a), we can infer that these elements may be cars, and that there is a parking lot on top of the building.

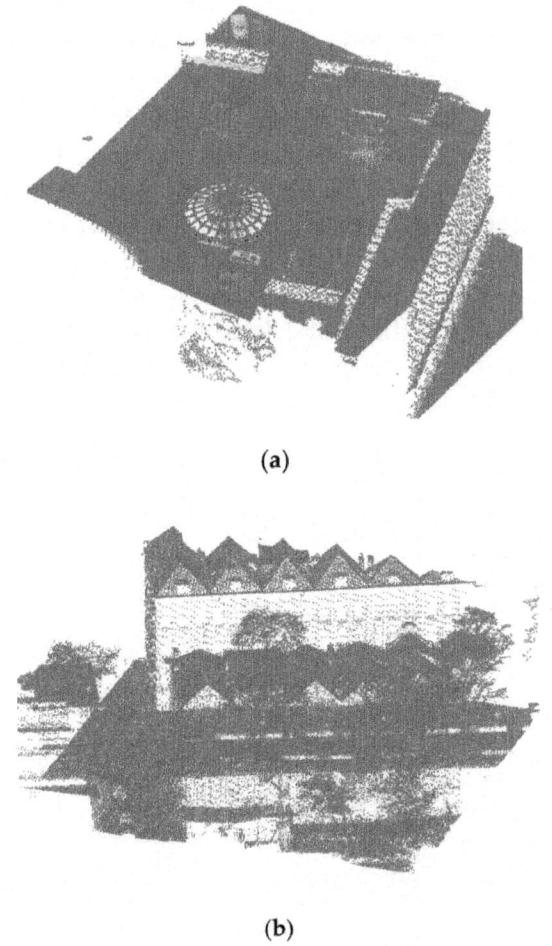

**(a)**

**(b)**

**Figure 7:** Connected components of changed objects are classified based on their context. **(a)** 2008 *vs.* 2010; **(b)** 2010 *vs.* 2012.

There are various types of objects that may be present on building roofs or attached to walls. Most changes to walls are caused by sun shades (which are open in one epoch but closed in the other epoch), stairs, flags, or vegetation near walls. Compared to the changes to roofs (e.g., a new dormer or the addition of a floor), changes near walls are less likely to be related to construction activities. For this reason, the second classification step is performed only for changes concerning roofs and roof elements.

Based on the size of changes and the underlying building structure, changes on a roof top are classified as dormers, cars or "other constructions". The difference between a dormer and "other constructions" above the roof line is that the latter involves relatively large changes compared to a dormer. The attributes used for the classification are area, height to the nearest roof, the normal vector direction of the nearest roof (which indicates a pitched or a flat roof) and the labels from the scene classification results.

Rules are defined to classify these components. Large areas of points labelled as changed indicate a change to a roof. Small changes occurring on a pitched roof are more likely to be newly built or removed dormers or chimneys. If changes occur on a flat roof, they are probably constructions above the roof line or cars. These rules are used in a decision-tree, and have been established as a rule-based classifier with some defined values according to the knowledge (see Figure 8). All defined values are based on the authors' knowledge and are explained in Table 3.

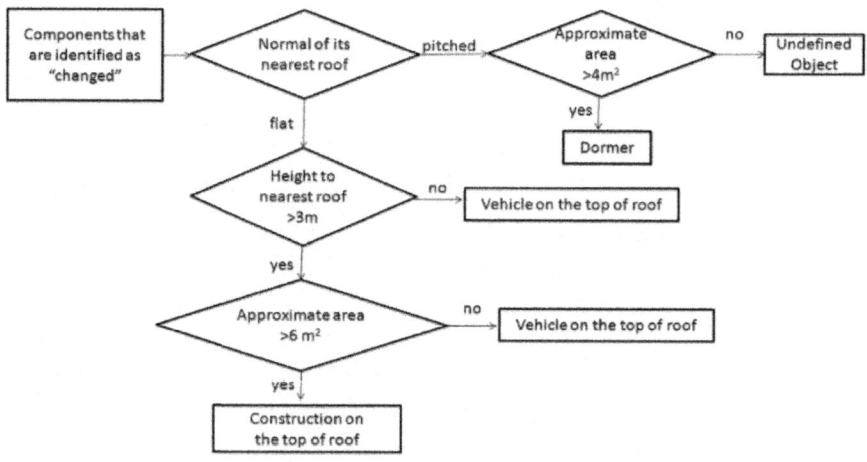

**Figure 8:** Rule-based classification of changes.

**Table 3:** Defined parameter values for the rule based classifier

| Attribute Name | | Value | | Explanation | |
| --- | --- | --- | --- | --- | --- |
| Approximate area | | | | | |
| On pitched roof | On flat roof | 4 m² | 6 m² | Minimum area of interest | Area of small vehicle |
| Height to the nearest roof | | 3 m | | The maximum height of a car | |
| Class label of point in the scene classification | | "roof"/"roof element"/"wall" | | For understanding the context | |

## Analysis of Changed Building Objects

After the points are identified as "changed" and classified, the results are points organized as connected components. These components have not been identified as building objects yet, and they may represent only a small part of an object. How points are organized as building objects and how the changed objects are analyzed is described in the following subsections.

### Changed Building Objects

The points that are identified as changed can be organized into building objects once we have labelled all the points with: (1) the type of object class from the change classification; (2) the labels "changed", "unchanged", or "unknown"; and (3) the signs from a difference map indicating a newly built or demolished structure. We distinguish buildings as being entirely changed or partly changed.

If there is a change to the roof of a building, all points that group together in a 2D connected component step will be labelled as an entirely changed building. If changes are made to building elements, the local 3D connected points will be only labelled as partly changed.

### Merging Objects at Tile Boundaries

So far, changes are detected within tiles of, for example, 50 m × 50 m. Typically, buildings are stored in different tiles, as shown in Figure 9a–c. As changed objects may occur on tile boundaries, a merging step is needed to detect a single object instead of two (*i.e.*, one in each neighbouring tile). To solve this problem, adjacent components are merged across tile boundaries, as described by Vosselman [25]. Figure 9d shows the components after they have been merged.

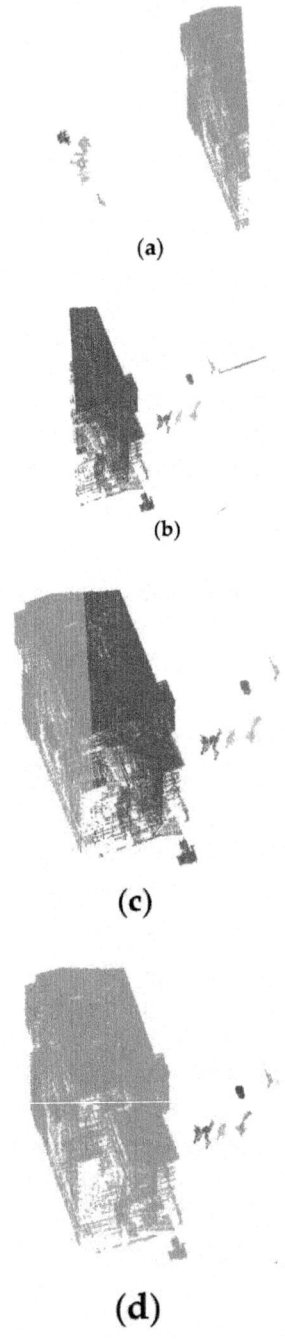

(a)

(b)

(c)

(d)

**Figure 9:** Objects are cut and stored in different tiles (**a**) and (**b**) during the processing of large data sets. Components have been merged in **(c)** and **(d)**. **(a)** building part in one tile; **(b)** building part in another tile; **(c)** the same building part belonging to different objects; **(d)** the same building part merged into the same object.

When building objects have been formed, minimum 3D bounding boxes are calculated for these objects. Their width, length, height, area, volume, and location of their center point are determined from their 3D bounding boxes. Among all the changes near buildings we found, there are some that are irrelevant for the municipality. Common examples are changes to extensive areas of flowers and shrubs, to fences, and railings in gardens on rooftops of buildings. Bounding boxes have been used to measure more precisely the size of a change. This measure of size is used to separate irrelevant changes ($<4$ m$^2$) from relevant ones. For changes $<4$ m$^2$ and a "length: width" ratio greater than 10, no 3D bounding boxes are calculated.

## DATA SETS

The research we present here originated from a request from the Municipality of Rotterdam (the Netherlands) to help in monitoring changes to buildings using ALS data. Three data sets, located in commercial and residential areas of Rotterdam, have been used. The data of the commercial area was scanned in years 2008 and 2010; data sets of the residential area were acquired in years 2010 and 2012 (Figure 10). The point densities for the data sets 2008, 2010, and 2012 are average 30 points/m$^2$, 35 points/m$^2$ and 40 points/m$^2$, respectively. The strip difference within an epoch and between different epochs had a maximum value of 10 cm. As the data sets already register well (with a strip difference less than 10 cm), no extra steps for registration were required. All data sets underwent scene classification and were then divided into tiles of 50 m × 50 m. Figure 11 shows four examples of the result from the scene classification step. Bridges and its cables are classified as building roof and roof elements in data set 2008 and 2010. Data set 2010 was obtained in the fall and the 2012 dataset was obtained in the fall. Therefore, tree points are more dense in the 2012 data set.

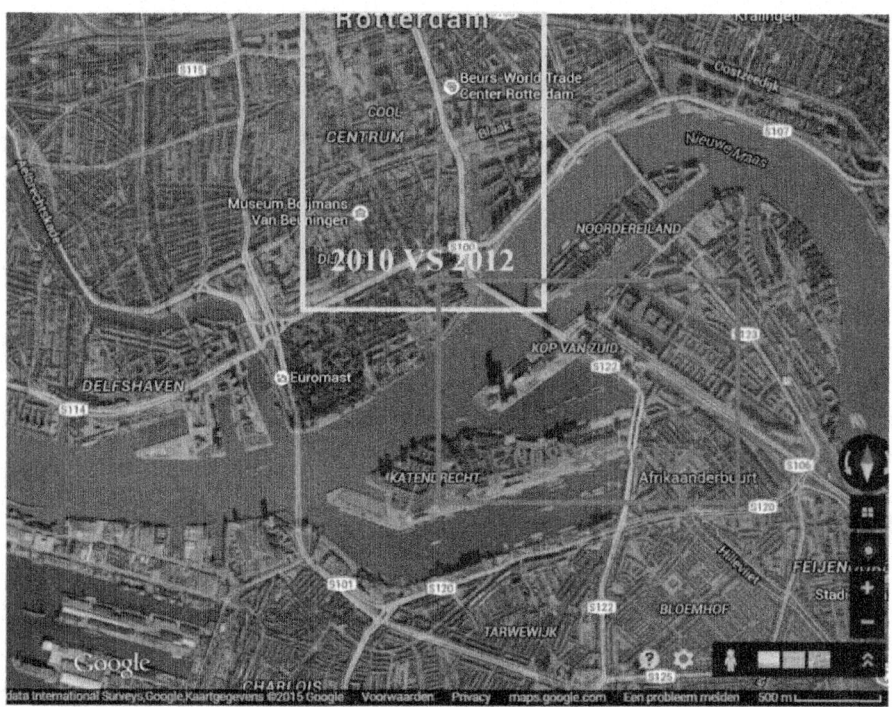

**Figure 10:** Locations of test areas and their data sets. (From Google earth and the coordinates are approximately measured from the map).

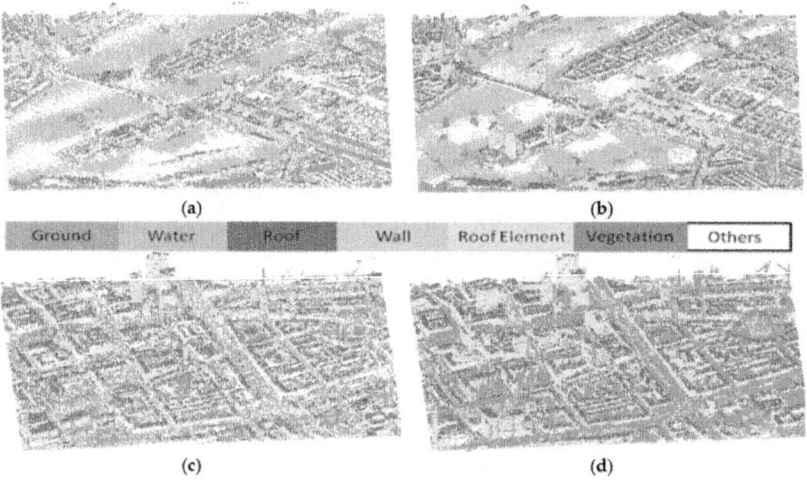

**Figure 11:** The scene classification results for the test areas. (a) 2008—commercial area; (b) 2010—commercial area; (c) 2010—residential area; (d) 2012—residential area.

# RESULT AND DISCUSSION

Our change detection procedure comprises change detection, classification of any change detected, and object-based analysis. This procedure includes generation of surface difference maps and change identification. The results for each step in the procedure are described below. As the compared data sets are merged into one with different epoch numbers, all the results will be shown in a merged version.

## Detecting Change

Results of the change detection step are shown and the causes of errors in detecting change to buildings are discussed here. The true positives (TP), false positives (FP), and false negatives (FN) are counted "per change" by three people after visual inspection of the compared data sets, the corresponding surface difference map, and our change detection results on a computer screen. To provide some context, an area with 10 m × 10 m around the detected change is shown, together with a high resolution aerial image of the area. The experts indicate whether the detected change is a change to a relevant object. The difference map is also used to verify whether all geometric differences resulted in a detected change. If it is a relevant changed object and we detected it correctly in our result, the change is counted as true positive. If we found a changed object in our result but three experts considered it not a change, the change is a false negative, and if the experts found some changes that we did not detect in the result, the change is a false positive. The completeness and correctness are calculated with the following equations:

$$completeness = \frac{TP}{TP + FN} \tag{1}$$

$$correctness = \frac{TP}{TP + FP} \tag{2}$$

It is notable that one large building change is counted as one change, the same as a small change, regardless of the number of points within the change. In order to know how the scene classification result affects the change detection result, we also calculated the percentage of errors that are caused by the scene classification result.

## Surface Difference Mapping

A surface difference map is generated from original data sets, regardless of the scene classification results. Some difference maps have already been shown in Figure 1 (Section 3.1) and, as discussed, the point density does not affect the surface difference map. More difference maps can be seen in Figure 12. The surface difference map gives only clues as to where changes may have occurred in the data sets. To improve their display, all the data sets have been merged into one. Different colors show different values of the differences between the compared data sets.

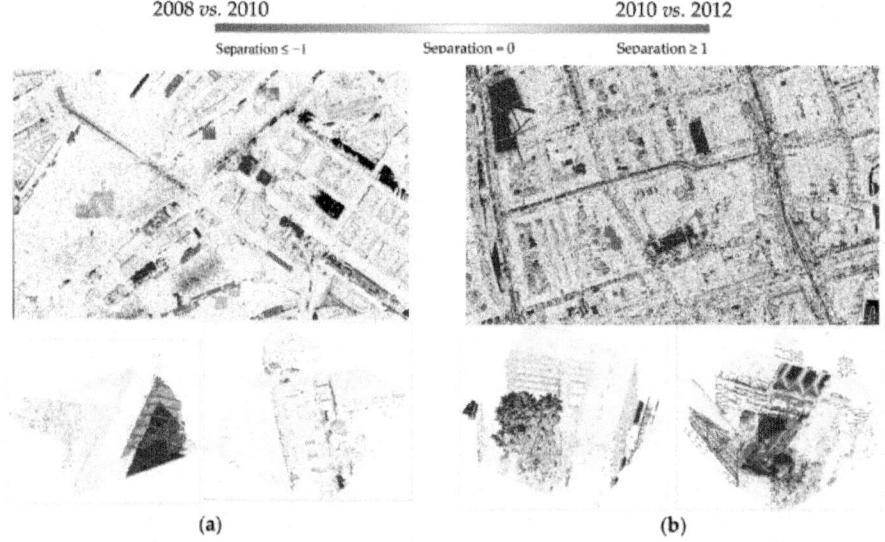

(a)                                                          (b)

**Figure 12:** Surface difference maps for **(a)** 2008 *vs.* 2010 and **(b)** 2010 *vs.* 2012. Maps in the first column are the overall visualization of the test area, and maps in the second column are detailed visualization from some tiles.

In Figure 12, significant differences can be directly observed on the surface difference maps where the colors become deeper. Red points appearing in the middle of water in Figure 12a indicate large differences in water surfaces. This large difference is caused by lack of points on the water in one of the epochs, while points were recorded in the other epoch. These difference maps are the inputs for the change identification step.

## Change Detection Results

Using the surface difference map, we labelled the data sets as "changed", "unchanged", and "unknown". Figure 13shows some results of both types of change (an entire building change (a), as well as changes to building elements, e.g., changes (b) and (c)). The interpretations "demolished" and "new" can be decided from the sign of the surface difference value in the difference map. For example, the points seen in epoch 2008 but not in epoch 2010, are "demolished" objects. A small change is shown horizontally for clearer visualization in Figure 13c. As there are new points in epoch 2010, as well as demolished points beneath them in epoch 2008, we can infer that there are some extensions of the objects. To understand which objects caused the extension, the changes need to be classified.

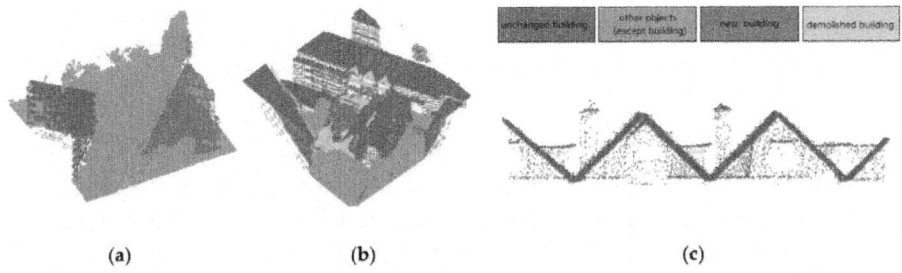

(a)                              (b)                              (c)

**Figure 13:** (a,b) Examples of large changes in buildings; (c) changes to dormer sizes in the merged data sets (removed dormers and extensions to dormers in height and length).

We selected 20 tiles with changes randomly from each test area to evaluate the performance of the change detection method. There are more false positives than false negatives, so completeness is greater than accuracy, *i.e.*, 50% (Table 4) of identified changes are not relevant. However, the vast majority of changes to buildings are actually detected, making it convenient to only look at the detected changes and ignore the irrelevant ones.

**Table 4:** The completeness and correctness of detected change

|  | True Positive | False Positive | False Negative | Correctness | Completeness |
|---|---|---|---|---|---|
| 2008 *vs*. 2010 | 34 | 34 | 3 | 50.00% | 91.89% |
| 2010 *vs*. 2012 | 35 | 30 | 6 | 53.86% | 83.33% |

As shown in Table 5, 40% of the errors (false positives and negatives) are due to incorrect scene classification in Test area 1, and 67% in Test area 2. In Test area 1, which is a commercial area of Rotterdam, the main propagation errors from scene classification are large and long balconies in walls, which are incorrectly classified as roofs. In Test area 2, a residential area, the error propagation mainly arose from trees that are so dense that they were wrongly classified as building roofs. So, changes in trees were incorrectly detected as changed building objects.

**Table 5:** Percentage of errors due to scene classification and errors due to the method of change detection used

| | False Positive | | False Negative | |
|---|---|---|---|---|
| | | 14 | | 1 | Errors due to the scene classification error 40% |
| 2008 *vs.* 2010 (Test area 1) | 34 | 20 | 3 | 2 | Errors due to limitations of the change detection method 60% |
| | | 18 | | 6 | Errors due to the scene classification error 67% |
| 2010 *vs.* 2012 (Test area 2) | 30 | 12 | 6 | 0 | Errors due to limitations of the change detection method 33% |

Note that in Table 5 there are two false negatives (relevant changes that are not detected) due to our method in Test area 1. These two relevant changes are detected in the change detection method, but the 3D bounding boxes are not generated. The reason for this is that we applied a threshold value for the ratio of the length to width to exclude some thin and long objects such as fences and railings during the change quantification stage. These two relevant changes were roofs that are long and narrow, so 3D bounding boxes were not generated for them. The advantage of using raw lidar data, instead of using DSM or image, is that lidar data can penetrate trees and the information under trees can be preserved. If a change happens under trees, DSM or image may not detect the change. One such case is shown in figure (i) in Section 5.2.1.

## *Change Detection Error Due to Scene Classification Error*

The points on the surface difference map that are labelled as part of a building are selected for the building change-detection step. When buildings are not correctly classified in the scene classification in any one of the epochs, it will influence the quality of building change detection. Errors in scene classification give rise to false positive and false negative results.

False positives occur when a change has occurred in the areas where non-building objects have been incorrectly classified as "buildings". In the yellow rectangles shown in Figure 14, there is an error in the scene classification of one of the epochs such that cars are incorrectly classified (in the top row) as a building roof; the water surface is also incorrectly classified as a building roof (in the bottom row). As there are changes in these areas, and one of them is incorrectly classified as part of a building, the changes are confirmed as a "change to a building". Although these represent real changes that have occurred, they are not changes to buildings. In fact, most false positives occur with trees, when a tree is incorrectly classified as a building roof in one of the epochs.

False negatives occur when a change is confirmed but this has not been classified as being part of a building in the data sets being compared: the change will not be signalled as a "change to a building". Some examples of false negatives resulting from scene classification errors are shown in Figure 15. If one of the data sets is correctly classified, false negatives can be avoided.

In addition, cranes, bridges, *etc.*, are sometimes classified as being part of a building. Nevertheless, we did not analyze changes in these objects because they did not fall within the scope of our research and we had not defined their features. Consequently, errors related to these objects have been ignored and not discussed in this paper.

## *Errors in Change Detection Due to Our Algorithm*

In addition to errors propagated from scene classification results, other errors arise that are caused by limitations of the change detection algorithm used. From visual inspection, we could conclude that no changes were missed (false negatives) due to our change detection method; some unchanged walls were, however, classified as changed (false positives). Figure 16a shows the incorrect classification of occluded walls as "changed". We assumed that if no change to the roof of a building had occurred, the attached walls would not change either, even if they had a large difference value. Under this assumption, all wall points in the left-hand image of Figure 16a should be labelled "unchanged". However, there are some projections—balconies, sun shades, *etc.*—on the walls that are far away from the roof, and these projections will be identified as "changed" because they have a large difference value and there was no unchanged roof found in their 2D neighborhood.

Figure 16b shows an example of what should be "unknown" points being incorrectly labelled as "unchanged". The lack of data for the roofs in one epoch was caused by a water layer that absorbed the laser signals. In the other epoch, the point distribution is rather regular. We expected that the entire area for which data are missing would be classified as "unknown". However, only the central part of the area was classed as "unknown" (light blue). The reason is because only the central parts of these areas in the merged data sets have a large difference value (>1.0 m radius). In other parts there are points from another epoch with a difference value less than 10 cm. Finally, these errors are not important because they did not influence the identification of the "changed" areas, they only resulted in a mix up between the designation "unchanged" and "unknown".

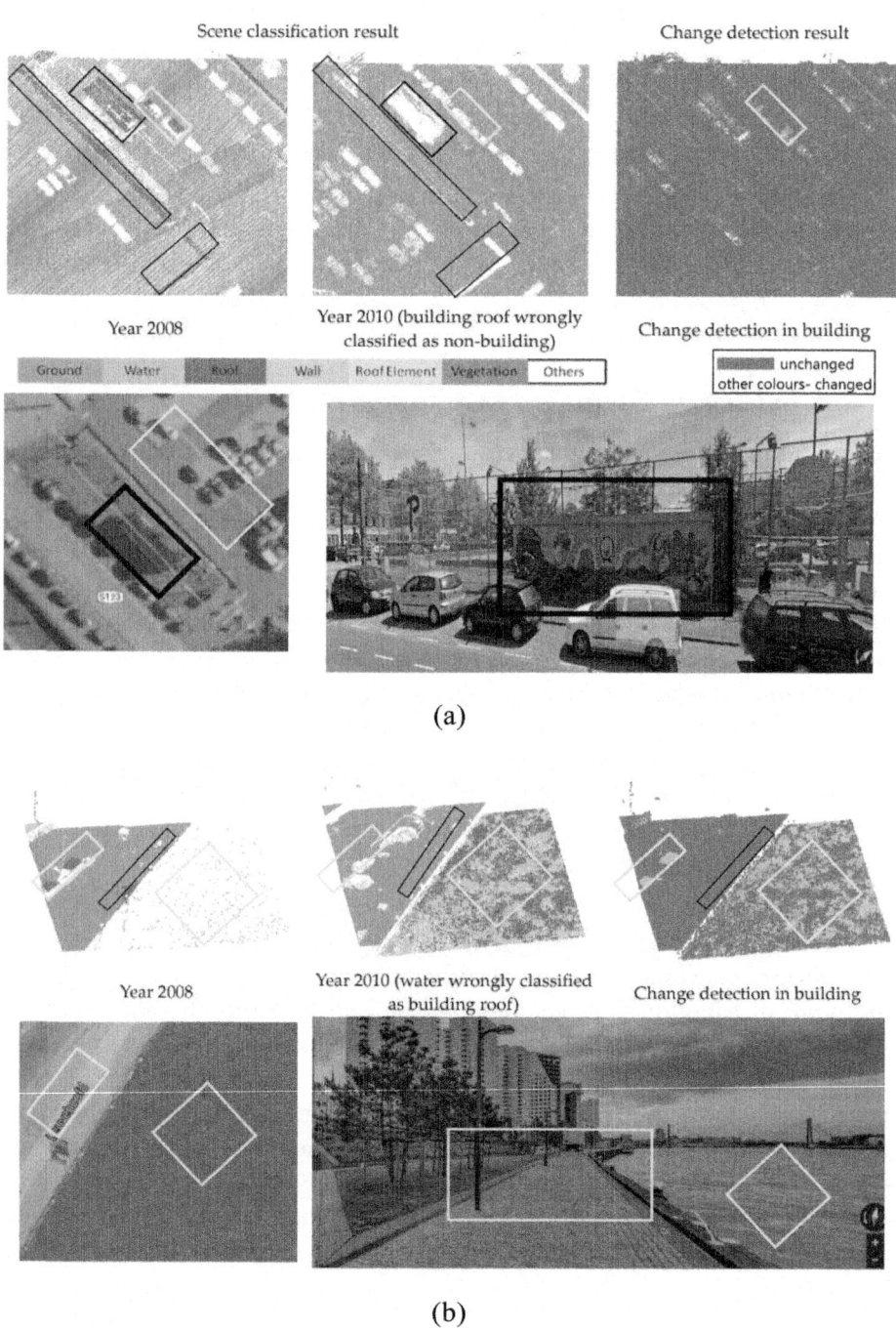

**Figure 14:** Scene classification errors that have no influence on the change detection

results (black rectangles), and ones that have a negative influence (yellow rectangles).

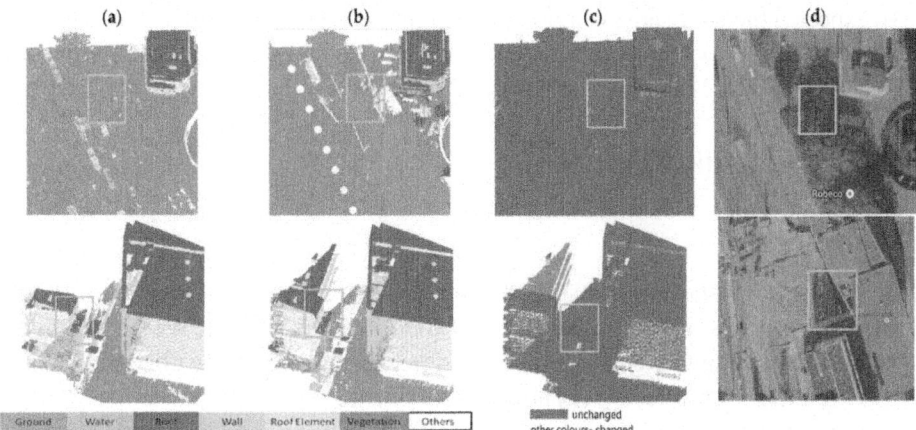

**Figure 15:** Examples of false negatives caused by scene classification errors. Because a building has been classified as an "undefined object" in the 2012 data set, changes to this building will not be detected (Column 3). **(a)** Data set 2010; **(b)** Data set 2012; **(c)** Change detection in building (merged); **(d)** Image from Google maps.

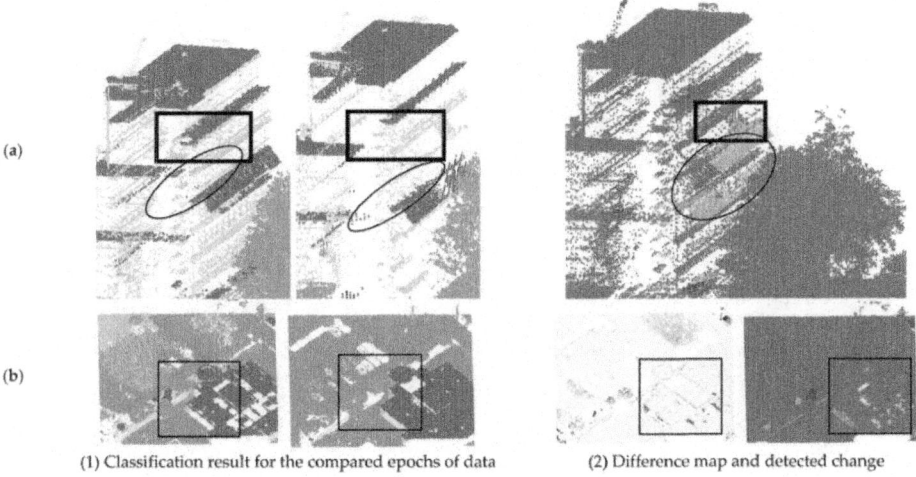

**Figure 16:** Errors due to our change detection algorithm. Colors in this figure represent the same objects as in Figure 15. In Figure 16, group (1a) shows some errors in the ellipse that are due to scene classification errors, and some in the box are due to lack of the contextual information (wall points far away from the roof). Group (1b) shows unknown points labelled as unchanged. As seen in the classification result of Group

(1b, left-hand), there is a lack of data (some holes) due to limited reflection (red) in the left-hand image of the roof, but the roof in the right-hand image is quite well covered. We expected the entire area where data is lacking to be labelled "unknown", but in the result in (2) only the central parts of the "gaps" are "unknown" (light blue), while other parts are labelled "unchanged" (brown).

## Change Classification

### *Results*

The change classification results for the two test areas are visualized in Figure 17; for examples with a higher level of detail see Figure 18. We chose several sites where changes were successfully detected and properly classified. These sites included examples of: (a) newly built dormers on roofs; (b) lack of data for roofs in one epoch (unknown) and lack of data for walls and ground because of occlusion; (c) undefined changes on roofs; (d) newly built constructions above roofs; (e) cars parked on top of buildings; (f) newly built and demolished buildings (two examples in one image); (g) add-on building constructions; and (h) insulation layers added to roofs.

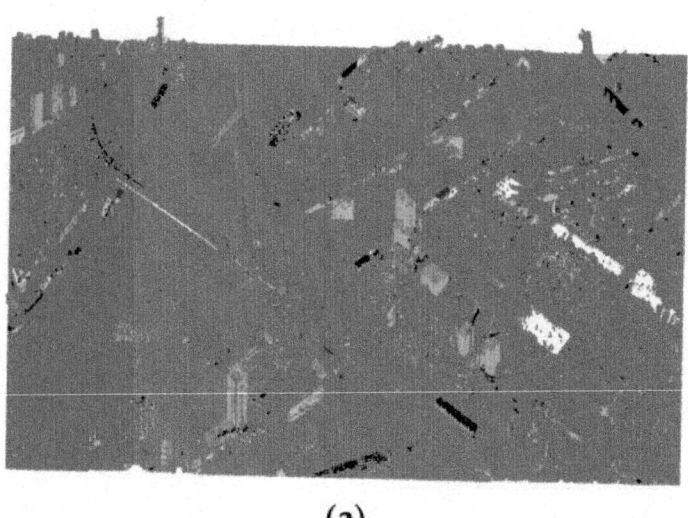

**(a)**

(b)

**Figure 17:** Classification of change in the test areas. (**a**) 2008 *vs.* 2010, (**b**) 2010 *vs.*2012.

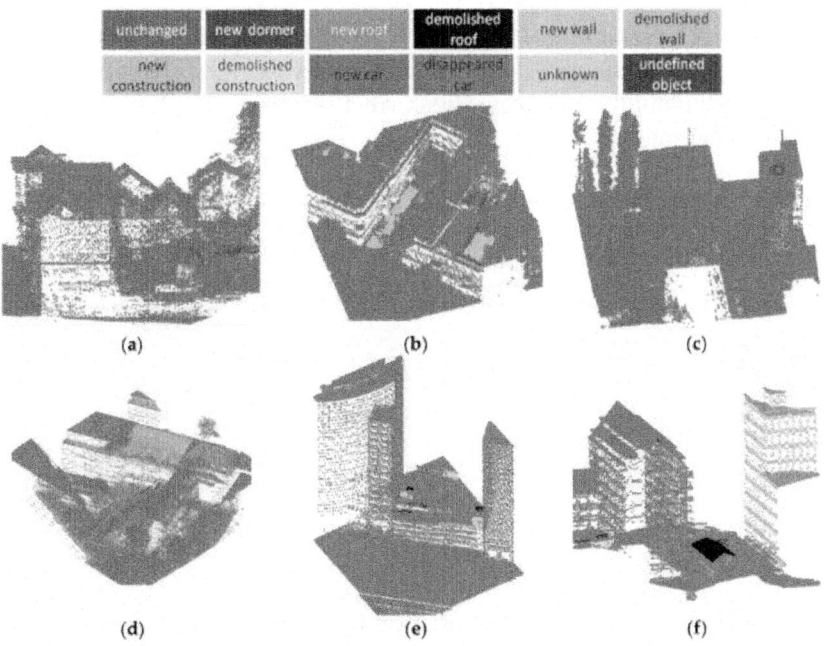

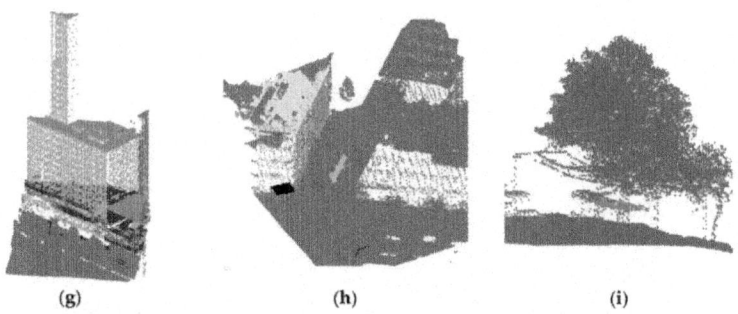

(g)                          (h)                          (i)

**Figure 18:** Successfully detected and classified examples of change. (a) newly built dormers on roofs, (b) lack of data for roofs in one epoch (unknown) and lack of data for walls and ground because of occlusion, (c) undefined changes on roofs, (d) newly built constructions above roofs, (e) cars parked on top of buildings, (f) newly built and demolished buildings (two examples in one image), (g) add-on building constructions, (h) insulation layers added to roofs. (i) new building roof under trees.

Some change detection errors are also caused by errors in scene classification—in addition to those arising from the limitations of our algorithm. These are discussed in Subsection 5.1.2. More false positives are observed, lowering the accuracy, which is analyzed in Section 5.3.

## Error Analysis

We randomly chose another 20 tiles with different changes from each study area to cover all types of changes on roofs. Error analysis of the change classification was done by manually assessing the correctness and completeness of changes detected (see Table 6).

**Table 6:** Completeness and correctness of the changed objects

| Label | 2008 vs. 2010 | | | 2010 vs. 2012 | | |
|---|---|---|---|---|---|---|
| | Correctness | Completeness | Number of Objects | Correctness | Completeness | Number of Objects |
| Undefined object | 91% | 35% | 29 | 100% | 26% | 23 |
| New dormer | 6% | 100% | 1 | 80% | 24% | 17 |
| Demolished dormer | - | - | 0 | - | - | 0 |
| New car on roof top | 33% | 11% | 9 | 50% | 7% | 14 |
| Car no longer on rooftop | 100% | 100% | 1 | - | - | 0 |
| New construction on rooftop | 13% | 25% | 8 | 42% | 56% | 9 |
| Demolished construction on rooftop | 100% | 40% | 5 | - | 0% | 2 |
| New roof | 78% | 81% | 26 | 59% | 100% | 17 |
| Demolished roof | 91% | 100% | 20 | 90% | 100% | 18 |
| New wall | 100% | 100% | 8 | 100% | 100% | 7 |
| Demolished wall | 100% | 100% | 8 | 100% | 100% | 4 |

The accuracy of change classification for roofs and walls is high compared to other objects because changes to roofs and walls occur most often in buildings undergoing entire change. In addition to that, scene classification is more reliable for larger objects: changes to buildings are generally large and can easily be correctly classified. Small changes are not so easily separated into their correct change classifications. Most constructions and undefined objects on rooftops are incorrectly classified as new or demolished dormers, especially if they are near a pitched roof. We have assumed that dormers are normally located near the pitched roof, but often roofs are in fact a combination of flat and pitched roofs, and there are as many changes on these roofs, which are not changed dormers but incorrectly classified as dormers. It is also difficult, based only on their size, to distinguish large cars on a rooftop parking lot from building constructions. As a result, some of these large cars are classified as constructions. We conclude that, even if the changes are detected correctly, in highly complex urban areas it is hard to completely and correctly classify the detected changes using geometrical and relational rules.

## Object-Based Analysis

### *Results*

Minimum 3D bounding boxes are generated around the connected components to enable selection of relevant changes in the buildings; for some results see Figure 19. The location of the center point, the area, and the volume of the minimum 3D bounding box are calculated. Changes are shown in Figure 19a. Figure 19b,c give two examples of 3D bounding boxes: Figure 19b shows a building object that has undergone complete change; and Figure 19c shows a changed building element (newly built dormer). The bounding box in Figure 19c is larger than the dormer because of the influence of some outliers occurring in the same plane as the dormer.

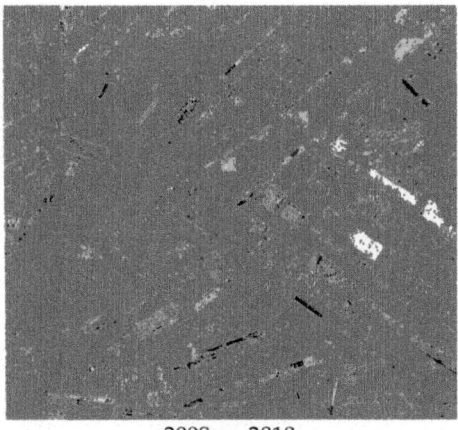
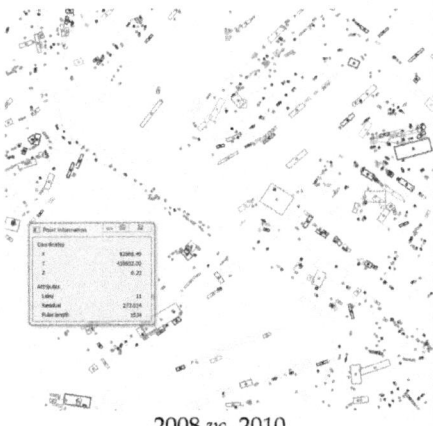

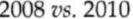
2008 *vs.* 2010                    2008 *vs.* 2010

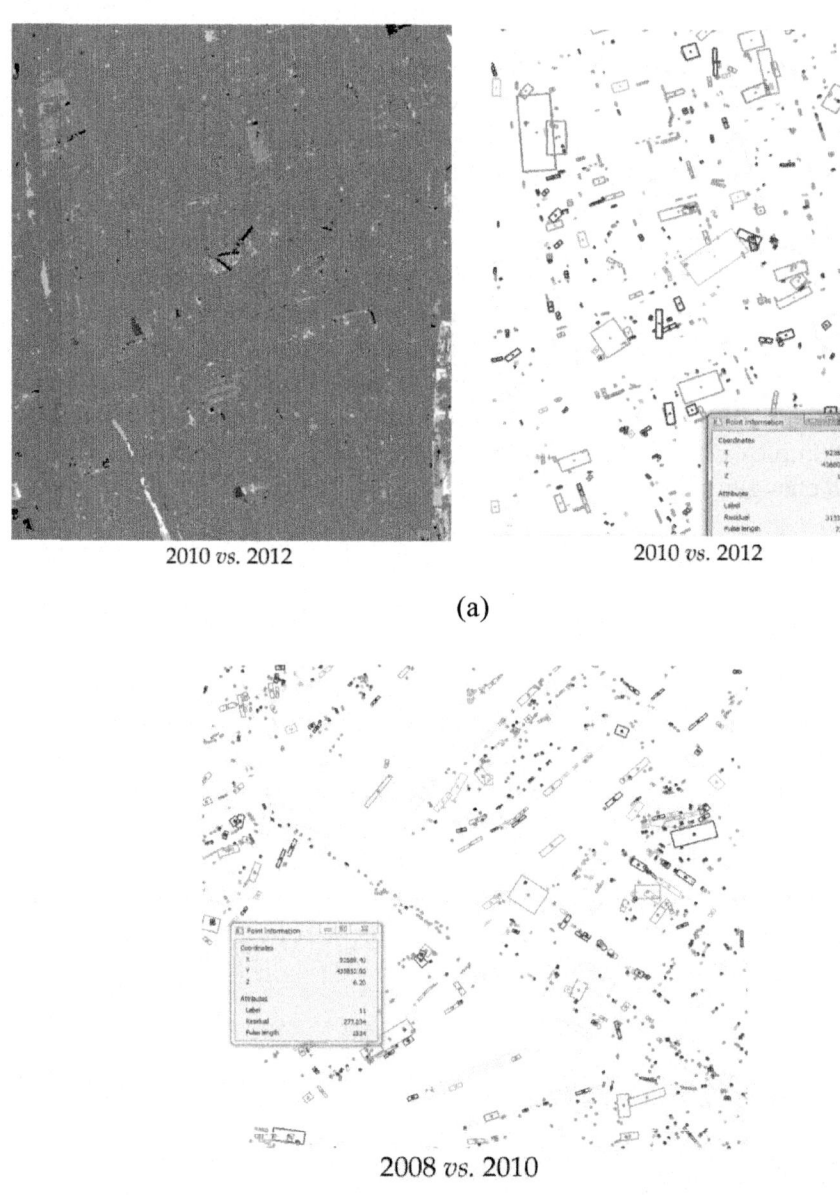

2010 *vs.* 2012                2010 *vs.* 2012

(a)

2008 *vs.* 2010

(b)

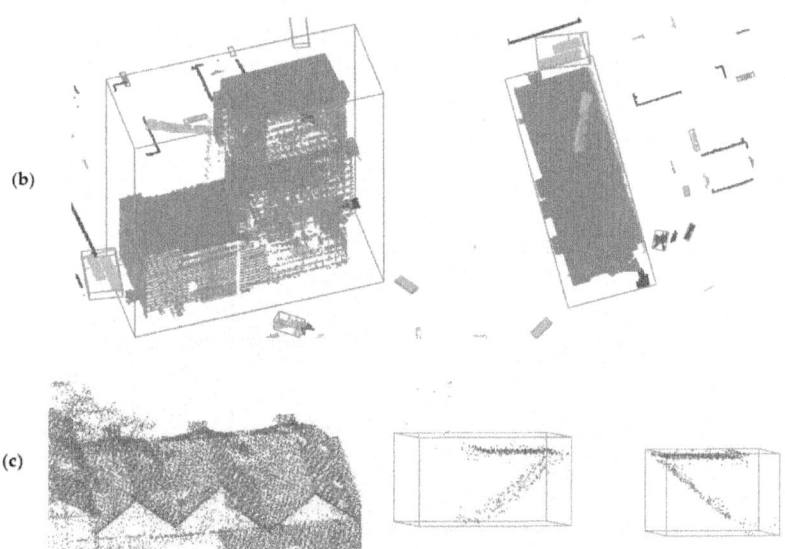

**Figure 19.** 3D bounding boxes and the center points of the relevant changes, together with the calculated area and volume; the label, area and the volume are shown in the point number window. In this figure, points without bounding boxes are irrelevant changes. (**a**) 3D bounding boxes for the two test areas; (**b**) an example of 3D bounding box for an entire new building; (**c**) an example of 3D bounding box for newly built dormers.

## Error Analysis

We used connected component labelling to form objects and their 3D bounding boxes to estimate their area and volumes. We found that connected component labelling failed to correctly form objects under two circumstances:

### (1) False positives in buildings

Sometimes points on walls or roofs are false positives. These points may belong to an unchanged wall that has been incorrectly detected as changed (see discussion in Subsection 5.1.4), or they can be sparse points of plants on a balcony or rooftop (irrelevant changes). If such points are close together, connected component labelling will form an object that is as big as an entire wall or even a complete building; see Figure 20a.

### (2) Small objects that are too close to each other

Small objects that are too close to each other will be connected together as one object. This often occurs with cars parked on rooftops of buildings and shelters at bus stops that are very close to each other. Figure 20b shows an example of cars parked on a rooftop. The consequence is an incorrect shape of detected change.

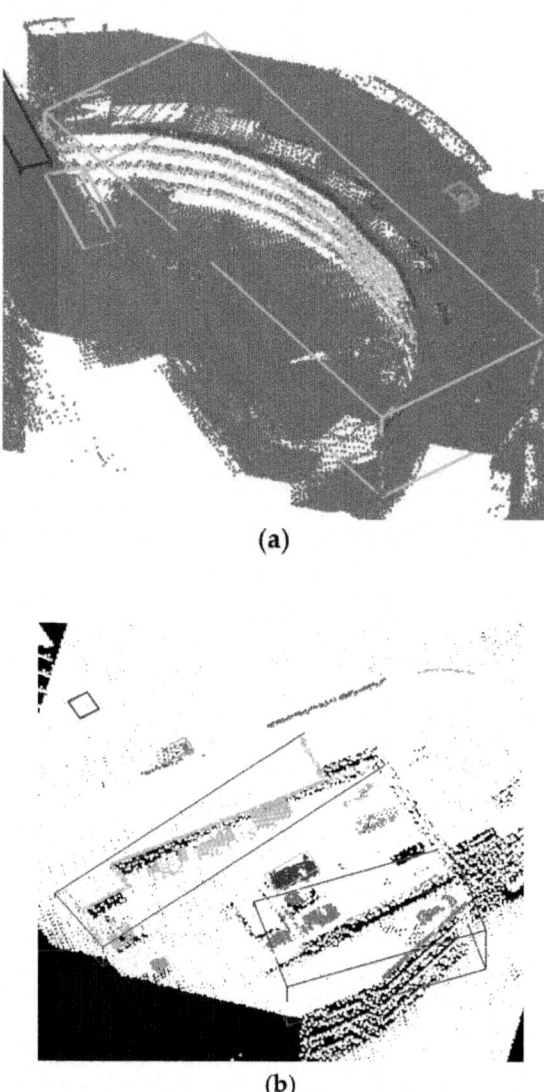

(a)

(b)

**Figure 20:** 3D bounding boxes of some wrong changes and irrelevant changes. (a) Incorrect changes in walls and irrelevant changes on roofs; (b) Cars that are close to each other and close to some fences are grouped to large objects. Yellow cubes are

show objects classified as new buildings. Blue cubes are new unknown objects. Green cubes are new constructions on top of the roof.

# CONCLUSIONS

In this paper, we present a method for detecting and classifying changes to buildings by using classified and well registered (strip difference <10 cm) laser data from several epochs. The analysis of our results leads us to draw several conclusions.

Provided distances between surfaces are greater than 10 cm and the area of change more than 4 $m^2$, both large and small changes can be automatically detected using a surface difference map and a rule-based change detection algorithm. Areas classified as "unknown" can be correctly identified in cases of occlusions and water reflection. The surface difference map is not affected by the point density of the compared data sets. Some of the other parameters, such as the number of points indicating the approximate area and radius for the surface growing method, should be adjusted with varying point densities.

Larger changes can be correctly assessed as belonging to a building provided that building has been correctly classified as such in the scene classification step for one of the data sets being compared. The accuracy of object recognition was evaluated by overlaying the 3D bounding boxes of the buildings for which change was detected on manually generated reference data. Our method detected 91% of actual changes in Test area 1 and 83% in Test area 2. Nearly half of the changes detected in objects were irrelevant changes.

About half of the false positives that occurred were caused by scene classification errors. The other false positives can be mostly attributed to our change detection algorithm. Mostly they are spurious changes in a wall that is at a larger distance to a roof, plants growing in a rooftop garden, containers, isolation layers, cleaning machines on the top of buildings, *etc.* Regarding classifying changes, we conclude that ,even if changes are identified correctly, in highly complex urban areas it is difficult to completely and accurately classify the smaller detected changes using geometrical and relational rules.

The classification result for the changed object shows that large changes affecting an entire building, for example its roof or a main wall, can be detected with a higher degree of accuracy than changes made to a building element, such as dormers or construction on top of the building.

Overall, our method detected 80-90% of changes to buildings. It is the first time that we tried to detect high-level detail changes in buildings using raw lidar data, so we did not compare our accuracy to other methods. Our method is, however, not yet capable of distinguishing small irrelevant changes to objects from relevant ones. A more accurate definition of a "changed building object" and its characteristics are required in order to better interpret differences between two data sets.

## ACKNOWLEDGMENTS

The authors would like to thank the Municipality of Rotterdam for providing the Lidar data sets.

## AUTHOR CONTRIBUTIONS

Sudan Xu designed the algorithms as described in this paper during her PhD research, and was responsible for the main organization and writing of the paper. Sander Oude Elberink and George Vosselman supervised Sudan during her PhD project contributed to paper writing.

## REFERENCES

1.    Xiao, W.; Vallet, B.; Paparoditis, N. Change detection in 3D point clouds acquired by mobile mapping system.*ISPRS Ann. Photogramm. Remote Sens. Spat. Inf. Sci.* 2013, *2*, 331–336.

2.    Xu, S.; Oude Elberink, S.; Vosselman, G. Multiple-entity based classification of airborne laser scanning data in urban areas. *ISPRS J. Photogramm. Remote Sens.* 2013, *88*, 1–15.

3.    Mas, J.-F. Monitoring land-cover changes: A comparison of change detection techniques. *Int. J. Remote Sens.*1999, *20*, 139–152.

4.    Byrne, G.F.; Crapper, P.F.; Mayo, K.K. Monitoring land-Cover change by principle component analysis of multitemporal landsat data. *Remote Sens. Environ.* 1980, *10*, 175–184.

5.    Sharma, B.; Rishabh, I.; Rakshit, S. Unsupervised change detection using RANSAC. In Proceedings of the IET International Conference on Visual Information Engineering, Bangalore, India, 26–28 September 2006; pp. 24–28.

6.    Yang, Z.; Qin, Q.; Zhang, Q. Change Detection in high spatial resolution images based on support vector machine. In Proceedings of the IEEE International Geoscience and Remote Sensing Symposium, Denver, CO, USA, 31 July–4 August 2006; pp. 225–228.

7.    Di, F.; Li, X.; Zhu, C. A new method in change detection of remote sensing image. In Proceedings of the 2nd International Congress IEEEExplore Digital Library, Image and Signal Processing (CISP' 09), Tianjin, China, 17–19 October 2009; pp. 1–4.

8.   Tanathong, S.; Rudahl, K.T.; Goldin, S.E. Object oriented change detection of buildings after a disaster. In Proceedings of the ASPRS 2009 Annual Conference, Baltimore, MD, USA, 9–13 March 2009. On CD-Rom.

9.   Kasetkasem, T.; Varshney, P.K. An Image Change detection algorithm based on Markov Random Field models.*IEEE Trans. Geosci. Remote Sens.* 2002, *40*, 1815–1823.

10.  Murakami, H.; Nakagawa, K.; Hasegawa, H.; Shibata, T.; Iwanami, E. Change detection of buildings using an airborne laser scanner. *ISPRS J. Photogramm. Remote Sens.* 1999, *54*, 148–152.

11.  Khoshelham, K.; Oude Elberink, S.J.; Xu, S. Segment based classification of damaged building roofs in aerial laser scanning data. *IEEE Geosci. Remote Sens. Lett.* 2013, *10*, 1258–1262.

12.  Gerke, M. Dense matching in high resolution oblique airborne images. *Int. Arch. Photogramm. Remote Sens. Spat. Inf. Sci.* 2009, *38*, 77–82.

13.  Vosselman, G.; Gorte, B.G.H.; Sithole, G. Change detection for updating medium scale maps using laser altimetry. *Int. Arch. Photogramm. Remote Sens. Spat. Inf. Sci.* 2004, *34*, 207–212.

14.  Vögtle, T.; Steinle, E. Detection and recognition of changes in building geometry derived from multi-temporal laser scanning data. *Int. Arch. Photogramm. Remote Sens. Spat. Inf. Sci.* 2004, *34*, 428–433.

15.  Rutzinger, M.; Ruf, B.; Hofle, B.; Vetter, M. Change detection of building footprints from airborne laser scanning acquired in short time intervals. *Int. Arch. Photogramm. Remote Sens.* 2010, *38*, 475–480.

16.  Choi, K.; Lee, I.; Kim, S. A feature based approach to automatic change detection from Lidar data in urban areas. *Int. Arch. Photogramm. Remote Sens. Spat. Inf. Sci.* 2009, *38*, 259–264.

17.  Matikainen, L.; Hyyppä, J.; Kaartinen, H. Automatic detection of changes from laser scanner and aerial image data for updating building maps. *Int. Arch. Photogramm. Remote Sens. Spat. Inf. Sci.* 2004, *35*, 434–439.

18.  Rottensteiner, F. Building change detection from digital surface models and multi-spectral images. *Int. Arch. Photogramm. Remote Sens. Spat. Inf. Sci.* 2007, *36*, 145–150.

19.  Champion, N.; Rottensteiner, F.; Matikainen, L.; Liang, X.; Hyyppä, J.; Olsen, B.P. A test of automatic building change detection approaches. *Int. Arch. Photogramm. Remote Sens. Spat. Inf. Sci.* 2009, *38*, 145–150.

20.  Chen, L.; Lin, L.; Cheng, H.; Lee, S. Change detection of building models from aerial images and Lidar data.*Int. Arch. Photogramm. Remote Sens.*

*Spat. Inf. Sci.* 2010, *38*, 121–126.

21.  Awrangjeb, M.; Fraser, C.S.; Lu, G. Building change detection from lidar point cloud data based on connected component analysis. *ISPRS Ann. Photogramm. Remote Sens.* 2015, *2*, 393–400.

22.  Hebel, M.; Arens, M.; Stilla, U. Change detection in urban areas by object-based analysis and on-the-fly comparison of multi-view ALS data. *ISPRS J. Photogramm. Remote Sens.* 2013, *86*, 52–64.

23.  Vosselman, G. Automated planimetric quality control in high accuracy airborne laser scanning surveys. *ISPRS J. Photogramm. Remote Sens.* 2012, *74*, 90–100.

24.  Vosselman, G.; Gorte, B.G.H.; Sithole, G.; Rabani, T. Recognising structure in laser scanner point clouds. *ISPRS Arch. Photogramm. Remote Sens. Spat. Inf. Sci.* 2004, *40*, 33–38.

25.  Vosselman, G. Point cloud segmentation for urban scene classification. *ISPRS Arch. Photogramm. Remote Sens. Spat. Inf. Sci.* 2013, *40*, 257–262.

# Chapter 10

## EARTH SHELTERS; A REVIEW OF ENERGY CONSERVATION PROPERTIES IN EARTH SHELTERED HOUSING

Akubue Jideofor Anselm[1]
[1]Architecture Department, University of Nigeria, Nigeria

## INTRODUCTION

Earth sheltering is an age long traditional practice. In modern times its benefits has prompted new definitions for its practice. With the potential thermal conservation qualities and physical characteristics of earth as a building mass, earth shelters can now be defined as structures built with the use of earth mass against building walls as external thermal mass, which reduces heat loss and maintains a steady indoor air temperature throughout the seasons. The popularity of earth sheltering was advanced mostly by research in energy conservation in residential housing. Originally conceived as dwellings developed by the utilization of caves within the traditional context, its evolution through technologies led to the construction of customized earth dwellings all across the globe. These structures in the past were built by people not schooled in any kind of formal architectural design or with identifiable building techniques rather they depended on the cover the very structure of the earth could provide them for purposes of shelter, warmth and security. Investigations into the traditional earth sheltered dwellings also identified sunken earth houses with characteristics that suggested potentials in passive building insulation which utilizes ground thermal inertia.

In the view of some researchers on earth supported housing, building underground provides energy savings by reducing the yearly heating and cooling loads in comparison with known conventional structures. Not only is the temperature difference between the exterior and interior reduced, but mostly because the building is also protected from the direct solar radiation [1]. One significant value of earth-sheltered housing and the reason for its evaluation is its potential energy savings when compared to conventional aboveground housing. This potential is based on several unique physical characteristics. The first of these characteristics is in the reduction of heat loss due to conduction through the building envelope because of the high density of the earth. According to [2], in an earth sheltered building even at very shallow depths and given normal environmental conditions, the ground temperatures seldom reaches the outdoor air temperatures in the heat of a normal summer day. This condition allows the conducting of less heat into the house due to the reduced temperature differential.

In the case of colder climates, it was noticed that during winters the rate of heat loss in bermed (earth supported) structure was less in comparison to that in on-grade structures. This indicates through results that the floor surface temperature increased by 3° C for a 2.0m deep bermed structure due to lower heat transfer from the building components to the ground, thus suggesting the presence of passive heat supply from the ground even at the extreme cold temperatures of winter [3]. This evidently contributes as a factor for energy saving in earth shelter buildings in cold climates.

Other characteristics include the reduction of air infiltration within the dwelling which is mainly surrounded by earth walls with very little surface area exposed to the outside air. These characteristics have been investigated in previous studies and the analysis on each location provides results and findings in terms of climatic effects, design styles and residential activities of the dwellers that bring about the unique energy saving value of these buildings.

Single unit earth sheltered houses are unique energy conservation ideas based on their earth contact characteristics as mentioned above. In order to achieve the maximum benefits from earth sheltered housing, its application could be examined also at an entirely community scale rather than simply at the scale of individual houses. One of the biggest challenges to the overall performance of earth sheltered housing would be the built conventional surroundings. While contemporary use of earth sheltering is confined to individual homes built on single plots of land or a small cluster of houses which will absolutely be affected by the surrounding conventional structures around, the traditional use encompassed entire communal design or villages that will stay within the same conditions the micro-environment provides. This communal development option is identified to be most effective as isolated pockets of earth sheltered houses do not really reach the scale needed for sustainable development [4]. Earth sheltered mass-housing may thence become the general concept for design and building with earth whereby entire communities are created, enjoying dual land use by locating all housing underground [5]. If a single case of earth sheltering is found to have significant advantages, these advantages can only increase in magnitude if applied to whole communities.

# FUNDAMENTALS OF EARTH SHELTERED HOUSING

The values of energy conservation in earth shelters are dependent on certain principles. These principles which form the ground rules for the design and construction of earth sheltered dwellings have been existent since prehistoric periods. Earth sheltered homes were primarily developed for shelter, warmth and security for the earliest human dwellers. Most of the recorded cases of these shelters are found extensively in areas like Asia and Northern Africa. In one of the earliest cases in Japan was discovered the oldest human habitation in a layer of earth about 600,000 years old in Kamitakamori, Miyagi Prefecture. Archaeologists from the Tohoku Paleolithic Institute, Tohoku Fukushi University and other institutes believed that the finding may be one of the oldest in the world. There are only a few remains of human dwelling structures from the early Paleolithic period in the world, as early humans such as the Peking-man lived in caves. Researchers believed the dwellings were built by primitive man who appeared some 1.6 million years ago and likely reached Japan 600,000 years ago at the latest, according to the archaeologists. The buildings could have been used as a place to rest, a lookout for hunting, a place to store hunting tools or to conduct religious rites.

In Tunisia, residents of Matmata were discovered to have lived in manmade caves for centuries (Figure 1). Here rooms were carved into the soft rock to create atrium houses that had several excavated rooms with up to 4 to 10 meter high and vaulted ceilings opening out onto a single sunken courtyard. The original objective for going below the ground in this case was to protect the inhabitants from the extremes of daytime North African heat and nighttime cold, typical of this desert region.

**Figure 1:** Aerial view of a typical Matmata earth shelter dwelling. Image by Tore Kjeilen

However through the years, more modern earth sheltered dwellings were revealed as studies on the earliest forms of human settlements progressed. In China, modern earth shelters habitats were discovered with histories that dated back to before 2000 B.C. This type of habitats were commonly called cave dwellings as they were strictly home units hewed out of the mountains. It is believed that underground housing preceded above ground housing in this area. Studies on these existing Chinese earth habitats presented analytical data on the climatic and topographical relationships to the unique design elements utilized to attain living comfort by the cave shelter dwellers. Such analysis as the rain, wind, sun and seasonal weather conditions that exist in these areas where these dwellings were located possibly necessitated the advantage of its existence in these locations [6]. Analysis on each location also provided results and findings in terms of climatic effects, design styles and residential activities of the dwellers. In the North-west of China, variety of these structures evolved, ranging from the cave dwelling units to the more advanced subterranean types. In the case of the traditional subterranean homes in China (called '*yao dong*'), rooms were dug into loose, silty soil to primarily combat the hot summers and bitterly cold winters. In the early 20th century the provinces of Shanxi, Jiansu and Henan still had traditional dwellers that faced with the need to preserve agricultural land and housing for their people, dug entire cities beneath their lands. Today, it is still believed that more than 10 million Chinese live underground, perhaps the largest number of troglodytes ever to inhabit a single region. The Shanxi homes (Figure 2, 3 and 4) were buried at depths of up to 10 meters with their underground homes built around courtyards. This atrium-style design offer ample sunlight as well as surface spaces for other activities.

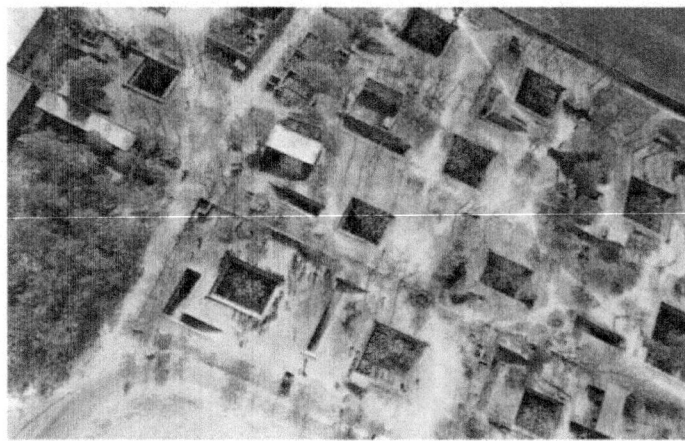

**Figure 2:** Aerial view of an earth shelter neighborhood in *Lian Jiazhuang*, Shanxi Province, North-western China

Research conducted in [6] also provided analytical data on climatic and topographical relationships to the structural design styles with single unit design solution, multi unit designs and finally urban planning initiatives on how to achieve a sunken city that exists beneath rather than above ground level as seen in Figure 2 below. Also fascinating in discovery included methods and techniques of ventilating the building units naturally. Such natural ventilation techniques are viewed today as ideas that advanced the notion of passive aeration of interiors which ultimately is a cost and energy efficient alternative to the whole process of earth sheltered housing.

(a)

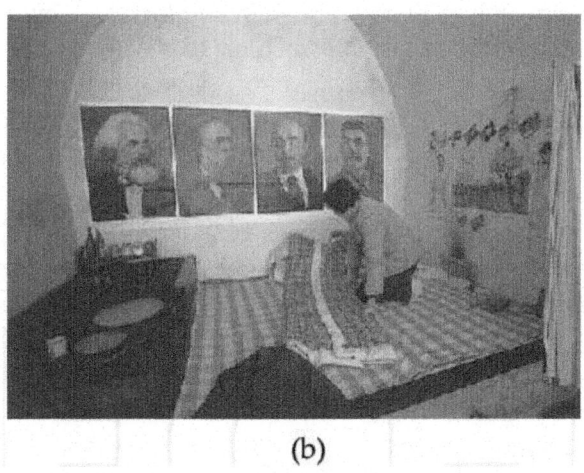

(b)

**Figure 3:** a) Courtyard view of an Atrium type subterranean earth shelter dwelling in *Lian Jiazhuang*, Shanxi Province. (b) Interior view of a typical room space. Image by Kevin Poh.

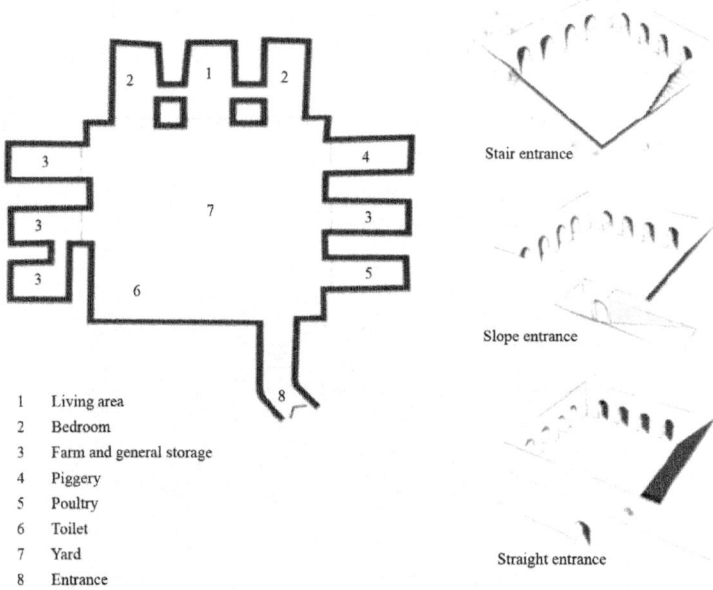

1   Living area
2   Bedroom
3   Farm and general storage
4   Piggery
5   Poultry
6   Toilet
7   Yard
8   Entrance

**Figure 4:** A typical earth shelter home layout in North-western China

With the challenges of global warming and fossil energy reduction, energy saving ideas has become an essential element in building designs and occupation. Since energy conservation is the practice of saving energy use without compromising occupant thermal comfort [7], building below the ground thence presents certain fundamentals that with the aid of research can significantly influence energy conservation efforts in modern housing. From reviews of the basic background of traditional earth sheltered housing, the fundamental objectives for building below the ground and significant energy conservation principles are listed as follows:

1.   Indoor temperature enhancement based on the natural principles of annual heat storage (PAHS) whereby the earth collects free solar heat all summer and cools passively while heating the earth around it, and keeping warm in winter by retrieving the stored heat from the soil in winters. This dual function presents a scenario that makes the practice of earth sheltered housing effective in both hot and cold climates.

2.   Huge temperature differential between the ground temperatures and the outdoor air temperatures. In this case the normal ground temperature seldom reaches the outdoor air temperatures in the heat of a normal hot day, thereby conducting less heat into the house due to the reduced temperature difference.

3.   Building protection from the direct solar radiation, thereby elimination the challenge of direct thermal load due to heat radiation through the building envelope.

Apart from the energy values which the subsurface climate of the earth provides, the other significant characters beneficial to earth shelters includes the major goal of recycling surface space by relocating functions to underground, by this earth shelters liberates valuable surface space for other functional uses and improves ground surface visual environment, open surfaces for landscaping and thus a more greener atmosphere.

# MODERN CONSTRUCTION TECHNIQUES AND DESIGN TYPOLOGY

The structural make up of a typical earth shelter house is made up of the supporting members and the compacted backfills in which case strength and composition can determine the ability to withstand overhead loads of moisture, dead and live loads, the distribution of which depend on the compaction strength of the backfill or supports.

However in modern designs, the supports are the parts of the house that brace against the side walls of soil and overlaying roof members that are made of backfills as in the case of underground homes. The design method and material choice will determine the resistance to failure of these structural members. In the traditional construction scenario where the earth-soil is used as building material; its strength is determined by the soil stability, which goes to improve the resistance to wind and in most cases rain erosion.

## Earth Shelter Structural Integrity

The structural make up of earth homes is mainly made up of the supporting members and the compacted backfills. As earlier mentioned, the strength and composition of the material used as backfill can determine the ability to withstand overhead loads. The supports are the parts of the house that brace against the side walls of soil and overlaying roof members that are made of backfills. The building design method and material choice will determine the resistance to failure of these structural members. In the case where the earth-soil is used as building material, its strength is determined by the soil stability, which goes to improved the resistance to wind and rain erosion. In most earth shelter construction the significant structural areas are the soil, walls and roof area. Apart from serving as a building material, the soil-walls of the shelter trench are regarded as the most valuable structural member of the Earth house structure. It provides the necessary support a normal wall gives in an ordinary house design. Nevertheless, not all soil types are efficient in use for earth sheltered house construction. From studies it is identified that the best soils are granular, such as sand and gravel. These soils compact well for bearing the weight of the construction materials and are very permeable, which means they allow water to drain quickly. The poorest soils are cohesive, like clay, which may expand when wet and has poor permeability. Soil tests, offered through professional testing services, can determine load-bearing capability of soils and possible settlements that may occur after construction. Study in [6] revealed certain traditional considerations for deciding the depth, thickness of mass and curvature of the support ceilings (vault) of the Chinese earth homes which can also be applied in modern day construction of earth shelters (figure 5).

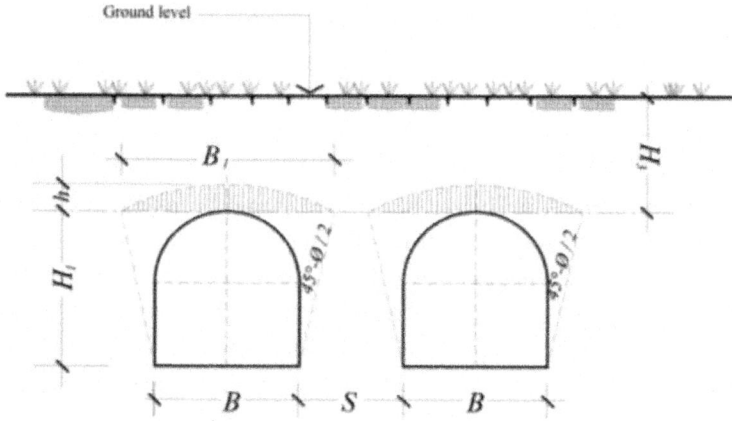

**Figure 5:** Structural consideration for a typical room space excavation in the Shanxi traditional earth shelters

$h = 1 \sim 2,\ \emptyset = 18°$.

$\frac{1}{2} B_1 = \frac{1}{2} B + H_1\, t\, g\, (45° - \emptyset/2)$

$= 3.5/2 + 3\, t\, g\, (45° - 18°/2)$

$\frac{1}{2} B_1 = 1.75 + 2.19 = 3.94\ m$

Then S = Thickness of Earth thermal mass wall

$H_3$ = Extent of depth clearance

Assuming B (room span) = 3.5m and $H_1$ (room height) = 3m

$H_3$ = Depth from ground surface to ceiling. This should be greater than $h$

The Dotted/shaded area indicates possible fault lines due to the pressure from the overlaying earth mass

Varieties of techniques have been used in the past for earth shelter wall construction. The construction materials for the walls of each type of structure will vary, depending on characteristics of the site, climate, soils, and design. However, general guidelines show that houses more deeply buried require stronger, more durable structural walls. Walls must provide a good surface for waterproofing and insulation to withstand the pressure and moisture of the surrounding ground. When soil is wet or frozen, the pressure on the walls and floors increases as pressure also increases with depth.

For the traditional earth supported homes built in the Chinese and Arid (dessert) climatic regions, there usually is no use for supporting walls as the naturally compressed soil structure already serves the function. However through recent research on improving the state of earth homes for most other climatic regions, the walls of Earth homes can be made of various materials ranging from Compressed Earth bricks to Concrete, while providing cavities and drainage patterns to aid damp proofing. In most earth home designs, the roof is usually the most challenging part of the entire structure. With recent ideas in ecology, the roof of earth shelters assume interesting landscaping functions. Especially for earth supported shelters which already posses the natural materials of earthen walls and members, the roof can also be finished to assume a natural finish too. Since the basic idea of this study is to discover techniques to achieve high performance as possible, the basic structural form for constructing the earth shelter roof is as follows:

1.   A frame strong enough to support the dead load brought by the soil overlay, rain, snow and ice loads where applicable.

2.   A solid deck built over the frame and a waterproof membrane installed on the deck prior to final earth cover.

3.   Treated soil backfill placed on the membrane (as the roof layer) and covered with a fine thick layer of soil. The roof will either grow a vegetation of its own or become a life garden depending on the appropriate type of maintenance.

Reinforced concrete is the most commonly used structural material in earth shelter construction. Products like Grancrete and Hycrete are becoming more readily available. They claim to be environmentally friendly and either reduce or eliminate the need for additional waterproofing. However, these are new products and have not been extensively used in earth shelter construction.

Some other unconventional approaches are also utilized in earth shelter construction. These techniques utilize recycled material of various forms and applications. One of such approaches is referred to as an Earth ship (figure 6). These houses are built to be self-contained and independent; their design allows occupants to grow food inside and to maintain their own water and solar electrical systems [8]. Some builders believe they have proven the design's ability to tap into the constant temperature of the earth and store additional energy from the sun in winter. These Earth ships carry out their environmentally conscious theme by employing unusual building materials in the form of recycled automobile tires filled with compacted earth for thermal mass and structure. While the tires form the major structural frames for the building, aluminum or tin cans are used for filling minor walls that are not load-bearing. Foam insulation can be applied to exposed exterior or interior walls and covered with stucco. Interior walls are also dry-walled giving it a conventional look.

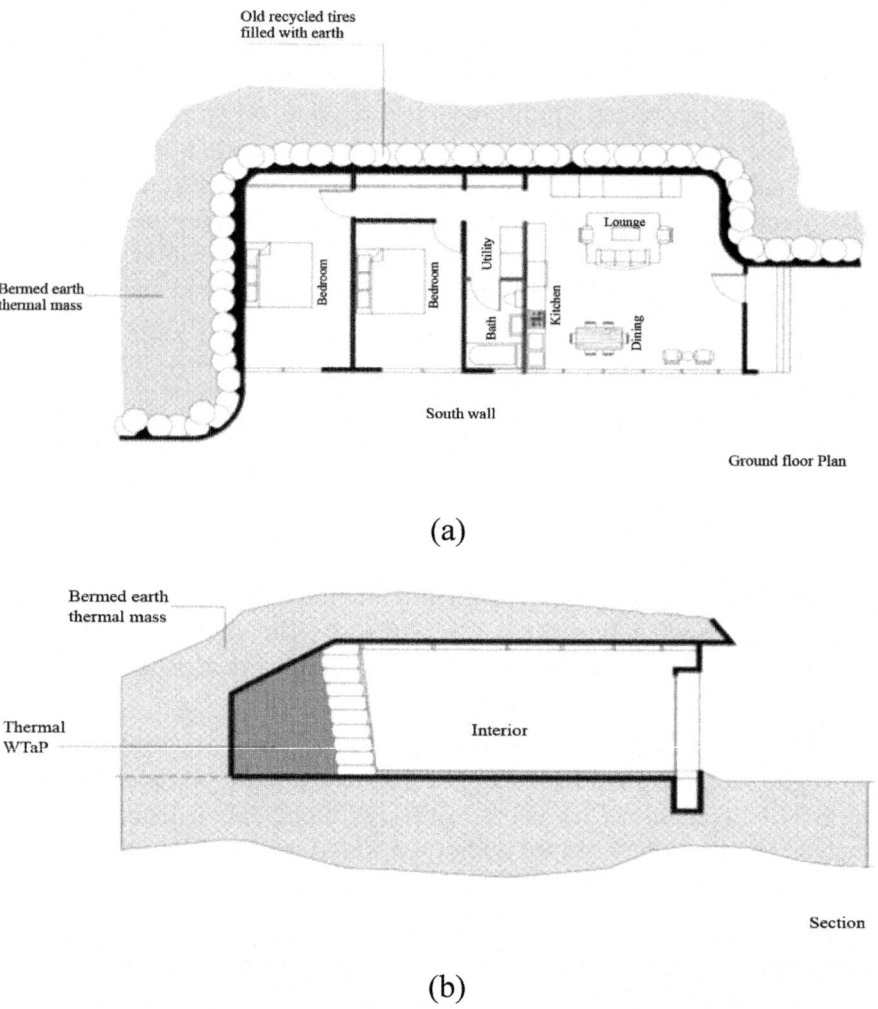

Figure 6: An Earth Ship design, using recycled materials

## Earth Shelter Construction Typology

Earth sheltered houses are often constructed with energy conservation and savings in mind. Though techniques of earth shelter construction have not yet become common knowledge, study into the most efficient application of the earth shelter principles reveals classifications of the major typologies that are utilized in the construction of earth houses. These major construction concepts are the Bermed or banked with earth type and the Envelope or True underground type. The energy conservation values of these typologies also vary depending on climate and physical challenges indigenous to each typology (table 1).

### *Bermed earth shelter*

In this type of construction, earth is piled up against exterior walls and heaped to incline downwards away from the house. The roof may, or may not be, fully earth covered, and windows/openings may occur on one or more sides of the shelter. Due to the building being above ground, fewer moisture problems are associated with earth berming in comparison to the fully underground construction. Other variations of bermed construction are the elevational and in-hill construction (figure 7). This type of construction is particularly appropriate for colder climates. With regards to energy efficiency in colder climates, all the living spaces may be arranged on the side of the house facing the equator. This provides maximum solar radiation to the most frequently used spaces like bedrooms, living rooms, and kitchen spaces [9]. Rooms that do not require natural daylight and extensive heating such as the bathroom, storage and utility rooms are typically located on the opposite in-hill side of the shelter. The compact configuration of this construction provides it with a greater ratio of earth cover to exposed wall thereby improving its energy performance benefits through the earth-contact principles.

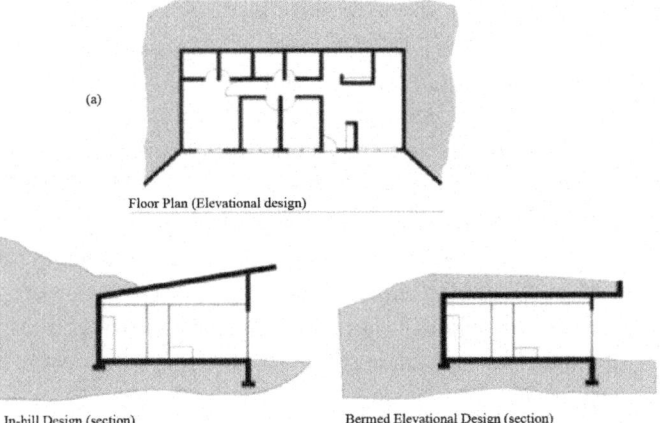

(a)

Floor Plan (Elevational design)

In-hill Design (section)                    Bermed Elevational Design (section)

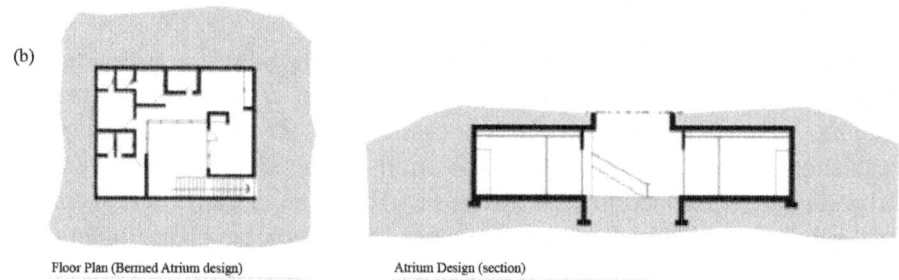

Floor Plan (Bermed Atrium design)          Atrium Design (section)

**Figure 7**: a) Elevational (beremed) and in-hill designs, (b) Atrium (bermed) design

However the case for both climates, the three major determinants for the building orientation remains the sun, wind and outside views. Proper orientation with respect to solar path and wind is significant for energy savings.

## Envelope or True underground earth shelter

In the true underground construction, the house is built completely below ground on a flat site, with the major living spaces surrounding a central outdoor courtyard or atrium. The windows and glass doors that are on the exposed walls facing the atrium provide light, solar heat, outside views, and access via a stairway from the ground level. The atrium effect offers the potential for natural ventilation. In the view of some researchers, this concept reduces the energy conservation properties in colder climates mostly due to the reduced solar exposure within the courtyard or atrium opening [9]. However recent studies in the area of soil temperature analysis with respect to energy conservation in earth shelters, provides information on the prospect of efficient underground earth shelter design. Such studies as in [10], provides mathematical method for predicting the long-term annual pattern of soil temperature variations as a function of depth and time for different soils and soil properties that are stable over time and depth. The likes of these studies were utilized by John Hait's [11] in his book on Passive Annual Heat Storage (PAHS) to advance the ideas of earth shelter housing. With the development of modern passive solar building design, during the 1970s and 1980s a number of techniques are developed to enabled thermally and moisture-protected soil to be used as an effective seasonal storage medium for space heating, with direct conduction as the heat return method. Other variations of the true underground typology are the Atrium/courtyard concept and the Penetrational type where earth covers the entire house, except where it is retained for windows and doors for cross-ventilation opportunities and access to natural light from more than one side of the house (figure 8).

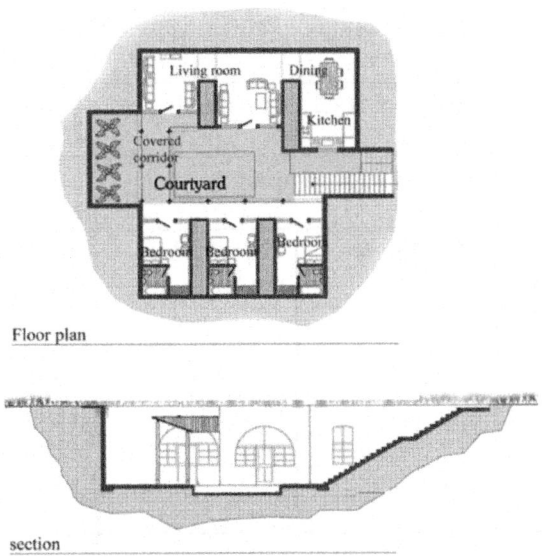

Floor plan

section

**Figure 8:** Underground earth shelter design

One of the most significant earth sheltered buildings in modern times is the Aloni House (figure 9). It was built in Antiparos Island in Greece and won the Greek Piranesi Award in 2009. The building epitomizes all that a modern time earth shelter represents. It combines all the design types mentioned above within a unique terrain. It also provided courtyard spaces with its landscape appearing to drift naturally into the courtyard thereby allowing for free solar penetration to the desired areas.

**Table 1:** Comparing efficiency values of the earth shelter building typology

| Factor | Earth shelter building type | |
|---|---|---|
| | *Bermed* | *Envelope/true underground* |
| Passive solar potential | Excellent | Less effective |
| Thermal stability | Less effective | Excellent |
| Natural lighting potential | Effective | Less effective |
| Wind protection | Less effective | Excellent |

| Noise protection | Less effective | | Excellent | |
|---|---|---|---|---|
| Visual convenience | Excellent (one directional view) | | Poor (allows only open sky view) | |
| Appropriate Climate | Effective for temperate | | Most effective for tropical | |
| Structural cost | *Modern design* | *Vernacular design* | *Modern design* | *Vernacular design* |
| | Intermediate | Less expensive | Most expensive | Least expensive |

**Figure 9:** Images of the Aloni House. (a) view from the hill top, (b) view from the top of the house, (c) opening leading to the courtyard, (d) the central courtyard, (e) interior view of the living room, (f) interior view from the kitchen. (Images by Julia Klimi)

## EVALUATION OF ENERGY CONSERVATION PRINCIPLES IN EARTH SHELTER SCHEMES

The most significant value of earth shelters and the basis for the exploitation of earth in energy saving building initiatives is its energy preservation potential. This is based on several unique physical characteristics of earth. As stated earlier, the dependability of earth in energy conservation designs is related to the natural principles of annual heat storage; huge temperature differential between the ground temperatures and the outdoor air temperatures and the insulation properties from direct solar radiation. In the cold climates, the significant property is the reduction of heat loss due to conduction through the building envelope. The amount of heat lost in this manner is a function of the thermal transmission coefficient (R-factor) of the envelope and the temperature difference between the inside of the envelope and the outside. While the R-factor for earth is substantially lower than that of other insulating materials, the large amount of earth inherent in earth sheltering can provide an overall R-factor comparable with more highly insulated structures [12].

According to investigations in [12], the temperature differential for conventional above ground structures is the difference between the outside air temperature and the interior temperature maintained for the comfort of its inhabitant. Under extreme conditions, this differential can be as much as 32°C. However, since the daily and seasonal fluctuations of temperature below the surface of the ground never equals that of the air above, therefore the deeper the temperature is taken, the less severe will be the variation. This reduced temperature differential results from the thermal storage capabilities of the soil which moderate extremes of temperature and create seasonal intervals, wherein energy from one season is transferred to the next season as in the principle of PAHS.

## Solar Radiation and Energy Conservation in Earth Sheltered Houses

It is common knowledge that the sun is one of the most significant determinants in energy efficient building design. The radiant energy from the sun can be used as both active and passive heat generators for a building. Generally in colder climates, the active solar receptor system is oriented directly to the south, whereas all passive solar collection methods are based on trapping the radiant energy of the sun which enters through the openings on the building envelope. In the case of earth sheltered houses, the best site orientation (in cold climates) is the south-side orientation which maximizes the presence of all of the window openings whereas the remaining sides of the building are completely earth covered. The use of passive solar collection in combination with other energy conservation values is a very desirable energy efficient concept in buildings since it does not involve the capital expense that an active solar collector does. Conversely, it is important to note that, while solar radiation is desirable in the heating season of cold climates, they are not as efficient in the cooling season of hotter climates. The effect of wind on the orientation of an earth sheltered structure is a serious energy consideration [13]. Since direct exposure to cold winter winds increases heat loss due to infiltration which consequently creates a wind chill effect, it is desirable to protect a building as much as possible from this exposure. In the north hemisphere the prevailing winter winds are from the northwest. Minimizing window and door openings on the north and west sides of the house in this region will enhance energy performance.

## Effects of Seasonal Thermal Storage Systems on Energy Conservation in Earth Sheltered Houses

A seasonal storage system can broadly be defined as one which stores energy in one season and delivers that energy in another season. Naturally for seasonal storage systems that function as solar thermal collectors, this means that energy is collected in periods of high radiation as is the case in summer seasons and delivered in winter seasons during periods of low radiation. However to further improve the efficiency of any of the seasonal thermal storage systems, very effective above-ground insulation or super insulation of the building structure is required to minimize heat-loss from the building, thereby improving the amount of heat that needs to be stored and used for space heating.

There are three major types of seasonal (annualized) storage systems that are classified as effective or beneficial to earth shelter buildings. These are:

## Low temperature Systems

This system utilizes the earth (soil) adjoining the building as a low-temperature seasonal heat store, thereby reaching temperatures similar to average annual air temperature while drawing upon the already stored heat for space heating. These systems can also be seen as an extension to the building design itself as the design involves some simple but significant differences when compared to conventional above ground buildings.

## Warm Temperature Inter-seasonal Heat System

This also uses soil to store heat, but utilizes active solar collection mechanisms in summer to heat up thermal banks (earth mass) in advance of the heating season. Warm temperature heat stores are generated from low-temperature stores in that solar collectors are used to capture surplus heat in summer and actively raise the temperature of large mass of soil so that heat extraction is made cheaper in winter.

## Passive Annual Heat Storage System (PAHS)

With the development of modern passive solar building design, during the 1970s and 1980s a number of techniques were developed that enabled thermally induced and moisture-protected soil to be used as an effective seasonal storage medium for space heating, with direct conduction as the heat return method. The concept of Passive Annual Heat Storage (PAHS) is such that solar heat is directly captured by the structure's spaces and surfaces in summer and then passively transferred through its floors, walls and roof into adjoining thermally-buffered soil by conduction. It is then passively returned to the building's spaces through conduction and radiation as those spaces cool in winter. This idea was originally introduced by John Hait [11]. It includes extensive use of natural heat flow methods, and the arrangement of building materials to direct this passive energy from the earth to the building, all without using equipment. PAHS is believed to be one of the most significant ideas for energy conservation in earth sheltered buildings.

## Concept of Passive Annual Heat Storage System (PAHS)

Globally, the earth receives electromagnetic radiation from the sun which is typically defined as short-wave radiation and emits it at longer wavelengths known typically as long-wave radiation. Figure 10below shows an analysis of the earth's shortwave and long-wave energy fluxes produced with details from [14]. This absorption and re-emission of radiation at the earth's surface level which forms a part of the heat transfer in the earth's planetary domain yields the idea for the principle of PAHS. When averaged globally and annually, about 49% of the solar radiation striking the earth and its atmosphere is absorbed at the surface (meaning that the atmosphere absorbs 20% of the incoming radiation and the remaining 31% is reflected back to space). This absorbed 49% of the solar radiation presents a premise for energy efficiency in building design. The concept of earth shelter design focuses fundamentally on the utilization of the absorbed/retained heat from this annual absorption and re-emission of radiation for indoor thermal environment control.

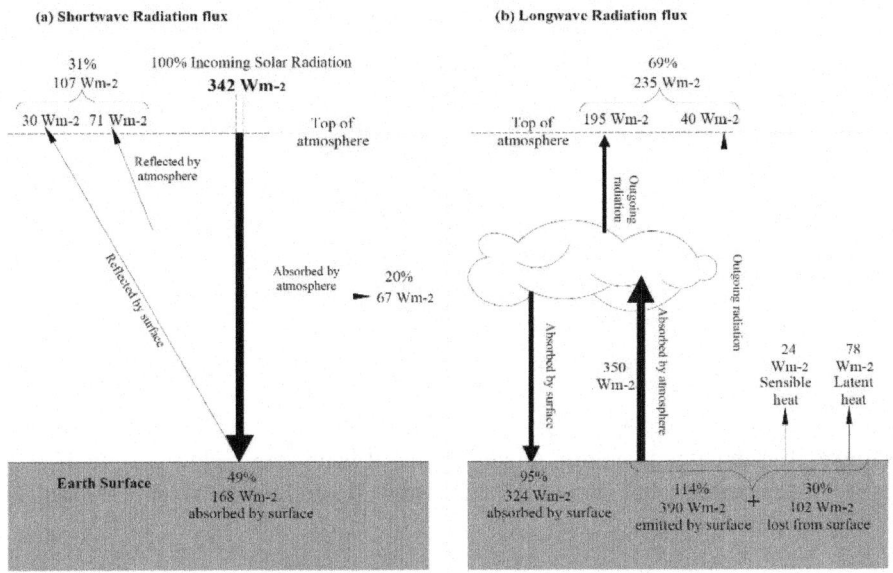

**Figure 10:** Earth's energy budget diagram showing the short-wave (a) and long-wave (b) energy fluxes

## Analysis of Soil Thermal Performance in Earth Shelter Designs

The thermal property of an earth-shelter soil is an essential factor in determining its performance against other conventional above-grond houses. Due to the relatively stable temperature of the soil, the earth shelter house in summer loses heat to the cool earth rather than gaining heat from the surrounding air, and in winter the relatively warm soil offers a much better temperature environment than the subzero air temperatures. This concept is clearly confirmed by examination of the daily and yearly soil temperature fluctuations at various depths. Daily fluctuations are virtually eliminated even at a depth of 20 cm of soil. At greater depths, soil temperature responds only to seasonal changes, and the temperature change occurs after considerable delay [15]. A reasonable level of soil study is necessary in order to facilitate the comparison of the energy needed for construction (soil excavation, dewatering and concrete works) with the energy to be saved in the long run, conditions related to the insulation efficiency of the soil [16]. However the expected efficiency varies with the soil type and its water content which in some cases may have a marked effect on the thermal properties of the soil. The figure below (Figure 11) presents a typical relationship between the annual air temperatures and corresponding temperature fluctuation below the ground surface.

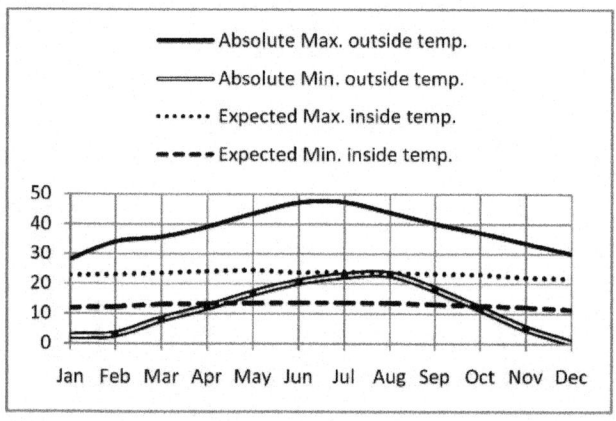

**Figure 11:** Annual temperature fluctuations in Riyadh from below zero to 48 ∘C and expected temperature fluctuation at 3.0m below ground level between 14∘C and 24∘C. (Data taken from [16])

In earth shelter houses, the overlaying thick earthen layer around much of the building effectively eliminates possibilities of infiltration through the building skin (as is the case in conventional above-ground houses). This can contribute significantly in reducing energy loss due to infiltration, except only through the exposed portions of the structure. Apart from the reduction of infiltration, studies identified that the application of thermal coupling of the earth-soil to the building wall places significant values to the thermal conditions of the earth shelter environment in winters. This process allows for improved thermal storage through the soil into the building walls. Since majority of modern earth shelters are built with concrete which possesses a large thermal storage capacity which can absorb the excess energy from the earth-soil, this absorbed heat is naturally released back into the building whenever the indoor air temperature is below that of the thermal mass. This thermal absorption and releasing process can provide essential heat energy required in the house for days without mechanical heating. The effect of this process is presented below (figure 12) in a thermal investigation study of a berm-type earth sheltered house in Missouri (US) covering a 4 day assessment period under a 6-hourly measurement interval [12].

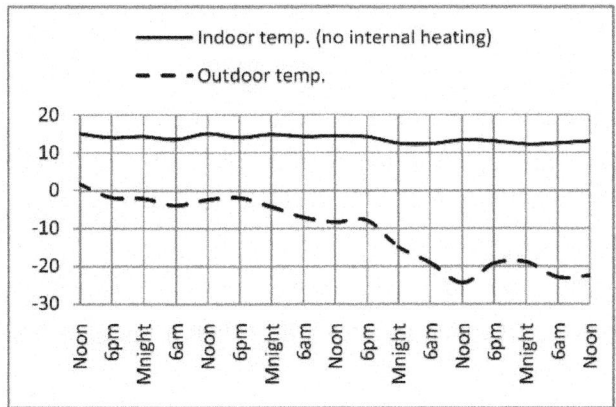

**Figure 12:** Temperature stability graph of an earth sheltered house in Missouri (Data taken from [12])

Determining the thermal performance of the soil for earth shelter construction involves assessing the long-term subsurface environment and above-ground temperature data. Consequently, this requires accurate environmental information on the boundary conditions, one of which is the temperature of the surrounding soil. For instance, in the case of a single basement study, a change in the mean annual ground temperature from 10∘C to 6 ∘C caused a 36% increase in heat loss [17]. Therefore, accurate data regarding diurnal and annual variation of soil temperatures at various depths is necessary to accurately predict the thermal performance of earth sheltered structures.

Study shows that actual data on soil temperatures is not usually abundant. However research has facilitated the evaluation of the underground climate in order to assess the suitability of earth sheltered structures. Algorithms for this calculation of the soil temperatures at various depths have already been developed based on existing field measurements in different regions of the world and by this, the annual pattern of soil temperatures at any depth can be accurately considered as a 'sine' wave about the annual average of the ground surface temperature. Accordingly, a mathematical method was developed to predict the long-term annual pattern of soil temperature variations as a function of depth and time for different soils and soil properties that are stable over time and depth [10]. This method is sufficiently accurate in the case certain thermal and physical characteristics are accurately estimated. The equation for estimating subsurface temperatures as a function of depth and day of the year is as follows (with the unit of cosine expressed in rad):

$$T_{(x,t)} = T_m - A_s e^{-x}\sqrt{\pi/365\alpha}\ \cos\left\{\frac{2\pi}{365}\left[t - t_0 - \left(\frac{x}{2}\right)\left(\sqrt{\frac{365}{\pi\alpha}}\right)\right]\right\}$$

(1)

Where:

$T_{(x,t)}$ = subsurface temperature at depth $x$(m) on day $t$ of the year (˚C),

$T_m$ = mean annual ground temperature (equal to steady state) (˚C), as the annual temperature amplitude at the surface ($x$= 0) (˚C),

$x$ = subsurface depth (m),

$t$ = the time of the year (days) where January 1 = 1 (numbers),

$t_0$= constant, corresponding to the day of minimum surface temperature (days),

$\alpha$ = the thermal diffusivity of the soil (m²/day)

Through this equation, the resulting temperature profile at different depths can now be graphed and compared with the annual average air temperatures. Following the evaluation of the subsurface climate, the calculated soil temperatures can then be used to calculate the heat flux through the building surfaces. The energy efficiency of a wall in contact with the earth at varying depths can thus be investigated for local climatic conditions. This can be done by simulating the heat transfer through a subsurface wall at varying depths using a computer program, and comparing the results with an above-ground wall using the same method. This procedure is a typical preliminary assessment method with minimal input required. The expected results from the simulations provides preliminary insight into the magnitude of reduction of heat flow that the building soil climate can provide in comparison to the above-grade climate and the analysis also provides a faster approach for determining the optimum depth placement for an earth sheltered building.

Although this theory seems rightly beneficial to the energy conservation concepts in earth shelter house construction, it is also right to consider other detrimental factors like the soil's heat and cooling losses due to normal thermal transmittance factors. Earth shelters are subjected to heat and cooling losses partly via the soil to the external air, via the soil to the groundwater below or directly to the groundwater. The quantity of loss is calculable in this case and the equation is generated in [18] as follows:

$$QT = Atotal \frac{(v_i - vOT)}{RAL} + \frac{v_i - vGW}{RGW}[W]$$

(2)

Where:

$\vartheta OT$= mean outside temperature

$\approx 0$ to $-5°C \approx (\vartheta e + 15K)$

ROT= Ri+ R$\lambda$A+ R$\lambda$B+ Re= equivalent resistance to thermal transmission room-outside air.

R$\lambda$A= equivalent resistance of the soil to thermal conductivity.

R$\lambda$B= resistance of building component to thermal conductivity.

RGW = Ri+ R$\lambda$B+ R$\lambda$s= equivalent resistance to thermal transmission room-groundwater.

R$\lambda$s= T/$\lambda$s = thermal conductivity resistance of soil to groundwater.

D = depth of groundwater

$\lambda$s= thermal conductivity coefficient of soil

$\approx 1.2$ W/mK

$\vartheta$GW = groundwater temperature = 10°C.

## Energy Conservation Values in Earth Shelter Design

Earth is a great moderator of temperature change. When warmed up, it can stay warm a long time without losing much of its heat [9]. Earth does not react as fast to temperature change as air does. This means that for instance if air surface temperatures ranges from -15°C to 35°C through the year (winter through summer), then about 3 meters below, the temperature of the earth will vary only between 10°C to 15°C. This short range in difference explains the ability of earth to maintain stable temperatures throughout the year. This is a significant energy conservation tendency in the case of reducing the load on home heating and air-conditioning systems. With regards to total operating cost (excluding estimates from heat-recovery systems), energy savings of up to 60% to 70% may be realized in residential scale structures within mid-temperate zones. Instances of this were presented in [19] from the energy cost studies undertaken in [20]. In this study, a conventional 135 sq m (9m x 15m) single level residence with a hypothetical subsurface structure of the same dimension was compared. With the use of climate data and energy rates of Denver metropolis in Colorado, the study establish that the underground house provides a 72% energy savings over the surface dwelling (Table 2, 3 and 4).

**Table 2:** Evaluation of rates of heat loss and gain in a typical above ground house and an earth sheltered house [20]

| Measured unit | Conventional surface house | Earth shelter house |
|---|---|---|
| Heat loss in winter (B.T.U. per hour) | 39,927 | 12,720 |
| Heat gain in summer (B.T.U. per hour) | 44,650 | 0 |

**Table 3:** Evaluation of annual energy consumption cost in a typical above ground house and an earth sheltered house [20]

| Measured unit | Conventional surface house | Earth shelter house |
|---|---|---|
| Winter: Gas (m³) Oil (gal.) electricity (kwh) | 2,656.9 m³ ($65.80) 710 ($129.90) 23,157 ($428.80) | 871.5 m³($27.60) 233 ($42.60) 7,596 ($191.10) |
| Summer Electricity (kwh) | 3,962 ($98.40) | 0 |

**Table 4:** Evaluation of annual cost of environmental control requirements in a typical above ground house and an earth sheltered house [20]

| Building type | Gas | Oil | Electricity |
|---|---|---|---|
| Above ground design (AGD) | ($395) | ($459) | ($758) |
| Earth sheltered design (ESD) | ($120) | ($135) | ($283) |
| Cost conservation comparison between ESD and AGD | 30% | 29% | 37% |

# SOIL SUITABILITY ANALYSIS FOR EARTH SHELTERED BUILDING CONSTRUCTION

As already discussed earlier, not all soil types are efficient in use for earth sheltered building construction. The choice of construction site is mainly determined by the soil type available in a given geographical area for issues of safety against landslides and other moisture originated hazards. Some types of soil are more suitable than others in the construction of sub-grade buildings. The strength of the soil must be determined for the proposed depth of building below ground level. Though may be desirous, excavations in a very strong soil may be difficult and in the case of rocky ground, may prove impossible. On the other hand, in very weak soils the excavations are easy. In the first two cases, the capital cost and the energy expenditures involved in construction need careful examination [21]. For the third case, however, the excavation may be difficult because high lateral earth pressure requires construction of heavy walls (retaining walls), preferably made of reinforced concrete, which implies increased capital costs and energy consumption. In modern earth sheltered home construction, compaction and permeability values are the most essential standards considered in the backfill process when building a berm or elevational type construction. This is mostly due to the dangers of soil drainage. It has been noted earlier that soil-water content has distinctive effect on the thermal properties of the soil hence may affect the overall energy performance of earth-homes. Choosing a site where the water will naturally drain away from the building is the best way to avoid water pressure against underground walls. In order to improve the energy performance of the earth-soil in temperate, humid or arid tropical scenarios, drainage systems must be designed to draw water away from the structure to reduce the frequency and length of time the water remains in contact with the building's exterior. Survey has identified that ideal sites are those of hilly or mountainous terrain. The partially buried (bermed-elevational) earth-sheltered home is identified as most suitable for maximizing passive-solar heating in cold climates, however since water tends to drain down the hill toward the building and off the roof toward the back of the home, it is advisable to build in highly water-permeable soils and to install water drainage systems around the perimeter of the buried walls. Hydrology discusses infiltration as the rate at which water passes into the soil. This is also affected by the ratio of macro to micro-pores of the soil in question. The more macro-pores a soil has the easier it is for water to soak into it and drain away. Soils with coarse particles like sand or gravel or nutty or block soil structures have a high proportion of macro-pores and as a result have high infiltration rates. Soils such as clays have a high proportion of micro-pores and therefore have low infiltration rates. Figure 14 below illustrates different infiltration rates based on soil structure and texture [22].

Through the analysis below, it could be said that a good earth home design site with natural drainage also requires permeable soils. The most permeable soils as identified above are the granular type which consists of a fair amount of sand or gravel while soils with high clay content are less permeable as they expand and contract as moisture levels fluctuate. Nonetheless, it is advisable to perform percolation tests on the construction site›s soil to determine the earth shelter soil permeability before construction.

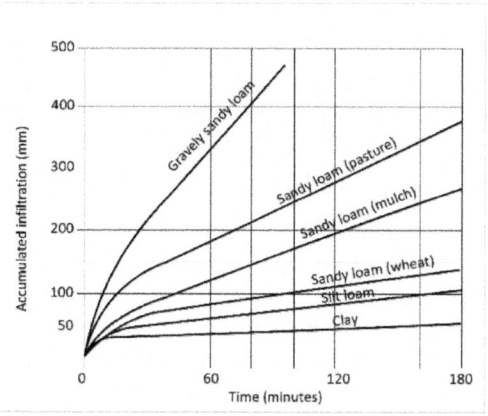

**Figure 13:** Infiltration curves for different soil textures

## THERMAL INTEGRITY ANALYSIS OF EARTH SHELTERED HOUSES

Thermal integrity factor (TIF) is a combined system for evaluating and comparing the energy performance values of building types. It is expressed in units which allow for direct comparison among such criterion as heating, ventilating, and air conditioning systems as well as the effect of various climatic conditions on different housing types. The standard unit for measuring thermal integrity values is $Btu/ft^2$ per degree day of the provided space condition. A TIF of 7.5 $Btu/ft^2$ per heating degree day is considered as representative of a baseline-factor for moderately insulated houses [23], while values in the ranges of 0.6 to 1.1 $Btu/ft^2$ per heating degree day are predicted for super-insulated houses [24]. Early indication of the performance of earth sheltered buildings against the conventional above-ground ones were recorded as far back as the late 1970s and 80s. Measurements were conducted on existing earth sheltered houses in some US cities. In one of the houses located in South Dakota which was monitored during 1978 and 1979, it consumed about 28,000 $Btu/ft^2$ for

8144 heating degree days, which yields a TIF of 3.5 Btu/ft$^2$ per heating degree day. The report on this house went on to note that typical above-ground framed homes in the same location generally required about 10 to 12 Btu/ft$^2$ per heating-degree day. This displays a 70% difference in the TIF of these two homes in the same location. Figure 13 below shows the comparative energy consumption for the above-ground and earth sheltered homes. In some other cases, earth sheltered houses display TIFs of 0 (zero) Btu/ft$^2$ per heating degree day. Below (table 5) are the results of the TIFs for five different buildings in Minnesota all of which recorded TIFs of less than 4.0 [25].

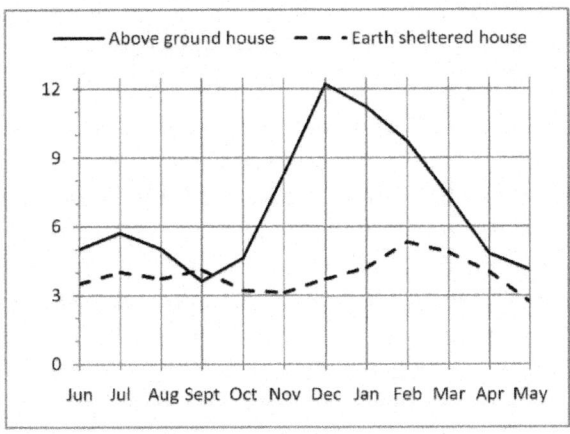

**Figure 14:** Comparison of monthly total energy usage in conventional above-ground and earth sheltered homes (taken from [12])

**Table 5:** Monthly thermal integrity factor for five Minnesota earth-sheltered residences

| House | June 1980 | July 1980 | Aug. 1980 | Sept. 1980 | Oct. 1980 | Nov. 1980 | Dec. 1980 | Jan. 1981 | Feb. 1981 |
|---|---|---|---|---|---|---|---|---|---|
| Burns-ville | *nil* | *nil* | *nil* | 0.65 | 0.84 | *nil* | *nil* | *nil* | 2.03 |
| Camden | 0 | 0 | 0 | 0 | 0.89 | 1.20 | 2.65 | 1.92 | *nil* |
| Seward | 0 | 0 | 0 | 0 | 0 | 2.14 | 3.60 | 2.53 | 3.19 |
| Wild River | 0 | 0 | 0 | 0 | 0.19 | 2.05 | 1.08 | 0.91 | 1.27 |
| Willmar | *nil* | *nil* | *nil* | 2.28 | 2.34 | 1.23 | 2.72 | 2.01 | *nil* |

# CONCLUSION

In this study, the following factors were analyzed in the hope of throwing light into the common questions that arise in the discourse of earth sheltered housing:

- Energy conservation elements for earth shelter housing,
- Thermal integrity values,
- Techniques for maximizing the thermal loads necessary for comfort conditions in passively heated or passively cooled earth shelters,
- Soil suitability, depth of placement and design techniques that optimizes structural integrity in earth sheltered house construction.

This study also presented some of the valuable analysis and results in earth shelter building evaluations as premise for assessing the potentials of passively heated earth sheltered houses. This is achieved through a review of previous performance assessments of monitored conditions in existing earth sheltered buildings. Through this review, thermal integrity factors (TIF) of existing earth sheltered homes were identified, which when compared with other housing types, perform significantly better than conventional above-ground dwellings. It also looked at both summer and winter impacts of earth shelter house types utilizing the passive approach under the different climate conditions. This study identified that the thermal integrity value of passively heated earth sheltered house is comparable with other energy-efficient approaches such as super-insulated and passive solar constructions which are much better in energy conservation performance than the conventional above-ground constructions. It further presents the criteria for identifying the appropriate soil type (sub-grade materials) needed in building earth sheltered houses with passive thermal approach. These are categorized under thermal inertia properties, bearing capacity and drainage properties. Based on the available information to date, it can be said that earth sheltered houses maintain heating energy consumption that is lesser by up to 75%. This claim appears to be substantiated as earth sheltered house compared to conventional above-ground house presents a lesser calculated or monitored TIF.

Having looked through the benefits and potentials of earth and the overall understanding of its potential for energy conservation through earth-sheltered construction, it is hoped that this review contributes to the information available so far on means of assessing the performance of earth shelters and associated thermal properties that affects it. It is then possible for designers and planners in different regions to have access to a simple framework for assessing its efficiency at the initial planning stages. The resulting outputs can then be used for the heat transfer and energy conservation analysis within the building units. Results from this analysis will provides insight into the degree of passive heating and cooling or reduction in heat flow that the soil climate can provide as compared to the surface climate as well as suggesting parameters for depth placement of earth shelter buildings for more efficient results.

## REFERENCES

1.  P. Carpenter, 1994Sod It: An Introduction to Earth Sheltered Development in England and Wales, Coventry University, Coventry,.

2.  J. Carmody, R. Sterling, 1984Design considerations for underground buildings, Underground Space 8352362

3.  R. Kumar, S. Sachdevab, S. C. Kaushik, 2007Dynamic earth-contact building: A sustainable low-energy technology, Building and Environment 4224502460

4.  J. Dodd, 1993Earth sheltered settlements, a sustainable alternative, in: Proceedings of the Earth Shelter Conference, Coventry University 32636

5.  Moreland F.L,1975An alternative to suburbia, in: Proceedings of the Conference on Alternatives in Energy Conservation: The Use of Earth-covered Buildings, National Science Foundation, Fort Worth, TX,.

6.  G. S. Golany, 1983Earth Sheltered Habitat (History, Architecture and Urban Design), Van Nostrand Reinhold Company Inc., New York.

7.  Rahman M.M, Rasul M.G, Khan M.M.K,2010Energy conservation measures in an institutional building in sub-tropical climate in Australia. Applied Energy 8729943004

8.  M. Reynolds, 1991Earth-ship Systems and Components, Solar Survival Press

9.  M. B. Wells, 1975To Build without Destroying the Earth. Alternatives in Energy Conservation: The Use of Earth Covered Buildings. Washington, D.C. U.S. Government Printing office. 211232

10. K. Labs, 1979Underground building climate, Solar Age 41044

11. J. Hait, 1983Passive Annual Heat Storage: Improving the Design of Earth Shelters, Rocky Mountain Research Center

12. R. L. Wendt, 1982Earth-Sheltered Housing: An Evaluation of Energy-Conservation Potential. U.S. Department Of Energy, Oak Ridge Operations, TN. 818

13. Minnesota University1979The Underground Space Center, Earth Sheltered Housing Design. Minneapolis, Minnesota: Van Nostrand Reinhold Company. 20

14. G. Bonan, 2002Ecological Climatology: Concepts and Applications, Cambridge Press, United Kingdom

15. US Department of Housing and Urban Development,1980Earth Sheltered Housing" Code, Zoning, and Financing Issues

16. A. M. Khair-Din El, 1991Earth Sheltered Housing: An Approach to Energy Conservation in Hot Arid Areas Architecture and Planning Riyadh. 3318

17. Mitalas G.P,1982Basement Heat Loss Studies, DBR/NRC, Ottawa.

18. D. Klaus, 2003Advanced Building Systems: A Technical Guide for Architects and Engineers, Published for Architecture Basel, Boston, Berlin. 50

19. K. Labs, 1975The Architectural Use of Underground Space: Issues and Application. Master's Thesis/Washington University, May, Mechanicsville PA. 121

20. L. Harrison, 1975Is it time to go Underground? The Navy Civil Engineer, Fall. 2829

21. A. M. Henna, 1980Building Underground Alternatives. Miami published research for the Energy Conservation Conference, Florida.

22. R. G. Mc Laren, K. C. Cameron, 1990Soil science. An introduction to Properties and Management of New Zealand Soils. Oxford university press, Auckland.

23. D. Lewis, W. Fuller, 1979Solar Age, 31

24. W. A. Shurcliff, 1980Super-insulated Houses and Double-Envelope Houses, A Preliminary Survey of Principles and Practice, 2nd Ed., Cambridge, Mass. 6

25. L. F. Goldberg, C. A. Lane, 1981A Preliminary Experimental Energy Performance Assessment of Five Houses in the MHFA Earth-sheltered Housing Demonstration Program, University of Minnesota, Minneapolis. 10

# Chapter 11

# DAMAGE TO BUILDINGS IN LARGE SLOPE ROCK INSTABILITIES MONITORED WITH THE PSINSAR™ TECHNIQUE

Paolo Frattini[1], Giovanni B. Crosta[1], and Jacopo Allievi[2]

[1]Department of Earth and Environmental Sciences, Università degli Studi di Milano-Bicocca, p.zza della Scienza 4, 20126 Milan, Italy

[2]Tele-Rilevamento Europa T.R.E., Ripa di Porta Ticinese 79, 20143 Milan, Italy

## ABSTRACT

The slow movement of active deep-seated slope gravitational deformations (DSGSDs) and deep-seated rockslides can cause damage to structures and infrastructures. We use Permanent Scatterers Synthetic Aperture Radar Interferometry (PSInSAR™) displacement rate data for the analysis of DSGSD/rockslide activity and kinematics and for the analysis of damage to buildings. We surveyed the degree of damage to buildings directly in the field, and we tried to correlate it with the superficial displacement rate obtained by the PSInSAR™ technique at seven sites. Overall, we observe that the degree of damage increases with increasing displacement rate, but this trend shows a large dispersion that can be due to different causes, including: the uncertainty in the attribution of the degree of damage for buildings presenting wall coatings; the complexity of the deformation for large phenomena with different materials and subjected to differential behavior within the displaced mass; the absence of differential superficial movements in buildings, due to the large size of the investigated phenomena; and the different types of buildings and their position along the slope or relative to landslide portions.

## INTRODUCTION

Large slow-moving non-catastrophic slope rock instabilities represent an important geological risk. They can cause the deformation of structures and infrastructures (*i.e.*, dams, tunnels, railway tracks, buildings, [1–3]) and, due to the damage of rock masses, resulting in the decay of mechanical properties, secondary landslides can be triggered within their limits [1]. Large slope rock instabilities include both very large rockslides and deep-seated gravitational slope deformations (DSGSDs) [4,5], the latter being characterized by the involvement of entire valley flanks, the presence of gravitational morphostructures (e.g., large scarps, open or infilled trenches, downthrown blocks, ridge top depressions, grabens, double or multiple ridges and counterscarps) and the geomorphological evidence of slope deformation and displacements along individual structures and inherited tectonic features [1,6–10]. Although DSGSD has been considered until recently to be a class of relict phenomena inactive under present climatic conditions, geomorphological and geochronological evidence recently demonstrated that movements associated with large slides and DSGSD, although slow, can continue for long periods, producing large cumulative displacements [11–14]. Reactivation may also happen after long periods of quiescence or inactivity. Surface displacements typically range from a few millimeters to several centimeters per year and are commonly close to the detection precision of monitoring equipment [8,15].

Recently, SAR (Synthetic Aperture Radar) interferometry [16–20] has been demonstrated to be a suitable technique to monitor these movements [1,21–31]. [21,22] used the Permanent Scatterers technique (PSInSAR™, [20]) to study DSGSDs (*i.e.*, Varadega or Confinale-Cima di Saline), landslides (*i.e.*, Ruinon) and active scree slope (*i.e.*, Premadio area) in the Central Italian Alps. [1] studied some DSGSDs from a structural point of view by using PSInSAR™ datasets; they described six different DSGSDs (*i.e.*, Mt. Varadega, Mt. Resverde, Mt. Pesciola, Mt. Baita Meriggio, Mt. Legnoncino and Mt. Cortafò) to demonstrate that the PSInSAR™ technique could give significant results in DSGSD detection and monitoring. Besides satellite-based SAR interferometry, ground-based SAR has been also applied to the investigation and monitoring of Alpine rockslides [32]. Damage to structures and infrastructures induced by very slow-moving rockslides has been rarely investigated [33–36], with special focus on infrastructures that have experienced significant deformation [1–3].

In this paper, we analyze seven DSGSDs and rockslides located in the Central Italian Alps (Lombardy Region, Northern Italy). These landslides are extracted from DSGSD and large landside inventories recently created for the entire Alpine range [5,37–39]. The aim of the paper is to analyze large slope movements in conjunction with radar interferometry and damage data in order to investigate the state of the activity of such phenomena and to describe the resulting level of damage as a function of the ground surface rate of movement.

# GEOLOGICAL AND GEOMORPHOLOGIC SETTING

The study area lies in the Alpine sector of the Lombardy Region (northern Italy) (Figure 1), which is composed of three main structural units [40–43]: Southern Alps, the Penninic unit and the Austroalpine domain.

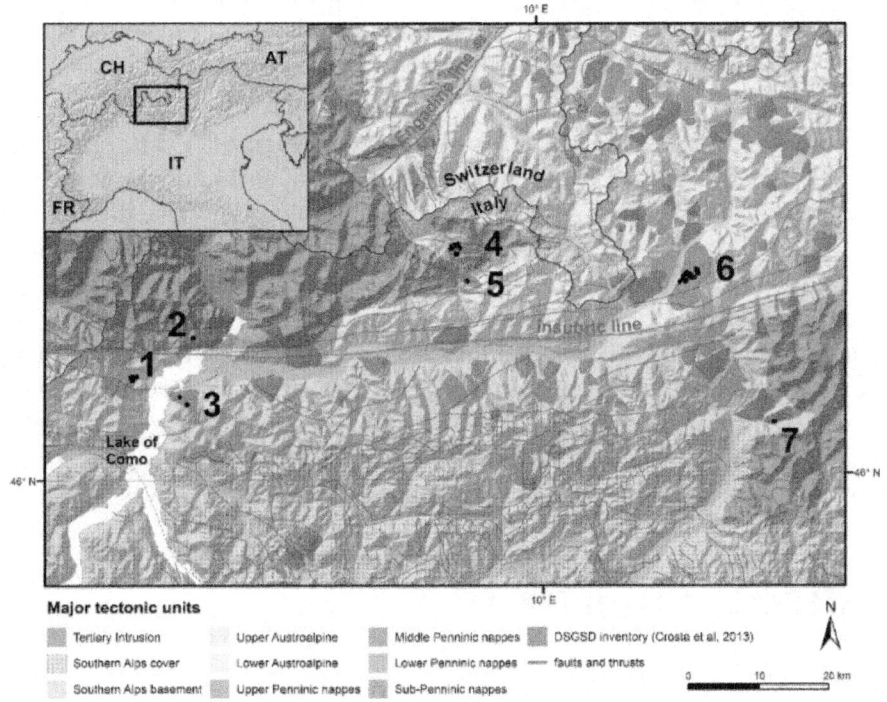

**Figure 1:** Location of the studied deep-seated gravitational slope deformations (DSGSDs) and rockslides: (1) Catasco rockslide; (2) Montalto rockslide; (3) Mt. Legnoncino DSGSD; (4) Lake Palù DSGSD; (5) Caspoggio DSGSD; (6) Mt Padrio-Varadega DSGSD; (7) Saviore DSGSD. Mapped major tectonic units and structures modified from [40,43], respectively. DSGSD polygons are from [5]. Black dots refer to surveyed buildings. CH: Switzerland; AT: Austria; FR: France; IT: Italy.

These units are separated by the Insubric line, representing a steeply north dipping and east-west trending fault zone. The Southern Alps represent the most recent part, interpreted as a fold-and-thrust system, where rocks can be divided into basement and sedimentary cover [41].

The units to the north of the Insubric line consist of the Austroalpine nappes to the east and the Penninic nappes to the west. Austroalpine units, although of similar paleogeographic provenance as the Southern Alps, consist of a completely rootless metamorphic basement and sedimentary cover that were detached from their lithosphere as early as the Cretaceous orogenesis

[42]. The Penninic units are of extremely heterogeneous paleogeographic provenance, including remnants of oceanic lithosphere (Malenco-Forno unit), as well as basement of the European margin (Adula, Tambò and Suretta units).

The Alpine territory is characterized by high mountains and deep valleys producing high relief energy. This morphology results from the combined action of geological structure, climate and its changes, causing a different action by glaciations and the fluvial system.

Seven slow-moving large slope instabilities have been analyzed in this paper (Figure 1). The Catasco rockslide (#1 in Figure 1), Mt. Legnoncino DSGSD (#3 in Figure 1) and Saviore DSGSD (#7 in Figure 1) occur in the basement of the Southern Alpine nappe. The lithology of these different sites is quite similar, mainly consisting of paragneiss and schist. The Montalto rockslide (#2 in Figure 1) belongs to the lower Penninic nappe, and its lithology is characterized by paragneiss. The Lake Palù DSGSD (#4 in Figure 1) and Caspoggio DSGSD (#5 in Figure 1) lie at the contact between the Upper Penninic nappe and the Austroalpine basement. Here, the Penninic units are composed of oceanic ophiolites and serpentines and the Austroalpine units of gneiss and schists. The Mt. Padrio-Varadega DSGSD (#6 inFigure 1) belongs to the Austroalpine basement, and the lithology consists of gneiss (Punta della Pietra Rossa Formation).

# MATERIAL AND METHODS

## PSInSAR™

PSInSAR™, Permanent Scatterers Synthetic Aperture Radar Interferometry, is an advanced interferometric technique developed at the end of 1990 by the SAR group of Milan Politecnico and T.R.E. (Tele-Rilevamento Europa) [17–20]. It is based on the processing of a long series of radar data acquired in the same geometry over the same area in order to single out those pixels, referred to as Permanent or Persistent Scatterers (PS), which have a "constant" electromagnetic behavior in all the radar images. This concept has been successively adopted by other researchers with similar PS processing tools [44–47]. If the scatterers correspond to objects whose reflectivity does not vary through time, temporal decorrelation is negligible and the average displacement rate can be determined with millimetric precision, removing the typical artefacts and noising affecting the traditional interferometric analysis (InSAR) [17–20,25]. The availability of long radar image archives, covering almost two decades, allows for obtaining ground displacement data since 1992, which is often not possible with more traditional methods, such as levelling and GPS surveys. PSInSAR™ displacements are measured along the satellite line of sight (LOS), which is the sensor to target direction, tilted at a θ angle to the vertical. Average displacement rate values can be both positive and negative. In the first case, the target approaches the sensor; in the second case, it moves away from the sensor. Due to the acquisition satellite geometry (the sensor flies along an orbit with an approximately N-S direction, acquiring a line of sight orthogonal to the orbit), InSAR measurements are not capable of detecting movements in the same N-S direction of the orbit. The combination between satellite orbit and Earth rotation allows the sensor to acquire data in two modes (Figure 2). The reference points are selected based on a statistical procedure that minimizes the standard deviation of measurements [48]. Later, the points are checked to verify if they are reliable stable points form a geological point of view.

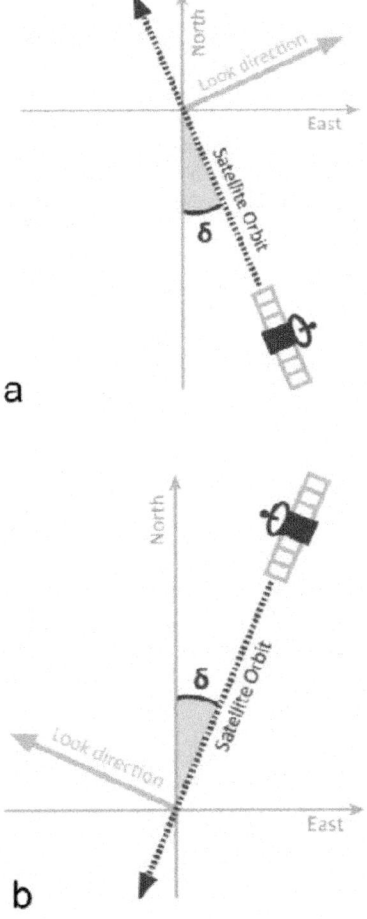

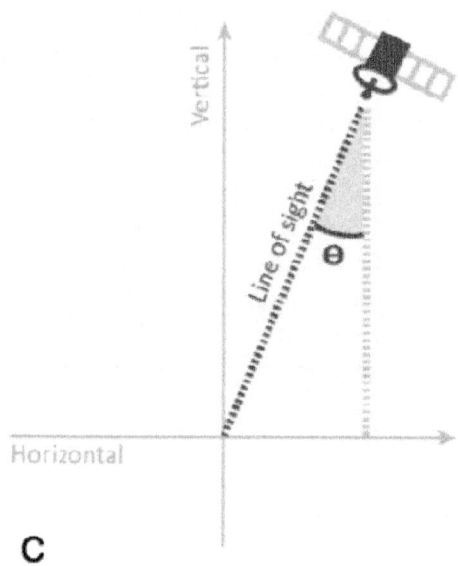

C

**Figure 2:** Satellite acquisition geometry: look direction and angle between the azimuth and north direction, δ, for (**a**) ascending and (**b**) descending modes; (**c**) local incidence angle, θ.

Different satellites detect points at regular time intervals (revisiting time): European Remote-Sensing satellites 1 and 2 (ERS1 and ERS2) every 35 days and Radarsat-1 (RSAT-S3) every 24 days. For our analysis, we used different datasets with ERS (1/2) and RSAT-S3 data and processed using the Standard PS Analysis (SPSA) processing engine [20].

**Table 1:** Synthetic Aperture Radar (SAR) datasets used for the analyses. D = descending; A= ascending; θ = local incidence angle of the center of the area of interest; δ = angle between the azimuth and north direction.

| Dataset Name | Satellite | Mode | Track | θ (°) | δ (°) | # of Scenes | Time Interval |
|---|---|---|---|---|---|---|---|
| LcED | ERS 1/2 | D | 208 | 23.09 | 11.99 | 80 | 05/16/1992–12/24/2002 |
|  |  |  | 480 | 23.11 | 12.50 | 82 | 04/30/1992–01/12/2003 |
| LED | ERS 1/2 | D | 208 | 23.09 | 11.99 | 81 | 05/16/1992–12/19/2000 |
| LRD | RSAT-S3 | D | 197 | 32.50 | 10.46 | 56 | 04/28/2003–06/18/2007 |
|  |  |  | 297 | 35.78 | 9.60 | 56 | 04/11/2003–06/01/2007 |
| LRA | RSAT-S3 | A | 147 | 34.49 | 11.51 | 59 | 03/07/2003–06/14/2007 |
|  |  |  | 247 | 32.60 | 12.15 | 59 | 04/07/2003–06/21/2007 |

**Table 2:** Slope geometry and information about the SAR datasets available at each site. $\beta$ = average slope gradient; $\bar{\alpha}$ = modal slope aspect; $\bar{\sigma}$ = average standard deviation of Permanent Scatterers Synthetic Aperture Radar Interferometry (PSInSAR™) displacement rates. Numbers are as in Figure 1

| # | Site | $\bar{\beta}$ (°) | $\bar{\alpha}$ (°) | Dataset Name | # Points | $\bar{\sigma}$ (mm/yr) |
|---|------|------|------|------|------|------|
| 1 | Catasco rockslide | 29 | 184 | LRA | 226 | 0.95 |
| | | | | LRD | 193 | 0.93 |
| 2 | Montalto rockslide | 28 | 190 | LRA | 19 | 0.96 |
| | | | | LRD | 17 | 0.95 |
| | | | | LcED | 1,050 | 0.90 |
| 3 | Mt. Legnoncino DSGSD | 29 | 340 | LRA | 159 | 0.96 |
| | | | | LRD | 345 | 0.94 |
| | | | | LED | 269 | 0.62 |
| 4 | Lake Palù DSGSD | 18 | 262 | LRA | 143 | 0.96 |
| | | | | LRD | 319 | 0.95 |
| | | | | LED | 514 | 0.58 |
| 5 | Caspoggio DSGSD | 26 | 312 | LRA | 859 | 1.06 |
| | | | | LRD | 786 | 0.94 |
| 6 | Mt. Padrio-Varadega DSGSD | 28 | 290 | LED | 648 | 0.56 |
| | | | | LRA | 172 | 1.34 |
| | | | | LRD | 1,948 | 1.17 |
| 7 | Saviore DSGSD | 175 | 25 | LRA | 349 | 1.37 |
| | | | | LRD | 623 | 1.17 |

LOS displacement data for each dataset have been converted to the direction of the maximum slope, assuming that the deformation is translational and parallel to the slope [25,49,50]. Although this assumption is not straightforward, we believe that the displacement vector along the slope improves the interpretation of deformation with respect to building damage. The precision (in terms of standard deviation) of PSInSAR™ displacements has been calculated at each site by averaging, within the instability area, the standard deviation of each single measurement (Table 2).

## Damage Survey

A field survey was performed to map the degree of damage of 182 buildings located on the studied sites. The surveyed buildings have been selected based on the distribution of PSInSAR™ points, to be sure that a reliable value of the displacement rate was available for each of them. For each building, we collected the GPS location and a photographic documentation of surveyed damages. If possible, interviews with local people have been conducted to reconstruct the damage history of the buildings.

For classification of the degree of damage (Table 3, Figure 3), we adopted a methodology derived from the European Macroseismic Scale [51]. For buildings showing damage intermediate between two classes, we assigned a half-class value (e.g., 2.5). Among the surveyed buildings, 75% are damaged (Figure 4) and about 20% of the buildings are severely damaged (damage class 3 to 4).

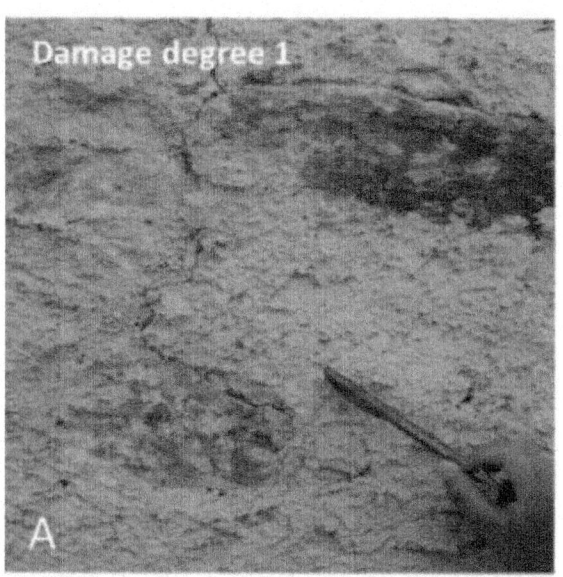

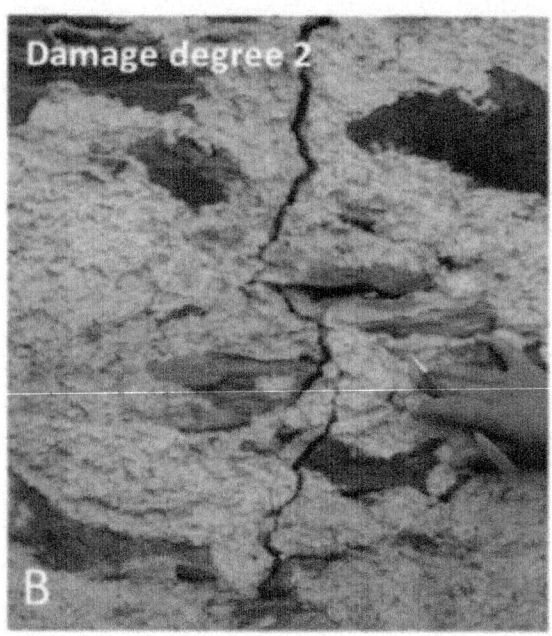

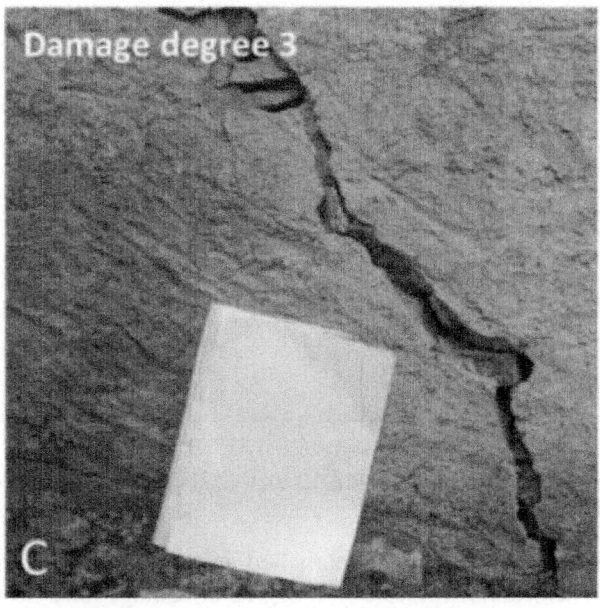

**Figure 3:** Examples of photographic documentation of damage reclassified according to the adopted classification (see Table 3).

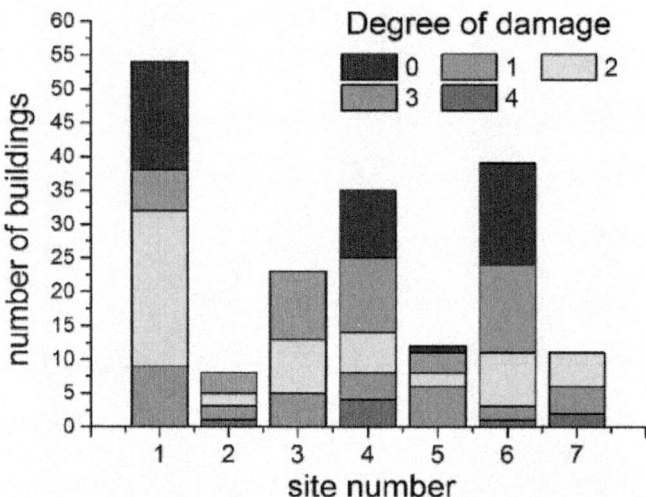

**Figure 4:** Frequency of buildings reclassified according to the proposed degree of damage scale. The original values of the degree of damage have been truncated. #1: Catasco rockslide (54 buildings), #2: Montalto rockslide (8), #3: Mt. Legnoncino DSGSD (23), #4: Lake Palù DSGSD (35), #5: Caspoggio DSGSD (12), #6: Mt. Padrio-Varadega DSGSD (39); #7: Saviore DSGSD (11).

**Table 3:** Building damage classification scheme, modified from [51]. Half-class values (e.g., 2.5) are also used to discriminate damage levels intermediate between two classes.

| Degree of Damage | | Description of Damage to Structures |
|---|---|---|
| 0 | None | No damage. |
| 1 | Negligible to slight | Hairline cracks in a few walls, falling of small pieces of plaster only. Falling of loose stone from the upper parts of buildings in very few cases. |
| 2 | Moderate | Cracks in many walls. Falling of large pieces of plaster. Partial collapse of chimneys. |
| 3 | Substantial to heavy | Large and extensive cracks in most of the walls. Roof tiles detached. Chimney fracture at the roofline; failure of individual non-structural elements. |
| 4 | Very heavy | Serious failure of walls; partial structural failure of roof and floors. |
| 5 | Destruction | Total or near total collapse. |

# ANALYSIS AND RESULTS

To understand the relationship between damage and the type and geometry of the instability, we prepared for each site a map displaying the available PSInSAR™ data, the surveyed buildings and the extent of the slope instability phenomenon. Moreover, we prepared a longitudinal swath profile of the displacement rate. To this aim, the profile along the longitudinal axis of the rock instability was subdivided into 100 m-long segments, from which polygons perpendicular to the down-slope direction and extending up to the lateral boundary of the instability have been created. By considering the PSs inside each landslide polygon, we calculated the average displacement rate and represented its trend in the plots. This approach for the construction of the displacement-rate profile was preferred with respect to the interpolation of the PSInSAR™ data, because it allows averaging over the entire width of the instability, which is an advantage where the PSInSAR™ targets are not evenly distributed within the rock instability. In the following, we present the results for each one of the sites obtained according to the above described methodology.

## Mt Legnoncino

The northern flank of Mt. Legnoncino (#3 in Figure 1) is affected by a 7 km² DSGSD moving toward Como Lake (Figure 5). PSInSAR™ data show displacement rates ranging from −2 to −12 mm/yr. The displacement time series show a nearly linear trend in time, both for the upper and the lower part of the slope (Figure 6). The spatial pattern of movement is in good agreement with structural lineaments and reactivated landslides, showing a progressive downslope decrease, which suggests an increase in the horizontal component close to the DSGSD toe [1].

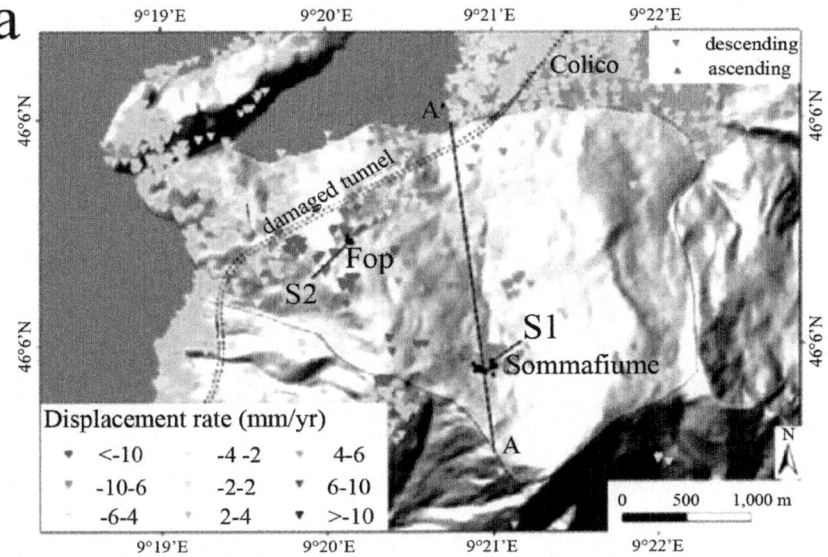

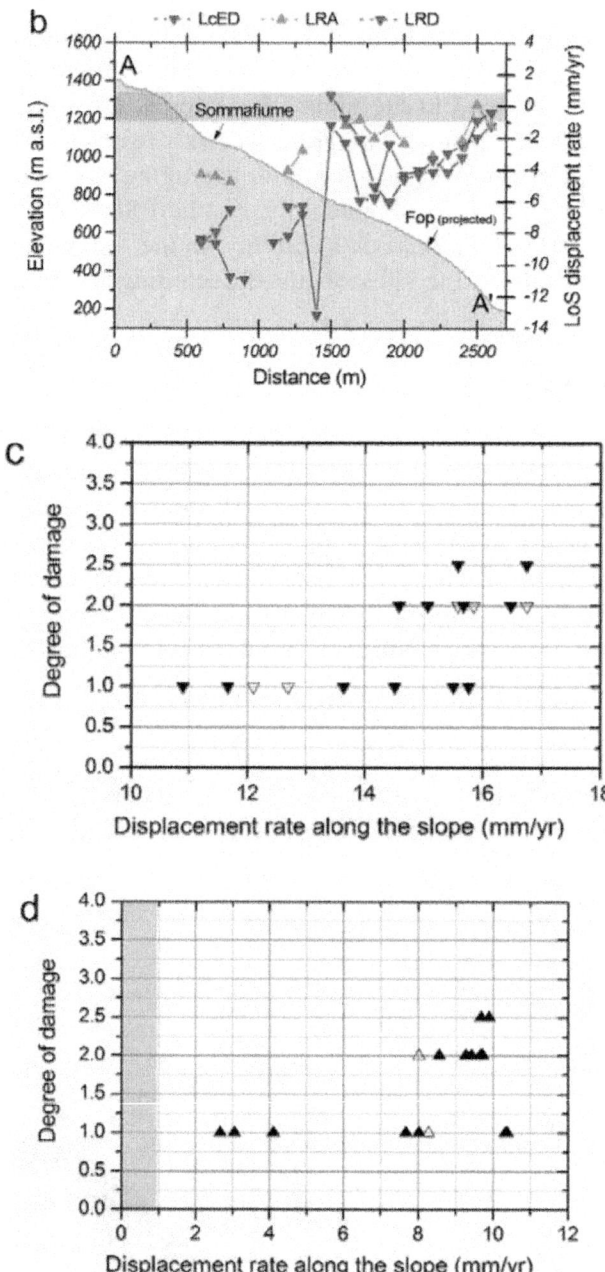

**Figure 5:** Displacement rate and damage to buildings for the Mt. Legnoncino DSGSD. **(a)** PSInSAR™ displacement rate map and location of the surveyed buildings (black polygons); **(b)** longitudinal swath profile of the displacement rate for available PS

datasets; (**c**) damage degree of buildings as a function of the displacement rate for the LRD dataset; (**d**) damage degree of buildings as a function of the displacement rate for the LRA dataset. Solid and open triangles in panels (c,d) refer to buildings with and without wall coating, respectively. The grey area in panels (b,d) represents the average standard deviation (Table 2). LOS, line of sight.

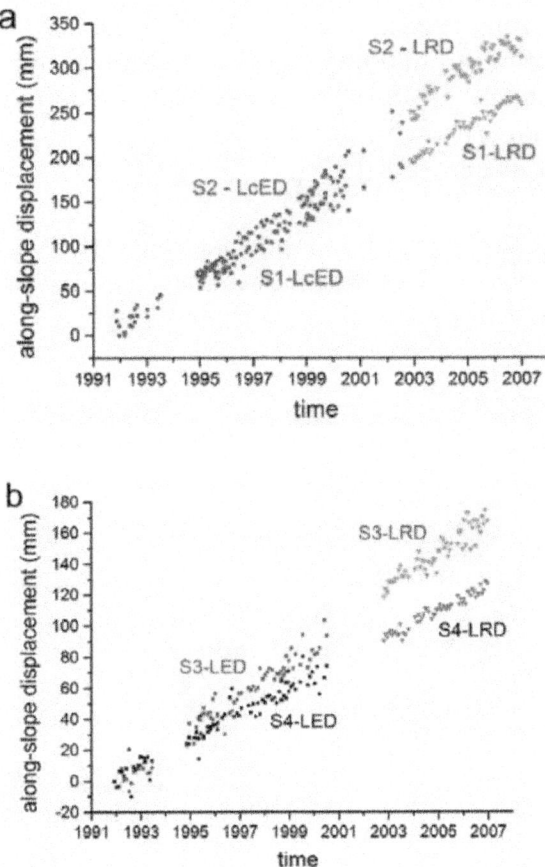

**Figure 6:** PSInSAR™ displacement time series for (**a**) Mt. Legnoncino DSGSD, sites S1 and S2 in Figure 5a, and (**b**) Mt. Padrio-Varadega DSGSD, sites in 1st figure in Section 4.3. The displacements monitored with ERS 1-2 satellites (LcED, LED datasets) were extrapolated to 2003 and added to the LRD displacements to show a continuous time history of the sites.

Due to the northward orientation of the slope, we can assume that PS data partially underestimate the actual movement, for the lack of capability to detect N–S-oriented movements, which are predominant in the DSGSD.

All the surveyed buildings show damages, but the degree of damage is mostly low or moderate. Both ascending and descending PSInSAR™ data seem to be correlated to the degree of damage (Figure 5c,d), as suggested by a general increasing trend of damage with displacement rate. Damages are also reported along the railway tunnel and along both tubes of the SS36 highway (Figure 5a). The latter required repairing of the tunnel concrete support because of large episodic displacements occurring in 2002 and 2012 and causing an approximate direct cost of 40 million euros. Unfortunately, these episodes are not detectable in the PSInSAR™ displacement time series, because of a lack of data during these periods (Figure 6).

### Caspoggio and Lake Palù DSGSDs

Caspoggio DSGSD (Figure 7, #5) and Lake Palù DSGSD (Figure 8, #4) show a significant ground surface displacement rate, up to 20 mm/yr in magnitude along the LOS.

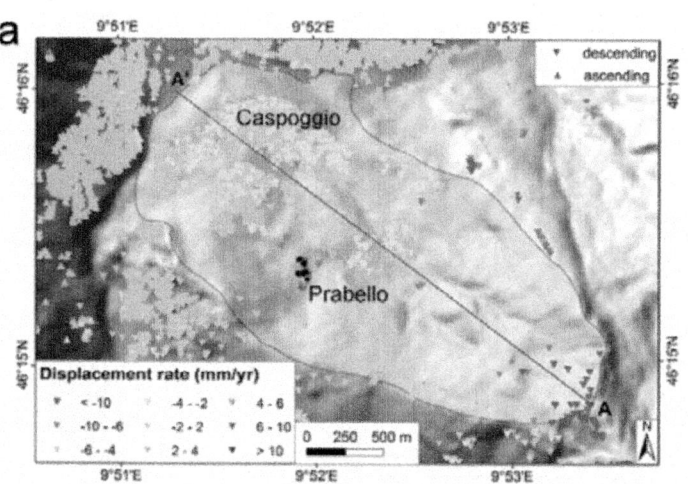

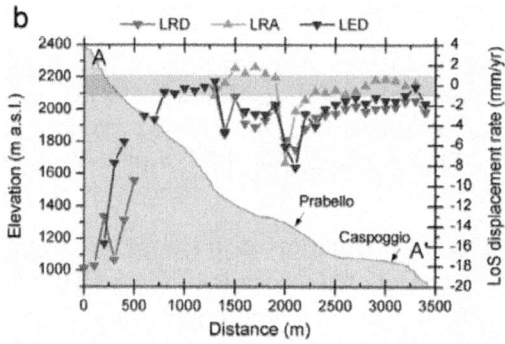

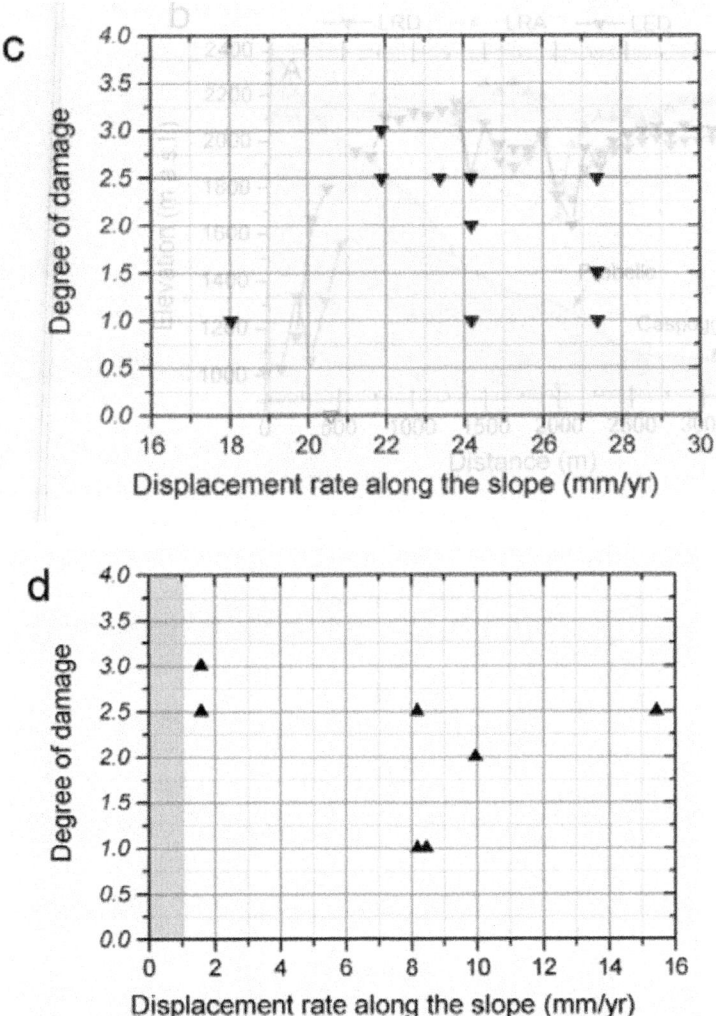

**Figure 7:** Displacement rate and damage to buildings for the Caspoggio DSGSD. (a) PSInSAR™ displacement rate map and location of the surveyed buildings (black polygons); (b) longitudinal swath profile of the displacement rate for available PS datasets; (c) damage degree of buildings as a function of the displacement rate for the LRD dataset; (d) damage degree of buildings as a function of the displacement rate for the LRA dataset. Solid and open triangles in panels (c,d) refer to buildings with and without wall coating, respectively. The grey area in panels (b,c) represents the average standard deviation (Table 2).

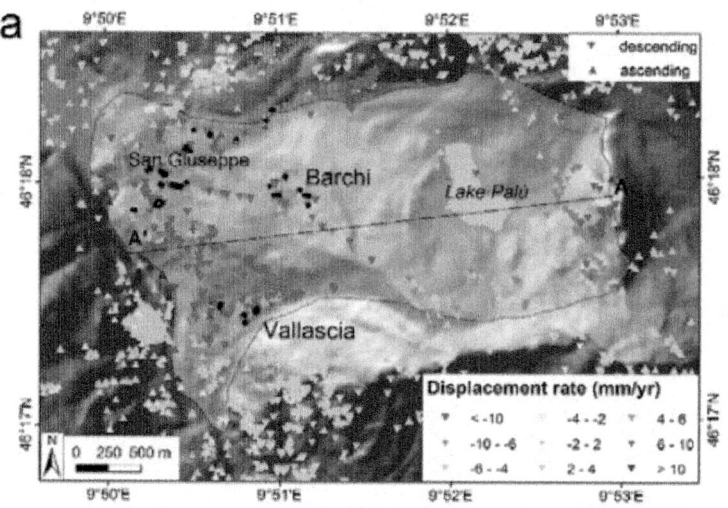

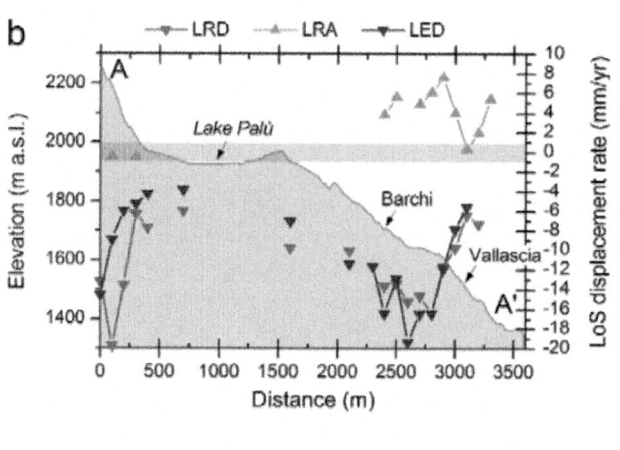

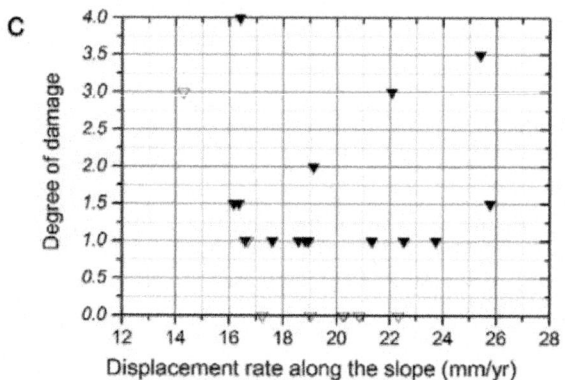

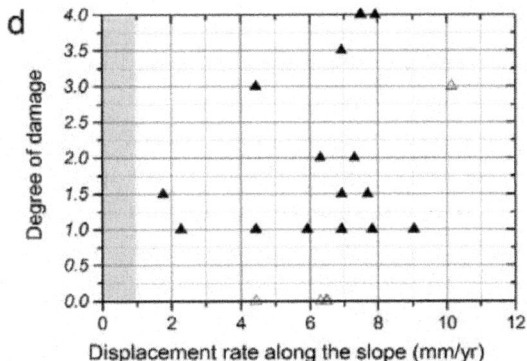

**Figure 8:** Displacement rate and damage to buildings for the Lake Palù DSGSD. (a) PSInSAR™ displacement rate map and location of the surveyed buildings (black polygons); (b) longitudinal swath profile of the displacement rate for available PSI datasets; (c) damage degree of buildings as a function of the displacement rate for the LRD dataset; (d) damage degree of buildings as a function of the displacement rate for the LRA dataset. Solid and open triangles in panels (c,d) refer to buildings with and without wall coating, respectively. The grey area in panels (b,d) represents the average standard deviation (Table 2).

For both DSGSDs, the displacement rate is larger in the upper part of the landslide, and decreases toward the toe of the DSGSD. Again, this behavior suggests a failure mechanism with a prominent vertical downward movement in the upper part and a more horizontal movement in the lower part of the landslide. This subcircular mechanism results in positive displacement rates in the lower slope sector, where the direction of the LOS is against the slope aspect (*i.e.*, the satellite has a frontal view of the landslide). This behavior is particularly clear for the Lake Palù DSGSD (Figure 8).

For both DSGSDs, we observe significant damage to buildings. Apparently, we cannot observe a trend of increasing damage with displacement rate. Damages in the Caspoggio DSGSD have been also reported for the hydroelectric penstock and the derivation tunnel located along the northern sector of the instability [52].

## Mt. Padrio Varadega DSGSD

The Padrio-Varadega DSGSD (Figure 9) is a large phenomenon, with differential internal movements due to large secondary landslides [21]. Overall, a progressive decrease of LOS displacement rate can be observed moving downslope, possibly due to a sub-circular or compound failure plane

[1,21]. The displacement time series show a linear behavior both for ERS 1–2 data and RSAT-S32 data (Figure 6). A good agreement is observed between geodetic and PSInSAR™ measurements for the hydroelectric power plant and the penstock located in the northern sector of the DSGSD (Figure 10) [1]. This provides a long-term validation of the PSInSAR™ technique, even if limited to a single sector of the DSGSD. For this study, we investigated buildings located in the central sector of the DSGSD, where the movements are also concentrated, because of the presence of secondary landslides within the main DSGSD. The ascending acquisition mode is unfavorable, due to the slope aspect, but shows a slightly positive displacement in the LOS direction, thus suggesting a sub-horizontal movement of the DSGSD toe.

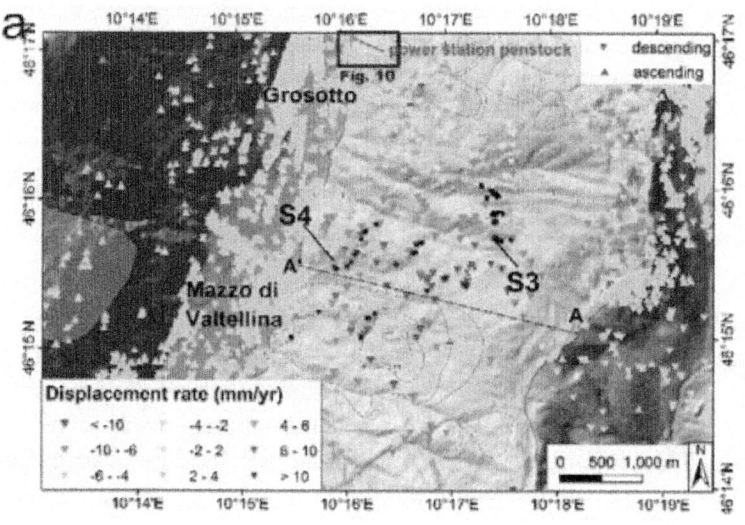

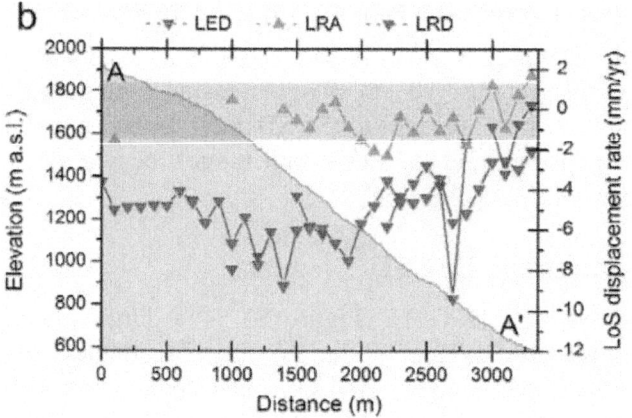

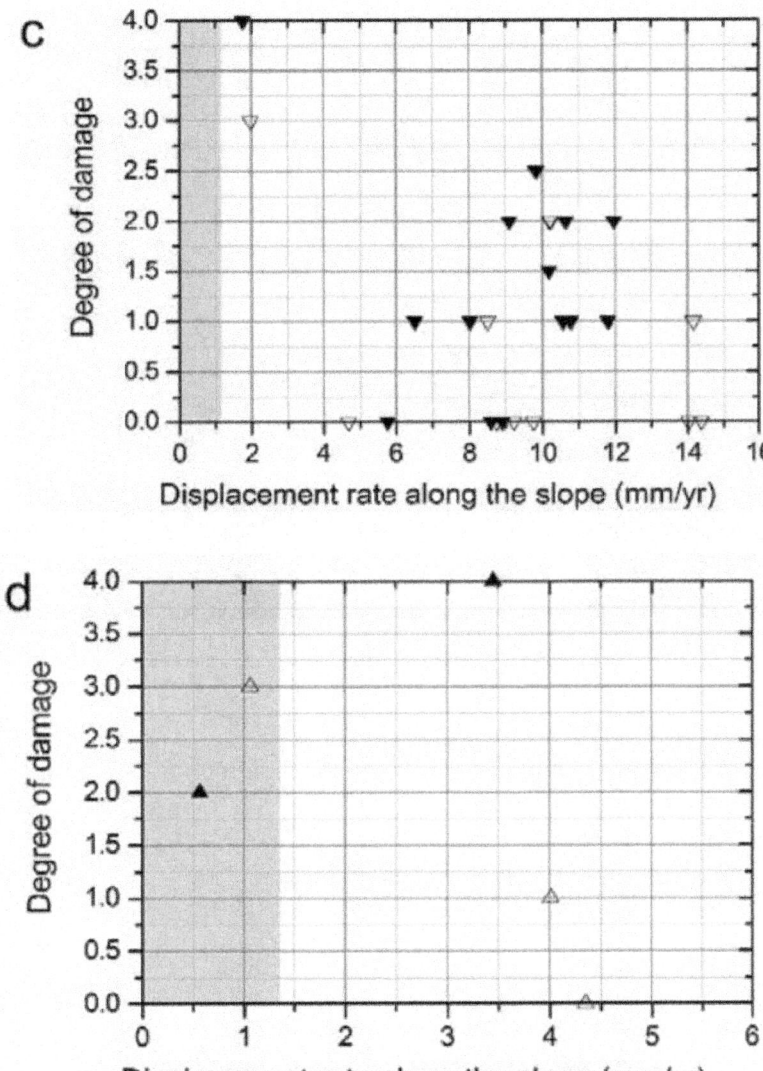

**Figure 9:** Displacement rate and damage to buildings for the Mt. Padrio-Varadega DSGSD. (a) PSInSAR™ displacement rate map and location of the surveyed buildings (black polygons); (b) longitudinal swath profile of the displacement rate for available PSI datasets; (c) damage degree of buildings as a function of the displacement rate for the LRD dataset; (d) damage degree of buildings as a function of the displacement rate for the LRA dataset. Solid and open triangles in panels (c,d) refer to buildings with and without wall coating, respectively. The grey area in panels (b,d) represents the average standard deviation (Table 2).

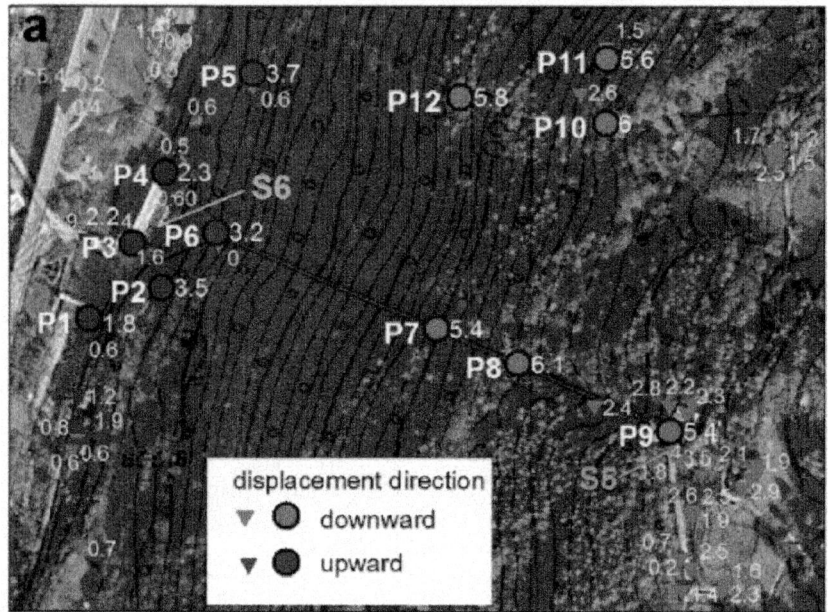

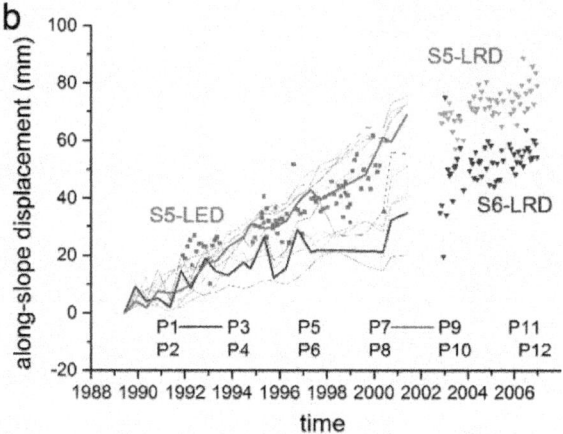

**Figure 10:** Comparison between geodetic and PSInSAR™ in the northern sector of the Mt. Padrio-Varadega DSGSD (see Figure 9a for location): **(a)** Displacement rate map, with geodetic monitoring (circles) and PSInSAR™ (triangles) points. The displacement rates reported in the figures are calculated along the slope direction. **(b)** Displacement time series for geodetic (lines) and PSInSAR™ (squares for ERS 1-2 and triangles for RSAT-S3) data. The PSInSAR™ displacements were shifted according to geodetic data to show a continuous time series.

This movement has been also observed by geodetic measurements at the hydropower station (Figures 9 and 10, [1] located on the alluvial deposits at the slope toe. Significant damage to residential buildings with a clear correlation with the displacement rate obtained by the descending PSInSAR™ data can be observed, with a few exceptions. In particular, the two buildings showing heavy damage and a small displacement rate in descending acquisition mode are located in the lower part of the slope, where the apparent direction of the movement becomes sub-horizontal and poorly visible in descending mode. In turn, these two buildings seem to be slightly correlated with positive displacement rates in ascending acquisition mode. The hydroelectric power plant underwent a complete stop for restoration and to allow for proper functioning of the turbines.

## Saviore DSGSD

Similar to Mt Padrio Varadega DSGSD, the Saviore DSGSD (Figure 11) is a complex phenomenon with secondary landslides affecting large part of the DSGSD [53]. Movements in the upper part of the slope are probably overestimated, due to the creeping of debris deposits moving faster than the DSGSD. In the southeastern sector of the DSGSD, it is possible to isolate movements associated with secondary rockslides, one of which has been recognized since the 1950s, due to damage to the Valle village.

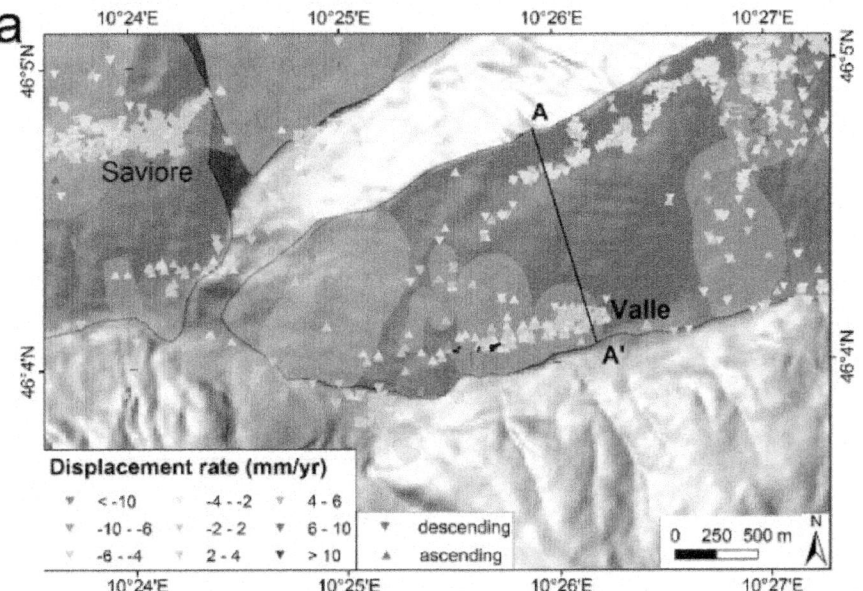

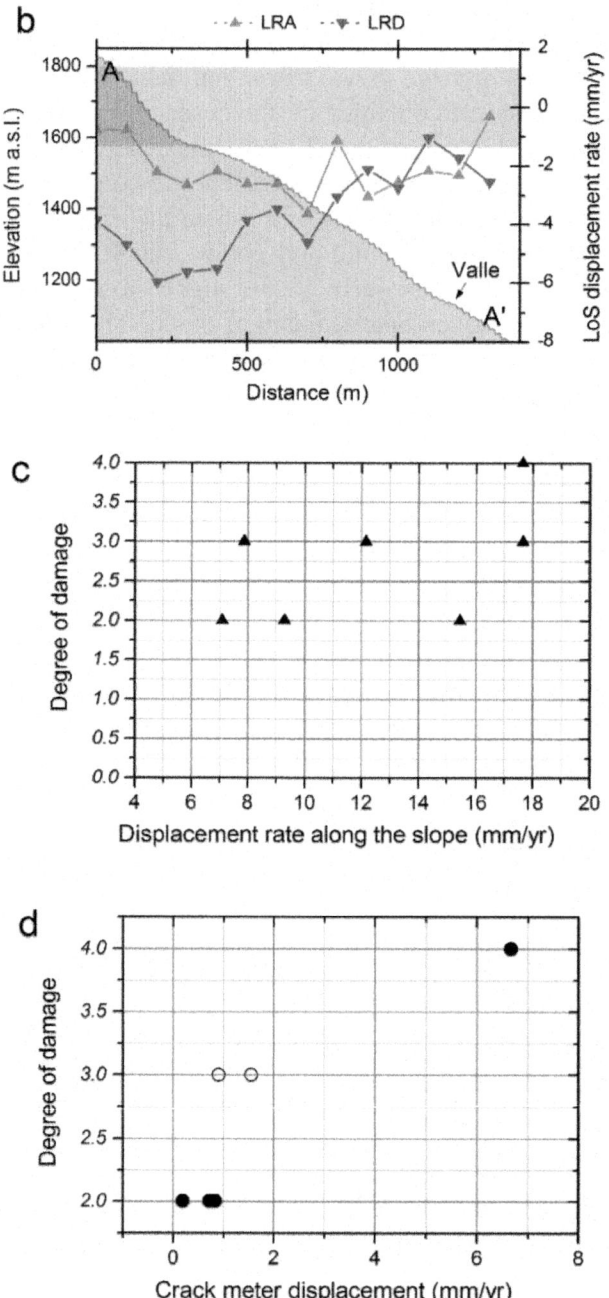

**Figure 11:** Displacement rate and damage to buildings for the active rock-slide inside the Saviore DSGSD. **(a)** PSInSAR™ displacement rate map and

location of the surveyed buildings (black polygons); (**b**) longitudinal swath profile of the displacement rate for available PSI datasets; (**c**) damage degree of buildings as a function of the displacement rate for the LRA dataset; (**d**) damage degree as a function of the crack meter displacement rate. Solid and open triangles in panel (c) refer to buildings with and without wall coating, respectively. Open circles in panel (d) refer to cemetery wall. The grey area in panel (b) represents the average standard deviation (Table 2).

This phenomenon has been monitored since 1987 with inclinometers, optical targets and crack meters installed on the most critical buildings [53,54]. Superficial monitoring data have been obtained by [54] and compared with displacements observed by PSInSAR™ data inside the rockslide. The latter appear to underestimate the actual landslide movements, due to the impossibility to fully characterize the N-S movements, which are predominant in the rockslide (Figure 11). However, we observe a good agreement of measured satellite displacement with the E-W and vertical components of the actual displacement vector (Figure 12) measured by a total station surveying a series of optical targets.

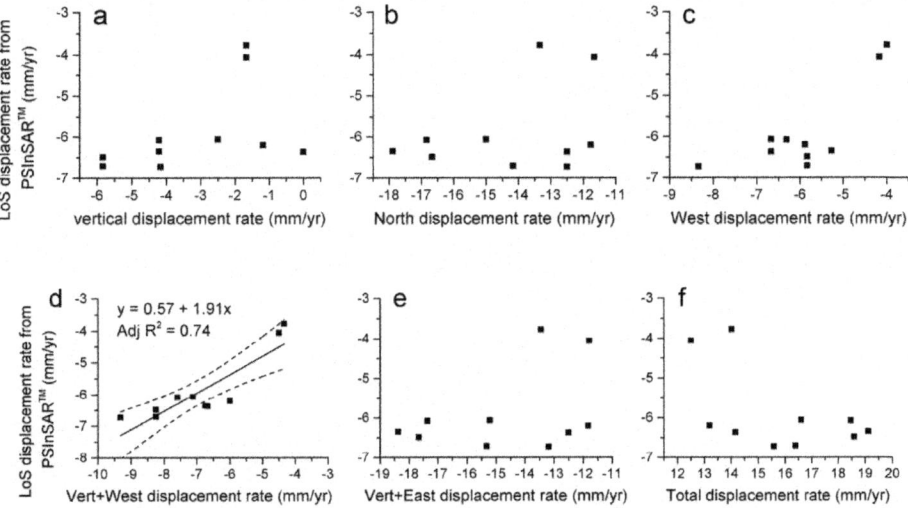

**Figure 12:** Comparison between PSInSAR™ LOS displacement rate (ascending) and displacement rates monitored by ground-based optical targets. Different components of optical target displacements vectors are used: (**a**) vertical component, (**b**) north component, (**c**) west component, (**d**) composed vertical and west component, (**e**) composed vertical and north component and (**f**) total vector. The best-fitting line in panel (d) was obtained by least square regression.

Damages to buildings are heavy in the western sector of the village, where buildings have been surveyed, and a very slight trend of increasing damage with displacement rate is observable. The rate of crack formation on the buildings, as monitored by crackmeters, has the same order of magnitude of observed PSInSAR™ displacements and shows a good correlation with the estimated degree of damage.

## Catasco and Montalto Rockslides

Catasco and Montalto landslides are smaller phenomena not classified as DSGSD (Figure 13). The instability that affects Catasco village (Figure 13a,c) is characterized by a large rockslide, which is partially active, with movements mostly localized in the eastern part of the village [55]. Notwithstanding the presence of long-term damage, the landslide complex has been only recently identified and mapped, and it was not studied systematically.

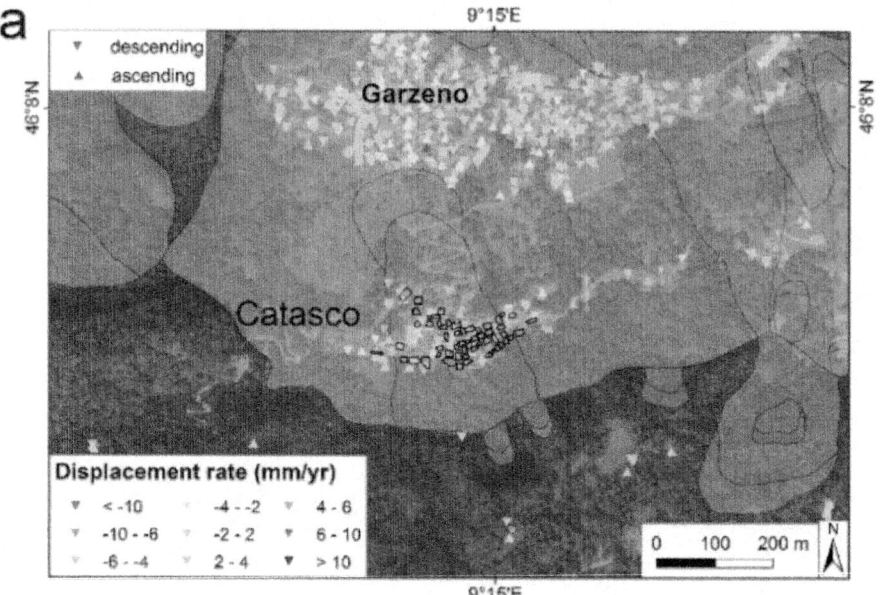

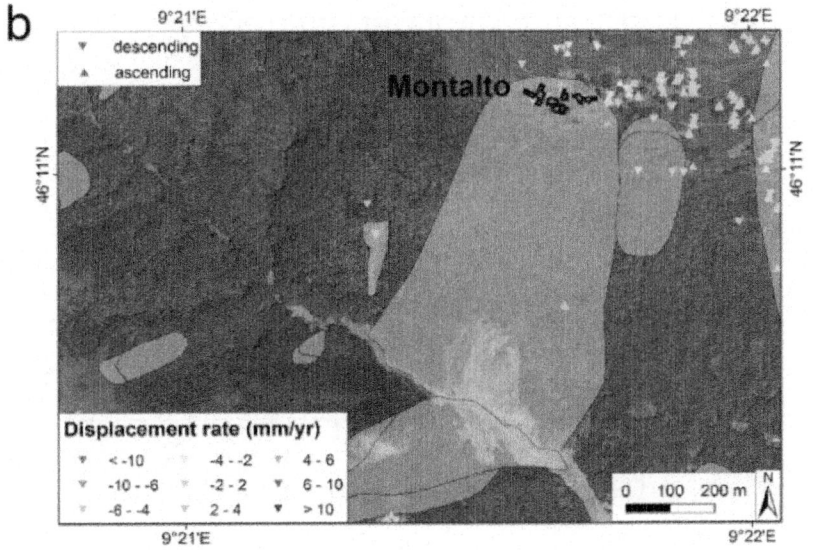

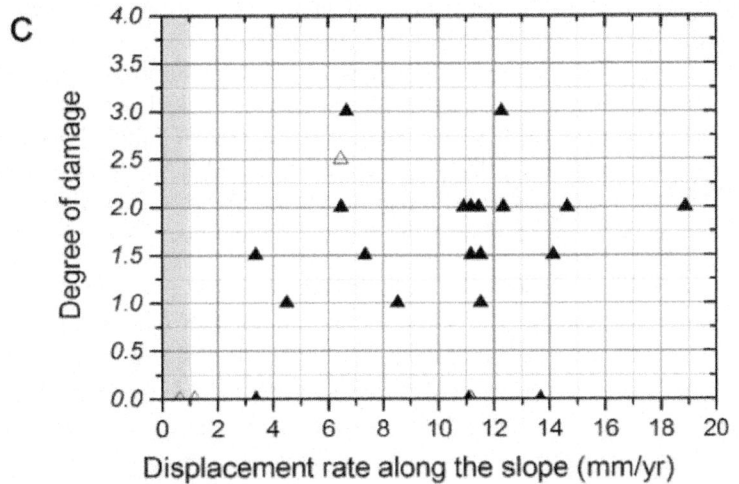

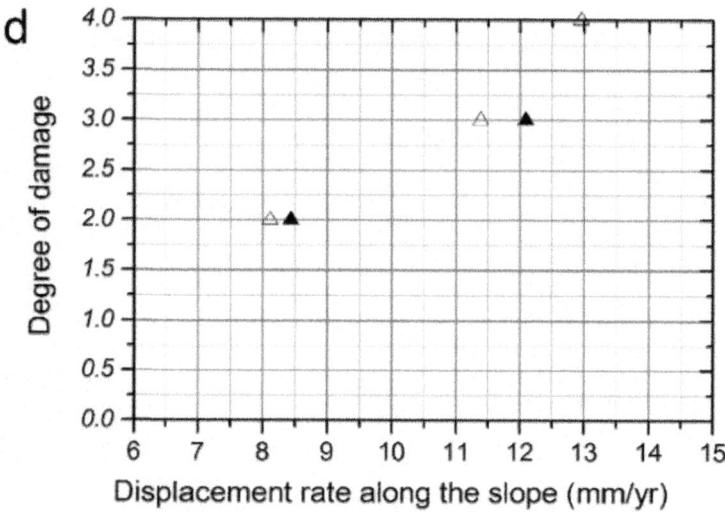

**Figure 13:** Displacement rate and damage to buildings for the Catasco and Montalto rockslides. (**a**) PSInSAR™ displacement rate map and location of the surveyed buildings for the Catasco rockslide (black polygons); (**b**) PSInSAR™ displacement rate map and location of the surveyed buildings for the Montalto rockslide (black polygons); (**c**) damage degree of buildings of the Catasco rockslide as a function of the displacement rate for the LRA dataset; (**d**) damage degree of buildings of the Montalto rockslide as a function of the displacement rate for the LRA dataset. Solid and open triangles in panels (c,d) refer to buildings with and without wall coating, respectively. The grey area in panel (c) represents the average standard deviation (Table 2).

Only in 2010 was on-site monitoring activity started in the Catasco village. The lower part of the village is characterized by a relatively shallow landslide (about 5 m deep) affecting the colluvial soil cover, while the upper part lies within a 30 m-deep rotational rockslide, as witnessed by inclinometer measurements available since 2010. The complex landslide behavior causes differential movements at the surface. These generate moderate-to-heavy damage to buildings, the latter related to buildings located in the lower part of the village, where the effect of shallow landsliding has been stronger.

The correlation between the degree of damage and the displacement rate is poor, probably due to the orientation of the slope, which is not optimal for the radar technique, and to the existence of different phenomena resulting in different deformations.

Montalto village (Figure 13b,d) lies at the head of a large rockslide, part of which has been very active since 1998, with a strong acceleration in 2002. The upper sector of the landslide shows a significant displacement in both ascending and descending modes (Figure 13b). Considering that the rockslide faces south, we can interpret the measured displacement as the vertical component of the movement associated with the landslide head. A possible horizontal component is not resolved by the satellites, but can be assumed as almost negligible with respect to the vertical component, due to the high slope gradient and the shape of the landslide. Damage to surveyed buildings is heavy, with a strong correlation between the degree of damage and the displacement rate (Figure 13d).

## DISCUSSION

The PSInSAR™ technique provides useful information for the analysis of DSGSDs and large rockslide characterized by low displacement rates. We showed in the analysis of the studied sites that PSInSAR™ could be used to describe the activity of slope instabilities and, also, the behavior and kinematics of the landslides. For instance, a progressive decrease of the displacement rate downslope could indicate a circular or compound failure mechanism. This analysis could benefit from the reconstruction of the actual displacement vector, at least in the E-W direction, by combining ascending- and descending-mode data [30,49]. Even if this reconstruction was not possible for all the studied sites, because of the scarcity of couples of ascending and descending data for the same points in space, we have shown the importance of considering both acquisition modes, especially for the example of Lake Palù DSGSD.

The PSInSAR™ data seems also suitable for the analysis of displacement rates associated with the damage of structures and infrastructures. Although this correlation is not always clear for the studied sites, the general trend of the data observed in box and whiskers plots (Figure 14) shows that the degree of damage is significantly correlated with the displacement rate. For instance, considering the 25th-percentile of the displacement rate (Figure 14b), we observe that this varies linearly between −5 and −10 mm/yr from a damage level of 0 to a damage level of 4. The distributions of the displacement rate values for each degree of damage (Figure 14) are very dispersed, due to a large uncertainty. This uncertainty mainly derives from the complexity of the slope instability behavior, sometimes from the difficulty of associating a unique damage value to a structure, and, finally, the different position of the structure within the landslide.

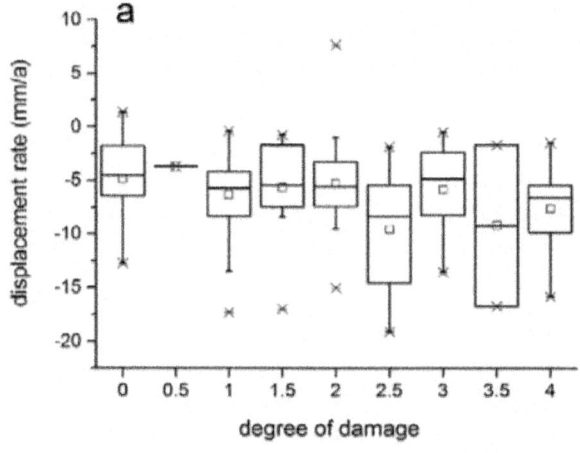

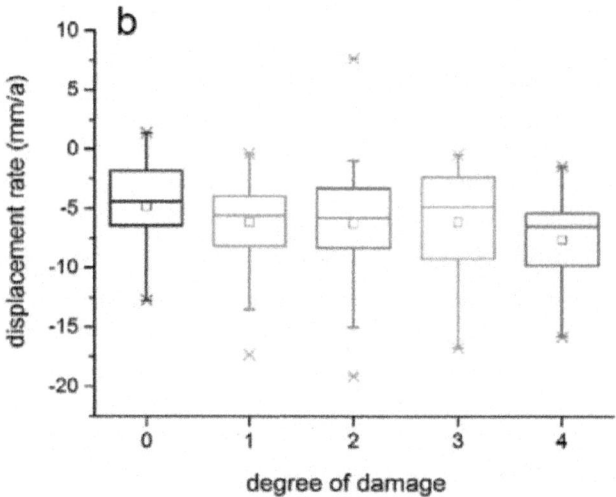

**Figure 14:** Box and whiskers plots of all displacement rate values for each degree of damage for all surveyed buildings at all the seven sites. (**a**) The original degree of damage values considering half degree intervals; (**b**) truncated values of the degree of damage. Box: 25th-, 50the- and 75th-percentile; whiskers: fifth- and 95th-percentile; square: mean.

The best correlation between the degree of damage and the displacement rate is observed for the Montalto rockslide. Although the number of surveyed buildings is low, we can argue that the good correlation is due to the fact that buildings are all located in the same area of the landslide, where the behavior is homogeneous. Moreover, the Montalto rockslide is the smallest phenomenon analyzed (0.32 km²), also showing the most simple kinematic behavior, which consists in the rotation of the whole slide mass.

The Mt. Legnoncino DSGSD also shows a good correlation between the degree of damage and the displacement rate, although some buildings show an anomalous degree of damage. In particular, a few buildings have a small degree of damage notwithstanding a large displacement rate. These buildings present recent wall coatings, possibly covering minor damages to the structure. This problem has been observed also at other sites, where walls of many buildings appeared to have been recently repaired and rendered. In those cases, the estimated degree of damage can be strongly underestimated. It must be stressed that some buildings have been built using different techniques and materials and at very different times. This can control the level of damage. Furthermore, in the case of slow moving landslides, old structures could have been subjected to a larger cumulated displacement or more acceleration events.

For the Lake Palù and Mt. Padrio-Varadega DSGSDs, we observe a poor correlation between the degree of damage and the displacement rate. Here, the surveyed buildings are widespread over the DSGSD, thus belonging to different sectors with different displacement rates and behavior. These buildings are subjected to a different style of movement (e.g., dominant vertical *vs.* horizontal movement), which induces different effects on the buildings and a different degree of damage.

For larger DSGSDs (Mt. Padrio-Varadega, Caspoggio and Legnoncino) with very deep failure surfaces, it is also likely that the movement of the phenomenon occurs without large ground surface differential movements, except along some specific morpho-structures, which can actually cause damage to building foundations and structures [26]. In this case, the movement of the structure can occur as a slight rigid translation of the structure, without any evident damage.

Finally, for the Saviore DSGSD and Catasco rockslide, the poor correlation between the degree of damage and the displacement rate can be attributed to the low capability of the PSInSAR™ technique to detect movements mainly directed toward north or south, thus parallel to the satellite orbit and perpendicular to the LOS.

## CONCLUSION

PSInSAR™ displacement data have been used for the analysis of deep-seated gravitational slope deformations (DSGSD) and large rockslide activity and kinematics in Alpine terrains. The main aim of the analysis was to investigate the correlation between the displacement rate and the degree of damage of buildings. In fact, no real effort has been done in the literature to verify the use of PSInSAR™ for building damage assessment and monitoring. The analysis of displacement data shows a continuous slope movement of DSGSD and rockslides with the presence of diffuse deformation along the slope. Displacement rates up to 20 mm/yr were calculated along the slope direction at most of the sites. In general, we observe an increase of the degree of damage when the displacement rate increases. For instance, the 25th-percentile of the measured displacement rates has been observed to increase linearly between −5 and −10 mm/yr for a damage level varying from null (damage level = 0) to very heavy (damage level = 4) (Figure 14b). However, this trend shows a large uncertainty, which can be due to different causes, such as:

- the uncertainty in the attribution of the degree of damage to recently renovated buildings;

- the complexity of the deformation for large phenomena, with differential behavior within the slope instability, due to reactivation of smaller events, which can locally increase the observed degree of damage, or the presence of debris at the surface;

- the possible absence of differential superficial movements causing damages to structures, due to the large size of the investigated phenomena; and

- the different behavior of buildings, depending on the type of structure, its age, the position along the slope and the occurrence of total/local recent reactivation/accelerations.

Future improvements of the PSI techniques could provide benefits to the analysis of large rock instabilities, especially due to the reduction of revisiting time and the improvement of resolution offered by other satellites (e.g., Cosmo SkyMed, TerraSAR-X, Sentinel). These developments will permit one to better define the instability behavior, to increase the PS density and the size of the structures database and to improve the possibility of reconstructing the actual displacement vector by combining ascending and descending data, referring to corresponding spatial positions.

## ACKNOWLEDGMENTS

We thank A. Ferretti from Tele-Rilevamento Europa and M. Ceriani from Regione Lombardia for making PSInSAR™ data available. The building damage survey was actively performed by E. D'Agostini and, for the Valle di Saviore, by M. Salvoni. The extraction of displacement swath profile was done by R. Ignagnaro.

# REFERENCES AND NOTES

1. Ambrosi, C.; Crosta, G.B. Large sackung along major tectonic features in the Central Alps. *Eng. Geol* 2006, *83*, 183–200.

2. MacFarlane, D.F. Observations and predictions of the behaviour of large, slow-moving landslides in schist, Clyde Dam reservoir, New Zealand. *Eng. Geol* 2008, *109*, 5–15.

3. Zangerl, C.; Eberhardt, E.; Perzlmaier, S. Kinematic behaviour and velocity characteristics of a complex deep-seated crystalline rockslide system in relation to its interaction with a dam reservoir. *Eng. Geol* 2010, *112*, 53–67.

4. Cruden, D.M.; Varnes, D.J. Landslide Types and Processes. In *Landslides: Investigation and Mitigation*; Turner, A.K., Shuster, R.L., Eds.; Transportation Research Board: Washington, DC, USA, 1996; pp. 36–75.

5. Crosta, G.B.; Agliardi, F.; Frattini, P. Deep seated gravitational slope deformations in the European Alps.*Tectonophysics* 2013, *605*, 13–33.

6. Zischinsky, U. On the Deformation of High Slopes. Proceedings of 1st Congress International Society for Rock Mechanics 2, Lisbon, Portugal, 25 September–1 October 1966; pp. 179–185.

7. Bovis, M.J. Rock-slope deformation at Affliction Creek, southern Coast Mountains, British Columbia. *Geol. Soc. Am. Bull* 1990, *93*, 804–812.

8. Varnes, D.J.; Radbruch-Hall, D.; Varnes, K.L.; Smith, W.K.; Savage, W.Z. *Measurement of Ridge-Spreading Movements (Sackungen) at Bald Eagle Mountain, Lake County, Colorado, 1975–1989*; US Geological Survey Open-File Report 90-543; US Geological Survey: Denver, CO, USA, 1990; p. 13.

9. Chigira, M. Long-term gravitational deformation of rock by mass rock creep. *Eng. Geol* 1992, *32*, 157–184.

10. Agliardi, F.; Crosta, G.; Zanchi, A. Structural constraints on deep-seated slope deformation kinematics. *Eng. Geol* 2001, *59*, 83–102.

11. Cruden, D.M.; Hu, X.Q. Exhaustion and steady-state models for predicting landslide hazards in the Canadian Rocky Mountains. *Geomorphology* 1993, *8*, 279–285.

12. Hippolyte, J.-C.; Brocard, G.; Tardy, M.; Nicoud, G.; Bourlès, D.; Braucher, R.; Ménard, G.; Souffaché, B. The recent fault scarps of the Western Alps (France): Tectonic surface ruptures or gravitational sackung scarps? A combined mapping, geomorphic, levelling, and 10Be dating approach. *Tectonophysics* 2006, *418*, 255–276.

13. Hippolyte, J.C.; Bourlès, D.; Braucher, R.; Carcaillet, J.; Léanni, L.; Arnold, M.; Aumaitre, G. Cosmogenic 10Be dating of a sackung and its faulted rock glaciers, in the Alps of Savoy (France). *Geomorphology* 2009, *108*, 312–320.

14. Bigot-Cormier, F.; Braucher, R.; Boulès, D.; Guglielmi, Y.; Dubar, M.; Stéphan, J.F. Chronological constraints on processes leading to large active landslides. *Earth. Planet. Sci. Lett* 2005, *235*, 141–150.

15. Bovis, M.J.; Evans, S.G. Extensive deformations of rock slopes in southern Coast Mountains, southwest British Columbia, Canada. *Eng. Geol* 1996, *44*, 163–182.

16. Curlander, J.C.; McDonough, R.N. *Synthetic Aperture Radar Systems and Signal Processing*; Wiley-Interscience: New York, NY, USA, 1991.

17. Ferretti, A.; Prati, C.; Rocca, F. Multibaseline InSAR DEM reconstruction: The wavelet approach. *IEEE Trans. Geosci. Remote Sens* 1999, *37*, 705–715.

18. Ferretti, A.; Prati, C.; Rocca, F. Nonlinear subsidence rate estimation using permanent scatterers in differential SAR interferometry. *IEEE Trans. Geosci. Remote Sens* 2000, *38*, 2202–2212.

19. Ferretti, A.; Prati, C.; Rocca, F. Multibaseline phase unwrapping for InSAR topography estimation. *Nuovo Cimento Della Soc. Ital. Fis. C* 2001, *124*, 159–176.

20. Ferretti, A.; Prati, C.; Rocca, F. Permanent scatterers in SAR Interferometry. *IEEE Trans. Geosci. Remote Sens* 2001, *39*, 8–20.

21. Allievi, J.; Ambrosi, C.; Ceriani, M.; Colesanti, C.; Crosta, G.B.; Ferretti, A.; Fossati, D. Monitoring Slow Mass Movements with the Permanent Scatterers technique. Proceedings of the IEEE International Geoscience and Remote Sensing Symposium (IGARSS'03), Toulouse, France, 21–25 July 2003; 1, pp. 215–217.

22. Colesanti, C.; Crosta, G.B.; Ferretti, A.; Ambrosi, C. Monitoring and assessing the state of activity of slope instabilities by the Permanent Scatterers Technique. *NATO Sci. Ser* 2006, *49*, 175–194.

23. Saroli, M.; Stramondo, S.; Moro, M.; Doumaz, F. Movements detection of deep seated gravitational slope deformations by means of InSAR data and photogeological interpretation: northern Sicily case study. *Terra Nova* 2005, *17*, 35–43.

24. Strozzi, T.; Farina, P.; Corsini, A.; Ambrosi, C.; Thüring, M.; Zilger, J.; Wiesmann, A.; Wegmüller, U.; Werner, C. Survey and monitoring of landslide displacements by means of L-band satellite SAR interferometry.

*Landslides* 1995, *2*, 193–201.

25. Colesanti, C.; Wasowski, J. Investigating landslides with space-borne Synthetic Aperture Radar (SAR) interferometry. *Eng. Geol* 2006, *88*, 173–199.

26. Osmundsen, P.T.; Henderson, I.; Lauknes, T.R.; Larsen, Y.; Redfield, T.F.; Dehls, J. Active normal fault control on landscape and rock-slope failure in northern Norway. *Geology* 2006, *37*, 135–138.

27. Strozzi, T.; Delaloye, R.; Kääb, A.; Ambrosi, C.; Perruchoud, E.; Wegmüller, U. Combined observations of rock mass movements using satellite SAR interferometry, differential GPS, airborne digital photogrammetry, and airborne photography interpretation. *J. Geophys. Res* 2010, *115*, F01014.

28. Calò, F.; Calcaterra, D.; Iodice, A.; Parise, M.; Ramondini, M. Assessing the activity of a large landslide in southern Italy by ground-monitoring and SAR interferometric techniques. *Int. J. Remote Sens* 2012, *33*, 3512–3530.

29. Del Ventisette, C.; Ciampalinik, A.; Manunta, M.; Calò, F.; Paglia, L.; Ardizzone, F.; Mondini, A.; Reichenbach, P.; Mateos, R.M.; Bianchini, S.; *et al.* Exploitation of large archives of ERS and ENVISAT C-band SAR data to characterize ground deformations. *Remote Sens* 2013, *5*, 3896–3917.

30. Tofani, T.; Raspini, F.; Catani, F.; Casagli, N. Persistent Scatterer Interferometry (PSI) technique for landslide characterization and monitoring. *Remote Sens* 2013, *5*, 1045–1065.

31. Strozzi, T.; Ambrosi, C.; Raetzo, H. Interpretation of aerial photographs and satellite SAR interferometry for the inventory of landslides. *Remote Sens* 2013, *5*, 2554–2570.

32. Tarchi, D.; Casagli, N.; Leva, D.; Moretti, S.; Sieber, A.J. Monitoring landslide displacements by using ground-based SAR interferometry: Application to the Ruinon landslide in the Italian Alps. *J. Geophys. Res* 2003, *108*, 2387–2401.

33. Moore, D.P.; Watson, A.D.; Martin, C.D. Deformation Mechanism of a Large Rockslide Inundated by a Reservoir. Proceedings of JTC Workshop on the Mechanics and Velocity of Large Landslides, Courmayeur, Italy, 25–28 September 2006.

34. Negulescu, C.; Foerster, E. Parametric studies and quantitative assessment of the vulnerability of a RC frame building exposed to differential settlements. *Nat. Hazard. Earth Syst. Sci* 2010, *10*, 1781–1792.

35. Mansour, M.F.; Morgenstern, N.R.; Martin, C.D. Expected damage from displacement of slow-moving slides.*Landslides* 2010, *8*, 117–131.

36. Fotopoulou, S.; Pitilakis, K. Vulnerability assessment of reinforced concrete buildings subjected to seismically triggered slow-moving earth slides. *Landslides* 2012.

37. Crosta, G.B.; Agliardi, F.; Frattini, P.; Zanchi, A. Alpine inventory of Deep-Seated Gravitational Slope Deformations. *Geophys. Res. Abstr.* 2008, *10*, EGU2008-A-02709.

38. Agliardi, F.; Crosta, G.B.; Frattini, P. Slow Rock-Slope Deformation. In *Landslides: Types, Mechanisms and Modeling*; Clague, J.J., Stead, D., Eds.; Cambridge University Press: Cambridge, UK, 2012; pp. 207–221.

39. Agliardi, F.; Crosta, G.B.; Frattini, P.; Malusà, M. Giant non-catastrophic landslides and the long-term exhumation of the European Alps. *Earth. Planet. Sci. Lett* 2013, *365*, 263–274.

40. Schmid, S.M.; Fogenschuh, B.; Kissling, E.; Schuster, R. Tectonic map and overall architecture of the Alpine orogen. *Eclogae Geol. Helv* 2004, *97*, 93–117.

41. Schönborn, G. Alpine tectonics and kinematic models of the central Southern Alps. *Mem. Sci. Geol* 1992, *44*, 229–393.

42. Froitzheim, N.; Schmid, S.M.; Conti, P. Repeated change from crustal shortening to orogen-parallel extension in the Austroalpine units of Graubünden. *Eclogae Geol. Helv* 1994, *87*, 559–612.

43. *Geologische Karte der Schweiz 1:500,000*; Institut für Geologie, Universität Bern, und Bundesamt für Wasser und Geologie: Bern, Switzerland, 2005.

44. Refice, A.; Bovenga, F.; Guerriero, L.; Wasowski, J. DInSAR Applications to Landslide Studies. Proceedings of the IEEE International Geoscience and Remote Sensing Symposium (IGARSS'01), Sydney, Australia, 9–13 July 2001; 1, pp. 144–146.

45. Werner, C.; Wegmuller, U.; Strozzi, T.; Wiesmann, A. Interferometric Point Target Analysis for Deformation Mapping. Proceedings of the IEEE International Geoscience and Remote Sensing Symposium (IGARSS'03), Toulouse, France, 21–25 July 2003; 7, pp. 4362–4364.

46. Hooper, A.; Zebker, H.; Segall, P.; Kampes, B. A new method for measuring deformation on Volcanoes and other natural terrains using InSAR persistent scatterers. *Geophys. Res. Lett* 2004, *31*, L23611.

47. Bovenga, F.; Nutricato, R.; Refice, A.; Wasowski, J. Application of multi-temporal differential interferometry to slope instability detection

in urban/peri-urban Areas. *Eng. Geol* 2006, *88*, 218–239.

48. Colesanti, C.; Ferretti, A.; Locatelli, R.; Novali, F.; Savio, G. Permanent Scatterers: Precision Assessment and Multi-Platform Analysis. Proceedings of IEEE Geoscience and Remote Sensing Symposium (IGARSS'03), Toulouse, France, 21–25 July 2003; pp. 1193–1195.

49. Cascini, L.; Fornaro, G.; Peduto, D. Advanced low- and full-resolution DInSAR map generation for slow-moving landslide analysis at different scales. *Eng. Geol* 2010, *112*, 29–42.

50. Cigna, F.; Bianchini, S.; Casagli, N. How to assess landslide activity and intensity with Persistent Scatterer Interferometry (PSI): the PSI-based matrix approach. *Landslides* 2012, *5*, 1–17.

51. Grunthal, G. *European Macroseismic Scale EMS-98*; Conseil de l'Europe, Cahiers du Centre Europeén de Geodynamique et du Seismoligie: Luxembourg, 1998; 15, p. 101. Available online: http://www.gfz-potsdam.de/en/research/organizational-units/departments-of-the-gfz/department-2/seismic-hazard-and-stress-field/products-and-services/ems-98/ (accessed on 24 September 2013).

52. Cossa, A. Analisi Dell'Evoluzione e Modellazione di Fenomeni di Instabilità di Versante Presso L'Impianto Idroelettrico di Lanzada (Valmalenco, SO). M.Sc. Thesis, University of Milano-Bicocca, Milan, Italy. 2006.

53. Istituto Sperimentale Modelli e Strutture (ISMES), *Studio per la Definizione dei Livelli di Soglia e Delle Procedure di Analisi dei Dati Strumentali Della Rete di Monitoraggio di Saviore dell'Adamello (BS)*; Technical Report No. 1; ISMES: Bergamo, Italy, 1999; unpublished.

54. Salvoni, M. Censimento e Monitoraggio Delle Lesiono Strutturali ed Analisi del Dissesto di Valle di Saviore ai Fini di Una Valutazione Quantitativa del Rischio. Bachelor's Thesis, University of Milano-Bicocca, Milan, Italy. 2007.

55. Nitti, D.O.; Bovenga, F.; Nutricato, R.; Rana, F.; D'Aprile, C.; Frattini, P.; Crosta, G.B.; Chiaradia, M.T.; Ober, G.; Candela, L. C- and X-band Multi-pass InSAR analysis over alpine areas (ITALY). *Proc. SPIE* 2010.

# CITATION

---

## CHAPTER 1

Jessica Giro-Paloma, Refat Al-Shannaq, Ana Inés Fernández, and Mohammed M. Farid, Preparation and Characterization of Microencapsulated Phase Change Materials for Use in Building Applications, doi:10.3390/ma9010011

## CHAPTER 2

Rongda Ye, Xiaoming Fang, Zhengguo Zhang, and Xuenong Gao, Preparation, Mechanical and Thermal Properties of Cement Board with Expanded Perlite Based Composite Phase Change Material for Improving Buildings Thermal Behavior, doi:10.3390/ma8115408.

## CHAPTER 3

Paul Joseph, and Svetlana Tretsiakova-McNally, Sustainable Non-Metallic Building Materials, doi:10.3390/su2020400

## CHAPTER 4

Antti Ruuska, and Tarja Häkkinen, Material Efficiency of Building Construction, doi:10.3390/buildings4030266

## CHAPTER 5

Sungwoo Lee, Sungho Tae, Seungjun Roh, and Taehyung Kim, Green Template for Life Cycle Assessment of Buildings Based on Building Information Modeling: Focus on Embodied Environmental Impact, doi:10.3390/

## CHAPTER 6

Kuang-Sheng Liu, Sung-Lin Hsueh, Wen-Chen Wu, and Yu-Lung Chen, A DFuzzy-DAHP Decision-Making Model for Evaluating Energy-Saving Design Strategies for Residential Buildings, doi:10.3390/en5114462

## CHAPTER 7

Miguel A. Gómez, Miguel A. Álvarez Feijoo, Roberto Comesaña, Pablo Eguía, José L. Míguez, and Jacobo Porteiro, CFD Simulation of a Concrete Cubicle to Analyze the Thermal Effect of Phase Change Materials in Buildings, doi:10.3390/en5072093.

## CHAPTER 8

Shaghayegh Mohammad and Andrew Shea, Performance Evaluation of Modern Building Thermal Envelope Designs in the Semi-Arid Continental Climate of Tehran, doi:10.3390/buildings3040674

## CHAPTER 9

Sudan Xu, George Vosselman, and Sander Oude Elberink, Detection and Classification of Changes in Buildings from Airborne Laser Scanning Data, doi:10.3390/rs71215867.

## CHAPTER 10

Akubue Jideofor Anselm (2012). Earth Shelters; a Review of Energy Conservation Properties in Earth Sheltered Housing, Energy Conservation, Dr. Azni Zain Ahmed (Ed.), ISBN: 978-953-51-0829-0, InTech, DOI: 10.5772/51873.

## CHAPTER 11

Paolo Frattini, Giovanni B. Crosta, and Jacopo Allievi, Damage to Buildings in Large Slope Rock Instabilities Monitored with the PSInSAR™ Technique, doi:10.3390/rs5104753

# INDEX